T0348551

PRACTICAL AND APPLIED
HYDROGEOLOGY

PRACTICAL AND APPLIED HYDROGEOLOGY

ZEKÂI ŞEN
Istanbul Technical University
Maslak 34469
Istanbul
Turkey

ELSEVIER

AMSTERDAM • BOSTON • WALTHAM • HEIDELBERG • LONDON • NEW YORK
OXFORD • PARIS • SAN DIEGO • SAN FRANCISCO • SINGAPORE • SYDNEY • TOKYO

Elsevier
Radarweg 29, PO Box 211, 1000 AE Amsterdam, The Netherlands
The Boulevard, Langford Lane, Kidlington, Oxford OX5 1GB, UK
225 Wyman Street, Waltham, MA 02451, USA

Copyright © 2015 Elsevier Inc. All rights reserved.

No part of this publication may be reproduced, stored in a retrieval system or transmitted in any form or by any means electronic, mechanical, photocopying, recording or otherwise without the prior written permission of the publisher.

Permissions may be sought directly from Elsevier's Science & Technology Rights Department in Oxford, UK: phone (+44) (0) 1865 843830; fax (+44) (0) 1865 853333; email: permissions@elsevier.com. Alternatively you can submit your request online by visiting the Elsevier web site at http://elsevier.com/locate/permissions, and selecting *Obtaining permission to use Elsevier material*.

Notices
Knowledge and best practice in this field are constantly changing. As new research and experience broaden our understanding, changes in research methods, professional practices, or medical treatment may become necessary.

Practitioners and researchers must always rely on their own experience and knowledge in evaluating and using any information, methods, compounds, or experiments described herein. In using such information or methods they should be mindful of their own safety and the safety of others, including parties for whom they have a professional responsibility.

To the fullest extent of the law, neither the Publisher nor the authors, contributors, or editors, assume any liability for any injury and/or damage to persons or property as a matter of products liability, negligence or otherwise, or from any use or operation of any methods, products, instructions, or ideas contained in the material herein.

Library of Congress Cataloging-in-Publication Data
A catalog record for this book is available from the Library of Congress
Sen, Zekai.
 Practical and applied hydrogeology / Zekai Sen, ITU Hydraulics Lab. Maslak, Istanbul, Turkey. – First edition.
  pages cm.
 1. Hydrogeology. I. Title.
 GB1003.2.S46 2014
 551.49–dc23
                                    2014023533

British Library Cataloguing in Publication Data
A catalogue record for this book is available from the British Library

ISBN: 978-0-12-800075-5

For information on all Elsevier publications
visit our website at http://store.elsevier.com

Working together
to grow libraries in
developing countries

www.elsevier.com • www.bookaid.org

Dedication

This book is dedicated to my newly born grandson.

Ali Zekâi Şen

With the hope that the content will be useful to anyone interested in hydrogeology and groundwater fundamentals.

Contents

5. Groundwater Quality

6. Groundwater Management

Preface

I had theoretical scientific information about the groundwater problems for more than 30 years, since then I came across with many practical questions for solutions in applications for which I found difficulties in answering and convincing geologists, hydrogeologists, groundwater hydrologists and engineers, who needed ready, simple, effective, and applicable practical solutions. The theoretical assumptions were criticized by clever students and field specialists, who would care for solutions with minimum level of mathematical background. I found that the practicality is possible through physical explanations, logical deductions, and rational formulations. In this manner, I also benefited from conflict of opinions and purposeful questions for practical goals toward final application stage in groundwater resources evaluation, assessment, and management.

Due to complicated subsurface geology, the science of hydrogeology becomes also complicated for groundwater occurrence, distribution, exploration, movement, quality variations, recharges, and managements. Most of those who are directly involved in fieldworks for hydrogeological applications are not concerned with the analytical or numerical solutions, which are highly mathematical. However, they are ready to grasp through the conceptual models, which are easy to explain, verbal in their characteristics, and simple in their mathematical bases. Since each mathematical analytical solution has linguistic fundamentals, valid logical rules and concepts, one can appreciate the mathematical

formulations as end products of complicated analytical derivations if these ingredients are provided. Although practicality of hydrogeological problems seems to be satisfied through the ready software, if the content and functionality of the program are not appreciated, then how could one write meaningful reports of interpretations? Each computer program involves a set of logical rules that are hidden in the analytical formulations and therefore the outputs from these softwares need meaningful interpretations, which are not possible without practical explanations. The purpose of this book is to present the principle logical, rational, and conceptual parts for each problem and then through simple calculus to arrive at the desired formulations under the light of a set of relevant assumptions.

When someone is confronted with a problem it is necessary to imagine related events, to plan and design the suitable configuration, and then to come out with a set of deductive statements. Since hydrogeological problems are very complicated depending on the subsurface geological setup, few or several assumptions are made for the simplification in order to arrive at the preliminary solutions. In the literature, there are numerous publications that expose analytical mathematical solutions in detail, but conceptual explanation of the models is not available as wide. In addition to imagination, especially experience provides further clarifications of questions toward the improvement of the existing methodologies. Logic and experience together indicate that nowhere in the world aquifers are homogeneous and isotropic and there are spatial variations in the aquifer parameters based on the control volume scale. In the groundwater movement toward the wells the control volume is the depression cone, which expands by time in a steadily slowing manner. During the expansion, control volume changes and different subsurface geological features enter the domain of influence and consequently the aquifer parameters should also vary spatially.

After the simple, practical, and conceptual principles, one can appreciate the key processes that play role in the groundwater movement and accordingly the parameters can be defined in a parsimonious manner. It is the main purpose to define the aquifer parameters for groundwater hydrogeological assessments in a region.

In order to render the practicability and applicability in hydrogeological projects, a set of case studies and examples are provided in all chapters after many sections. The most successful applied groundwater studies can be achieved with a team where theoretical, practical, and experienced staff incorporates for effective solutions. I could not complete this work without the patience, support, and assistance of my wife Fatma Şen.

Zekâi Şen
February 2014
Erenköy, Istanbul

1

Water Science Basic Information

1.1 HYDROLOGY (WATER SCIENCE) ELEMENTS

Hydrology is the science of water occurrence, movement and transport in nature. It gives weight toward the study of water in the Earth and is concerned with local circulations related to the atmosphere, lithosphere, biosphere, and hydrosphere leading to water movement, distribution, quality, and environmental aspects. Broadly, it deals also with the physical as well as chemical relationships. In general, it is concerned with natural events such as rainfall, runoff, drought, flood and runoff, groundwater

Copyright © 2015 Elsevier Inc. All rights reserved.

occurrences, their control, prediction, and management. On the application side, hydrology provides basic laws, equations, algorithms, procedures, and modeling of these events for the practical use of human comfort. It also covers the practical and field applications for water resources assessments with simple rational calculations leading toward proper managements. Hydrology related topics that reflect the content of this book are given in Figure 1.1 from engineering and earth sciences aspects point of view.

Surface water or groundwater studies require basic hydrological information as for rainfall assessment, evapotranspiration, infiltration, runoff, subsurface flow, and their modeling aspects for practical engineering, agricultural, irrigation, and hydrogeological applications.

Hydrogeology is the part of hydrology that deals with the occurrence, movement, and quality of water beneath the Earth's surface. Hydrogeology deals with water in complex subsurface environments, and therefore, its complexity as a science is more than surface hydrology. It is concerned with permeable geological formations or group of formations that bear water in saturation and bound to yield significant quantities through wells and springs. As will be explained in Chapter 2, void ratio and hydraulic parameters of these formations are among the most significant factors that reflect the water storing and transmitting properties of geological formations.

Quantification and practical uses of the water related topics in Figure 1.1 require scientific observations, measurements, investigations, and evaluations of various water balance components. Hence, a multitude of disciplines (geology, hydrogeology, hydrology, geophysics, geochemistry, hydraulics), and accordingly, different specialists are involved in any large-scale groundwater study.

There are specific scientists who are concerned individually with each one of the components. For instance, hydrologists pay a great attention in studying the physical occurrences of the source components to the groundwater system; geologists are mainly interested in the rock composition of the groundwater reservoir domain. The names groundwater hydrologist and geohydrologist are used synonymously and they are more interested in the source component of groundwater system and less worried about the geological composition of the reservoir. Their main research methods are mainly water abstraction modeling studies. Hydrogeologists are more concerned with the geological setup of the groundwater reservoirs with less emphasis on the source and abstraction except springs. Their research methods are mainly the field works in the forms of data as direct measurements for quantitative evaluations (pumping tests, piezometric level, joint and fracture measurements, etc.) and water

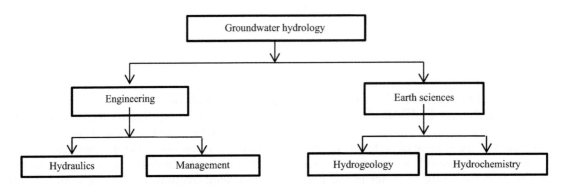

FIGURE 1.1 **Hydrology related topics.**

sample collections for the qualitative studies. At the end, with all these information availability, hydrogeologists prepare a detailed report for the groundwater reservoir concerning the possibility of exploitable groundwater storage, flow rates and quality classifications for domestic, agricultural and industrial usages. Finally, hydraulic engineers are concentrated on the engineering part of the system in Figure 1.1. Integrated groundwater management studies cannot be achieved without cooperation among the aforementioned specialists.

This book is prepared to guide the practical workers, and therefore, subject matter is not presented in a detailed theoretical manner. However, at places of relevance some detail is provided first linguistically and then the relevant mathematical expressions are derived.

1.2 HYDROLOGIC CYCLE

In general, hydrologic cycle is the combination of all possible waterways between the atmosphere, lithosphere, biosphere, and hydrosphere in addition to specific ways within each sphere. Living creatures are dependent on some gases, water, nutrients, and solids that are available in nature rather abundantly in finely balanced quantities for their survivals. The most precious commodities are the air (atmosphere), water (hydrosphere), earth (lithosphere), and energy (especially solar energy). From the geological records it seems that about 1.5 billion years ago free oxygen first appeared in the atmosphere in appreciable quantities (Harvey, 1982). The appearance of life was dependent essentially on the availability of oxygen but once sufficient amount was accumulated for green plants to develop, then photosynthesis process liberated more oxygen into the atmosphere.

In the atmosphere, water is in the vapor form and mixed with various gasses. Biosphere is the plant world that is entirely dependent on water. However, the lithosphere is the general name for the rock domain (geology) and it is the main environment that will be dealt with in this book from the water storage, movement, and quality and management points of view. Hydrosphere includes environments of sole water such as lakes, rivers, oceans, and groundwater reservoirs. The water in Earth circulates among these environments from the hydrosphere (oceans) to atmosphere and then to the lithosphere due to solar energy (SE) (Figure 1.2). The circulation includes complex and interdependent processes such as evaporation (EV), transpiration (TR), evapotranspiration (ET), plant water (PW), precipitation (PR), and groundwater recharge (RE).

The hydrological cycle in nature is assumed to start from free water bodies (oceans, seas, lakes, and rivers) through the evaporation process and closes on to itself after stages of cloud formation, rainfall, runoff, surface water, and groundwater storages (Figure 1.3).

One of the basic elements is the precipitation (rainfall) in humid (arid and semiarid) regions. There are many techniques in geography and meteorology disciplines for the assessments and evaluations of rainfall amounts, but its practical evaluation and application are within the surface hydrology domain. The part of hydrological cycle within the lithosphere that transforms the precipitation into the groundwater is shown in Figure 1.4.

Depending on the surface and subsurface geological setup some part of the precipitation (PR) crosses the Earth's surface through the infiltration (IN) process whereby the water moves downward due to the gravitational force. Some part of the infiltration water remains near in the subsurface as soil moisture, the remaining part advances deeper into the earth through percolation (PE) process, which ends up within the groundwater reservoir, where groundwater flow (GF) takes place. Part of the precipitation occurs as runoff (RO) on the Earth's surface. Groundwater reservoir is exploited through wells by abstraction (AB).

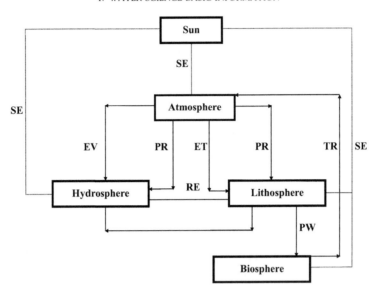

FIGURE 1.2 **Water environments.** SE, solar energy; EV, evaporation; TR, transpiration; ET, evapotranspiration; PW, plant water; PR, precipitation; RE, groundwater recharge. *(Modified from Şen (1995).)*

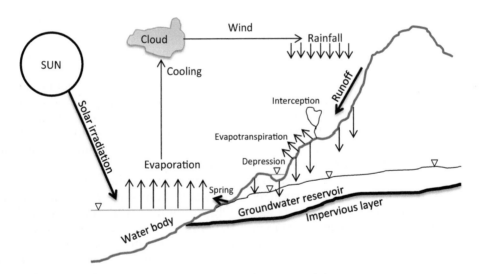

FIGURE 1.3 **Hydrological cycle components.**

1.3 RAINFALL

Rainfall is composed of water drops that reach the Earth surface from clouds after condensed atmospheric vapor. It is the source of all water resources, and especially, groundwater reservoirs are replenished after each storm rainfall. For practical applications, the quantity of water that reaches the Earth's surface from the clouds is important and it is recorded in

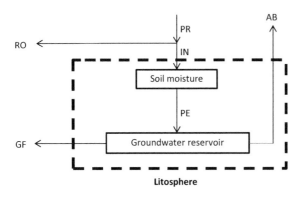

FIGURE 1.4 **Subsurface hydrologic cycle parts.** PR, precipitation; IN, infiltration; PE, percolation; GF, groundwater flow; RO, runoff; AB, abstraction.

height per duration per area. In practice, it is expressed frequently as mm/sec, which is the rainfall intensity. The change of intensity during the storm rainfall event provides significant information for many applications including groundwater recharge. The most important features can be summarized as follows.

1. Especially in arid and semiarid regions rainfall has temporally and spatially erratic behaviors.
2. Individual storm rainfall total can be very high in many cases; it may even far exceed the mean annual rainfall. Consequences of such rainfall events may lead to occasional floods and flash floods.
3. Rainfall intensities can be very high, and consequently groundwater recharge can be increased greatly either naturally or in arid and semiarid regions artificially (Chapter 4).
4. The amount of groundwater recharge decreases by the scaling effects of rainfall impact, but increases runoff transport capacity, and hence, debris flow, sedimentation, and deposition.
5. Due to the seasonal pattern of the rainfall, erosion, sediment, and groundwater recharge yields follow similar pattern, where the most valuable period for erosion, sediment, and

groundwater yield is the early part of the wet season when the rainfall is high but the vegetation has not grown sufficiently to protect the surface (Şen, 2008).
6. Weather patterns in many regions are most often under the effect of small-scale orographic and convective rainfall occurrences rather than occasional large-scale frontal rainfalls.

Spatial rainfall variability is directly related to the local and regional topography. At high elevations (escarpments, cliffs, or high mountains) orographic rainfall events occur. For instance, along the Red Sea coastal plain (Tihamah), there are sudden escarpments rising to 3000 m above mean sea level (MSL) over 50—150 km distance from the coastal area (Al-Sefry et al., 2004). Moisture laden air moves toward inlands, rises and cools in the meantime (Figure 1.5).

At any place the most significant treatment of the rainfall data in water resources, and particularly in groundwater and hydrogeology studies are given in the following points.

1. If time series of precipitation records are available at a station then one can search for meteorological wet (W) and dry (D) spell features (duration, intensity and magnitude), which play significant roles in flood,

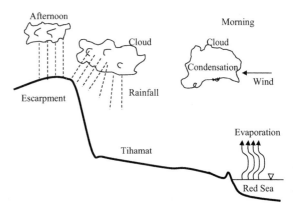

FIGURE 1.5 **Orographic rainfall mechanisms.**

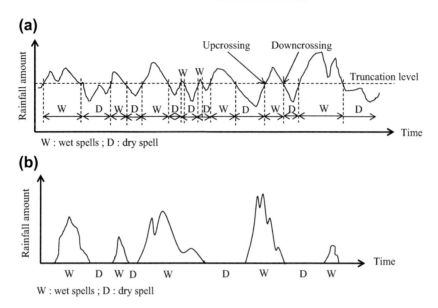

FIGURE 1.6 **Wet and dry spells.** (a) Humid region, (b) arid region.

drought, water management, and groundwater recharge studies and implementations (Figure 1.6). In humid regions, monthly and annual records may have W and D spells depending on a truncation level such as average or some fraction of it, demand, etc. (Şen, 1998a). In arid regions, there are actual W (rainy) and D (nonrainy) spells as in Figure 1.6(b) (Al-Yamani and Şen, 1997).

2. If a set of records are available at different stations within or in the vicinity of the study area then one can calculate mean areal precipitation (MAP) for the study area using classical methods as the arithmetic average, Thiessen polygon (Thiessen, 1911),

percentage weighted polygons (Şen, 1998b), isohyet maps by inverse square distance, or Kriging methodologies (Şen, 2009). During an effective study at the Saudi Geological Survey monthly isohyet maps are prepared for each month over the Wadi Fatimah catchment area along the Red Sea coastal area (Al-Sefry et al., 2004).

EXAMPLE 1.1 MEAN AREAL RAINFALL CALCULATION

During a storm rainfall the precipitation records, R, are given in Table 1.1 for a set of stations in and around a drainage basin (Figure 1.7).

TABLE 1.1 Rainfall Records

Station	A	B	C	D	E	F	G	H	I	J	K
R (mm)	8.7	17.3	18.3	17.7	21.7	23.7	24.9	34.3	29.3	33.2	35.5

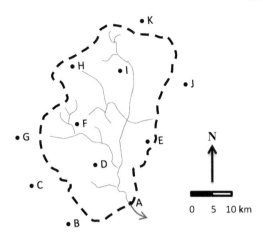

FIGURE 1.7 **Drainage basin areas.**

TABLE 1.2 Drainage Basin Stations

Station Number	Station Name	R_i (mm)
1	A	8.7
2	D	17.7
3	E	21.7
4	F	23.7
5	H	34.3
6	I	29.3

where N is the number of stations and R_i is the point rainfall amount at station number, i ($i = 1, 2, 3, 4, 5, 6$). The substitution of the station rainfall amounts from Table 1.2 into this last

$$\overline{R}_A = \frac{1}{6} \sum_{i=1}^{6} R_i = \frac{135.4}{6} = 22.6 \text{ mm}$$

1. Calculate the mean areal precipitation (MAP) over the region according to arithmetic average, Thiessen polygon, and isohyet methods (Note: Take isohyet curves at each 5 mm interval.).
2. Determine the areal variation of the rainfall height over this drainage basin and draw the rainfall height-area curve.

Solution 1.1

1. In practice, for MAP calculation most frequently used methodologies are the arithmetic mean, Thiessen polygon, and isohyet map approaches.

1.3.1 Arithmetic Mean Method

This method is based on all the stations that remain within the drainage basin and their arithmetic average represents MAP, $\overline{R}_A$. The stations inside the drainage basin are given in Table 1.2.

The general formulation of the arithmetic average methodology is as follows.

$$\overline{R}_A = \frac{1}{N} \sum_{i=1}^{N} R_i$$

1.3.2 Thiessen Method

The basis of this approach is to find weight for each station separately. The application of the Thiessen polygon methodology can be achieved after the execution of the following steps.

1. Connect the nearest stations by straight-lines, and hence triangularization is achieved over the drainage basin.
2. Draw the mid perpendicular lines to each triangular side and find separate polygon for each station.
3. Calculate the area, A_i, of each polygon according to the scale.
4. The Thiessen polygons for the given drainage basin are given in Figure 1.8.
5. The necessary calculations are given collectively in Table 1.3.

If i-th station has the polygonal area, A_i, then the general formulation of MAP for the Thiessen method can be written as follows.

$$\overline{R}_T = \frac{1}{A} \sum_{i=1}^{N} R_i A_i$$

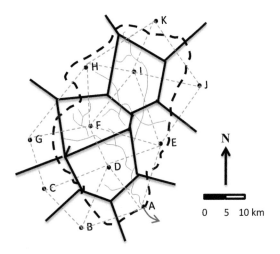

FIGURE 1.8 **Thiessen polygons.**

TABLE 1.3   Thiessen Polygon Calculations

Station	R_i (mm)	A_i (km^2)	R_iA_i
A	8.7	233.10	2027.97
B	17.8	644.91	11,479.40
C	18.3	481.74	8815.84
D	17.7	186.48	3300.70
E	21.7	85.47	1854.70
F	23.7	828.80	19,642.56
G	24.9	16,058	3998.44
H	34.3	297.85	10,216.26
I	29.3	903.91	26,484.56
J	33.2	297.85	9888.62
K	35.5	248.64	8826.72
	Total	4369.33	106,535.77

Herein, R_i is the rainfall amount at station i with its polygonal area, A_i; A is the total area, and N is the number of polygons. In this case stations inside and outside the basin contribute to the MAP calculation. The substitution of necessary values into the last expression leads to MAP height for the given drainage basin as,

$$\overline{R}_T = \frac{106,535.77}{4369.33} = 24.4 \text{ mm}$$

1.3.3 Isohyet Method

This is also another area weighted average method for MAP calculation. In this approach similar to the Thiessen method there are areal weights, but instead of polygons this time equal rainfall lines (isohyets) are drawn as in Figure 1.9 and subareas are considered as the area, A_i, that remains between two successive isohyets within the basin.

The rainfall amount over each subarea is taken as the arithmetic average of the two neighboring isohyets, R_i. The subareas and rainfall amounts over each area are given in Table 1.4.

It is possible to calculate the MAP over the drainage basin according to the same formulation as for the Thiessen method but with different meanings of each term as explained above. Hence, the substitution of the relevant values from Table 1.4 into the valid formulation yields

$$\overline{R}_I = \frac{\sum_{i=1}^{N} R_iA_i}{A} = \frac{105,264.08}{4369.33} = 24.1 \text{ mm}$$

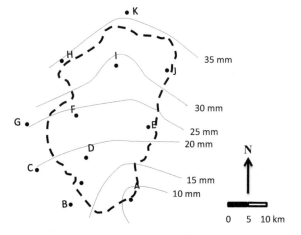

FIGURE 1.9 **Equal rainfall lines (Isohyets).**

TABLE 1.4 Isohyet Method Calculations

Isohyets	R_i (mm)	A_i (km²)	R_iA_i
35−30	32.5	1103.34	35,858.55
30−25	27.5	818.44	22,507.10
25−20	22.5	1186.22	26,689.95
20−15	17.5	924.63	16,181.03
15−10	12.5	300.44	3755.50
<10	7.5	36.26	271.95
	Total	4369.33	105,264.08

1. The areal distribution of the storm rainfall is given in Table 1.5. These values help to plot average rainfall height-area curve as in Figure 1.10. It is obvious from this figure that as the area, A, increases the MAP height, $\overline{R}_I$, decreases.
2. Rainfall intensity is among the most dominant characteristics in surface water, flood, and groundwater recharge calculations. In practice, rainfall intensity-duration-frequency (IDF) curves are necessary in almost all water resources studies. Such curves can be obtained only when a set of single storm rainfall records is available. If one imagines a single storm rainfall event from its start to the end in terms of cumulative rainfall (CR) amounts, we will logically come out with a

curve similar to the one shown in Figure 1.11, where the CR amount steadily increases until the end of the storm.

At any time during the storm rainfall the slope of the CR curve indicates the intensity. In practice, it can be calculated at a set of time subintervals as 5-min, 10-min, 20-min, 30-min, 1-hour, 3-hour, 5-hour, 8-hour, 10-hour and 12-hour. Logically, one can deduce from a set of similar figures a nonlinear inversely proportional relationship between the maximum intensity and storm duration. In practice, the important question is "what the relationship is between the frequency of storm occurrence and rainfall intensity?" The frequency of storm occurrence implies a single worst case in a certain time duration. In practical applications, different design durations, D, (2-year, 5-year, 10-year, 24-year, 50-year, 100-year, 200-year, and 500-year) are taken into consideration For instance, 10-year means the occurrence of most dangerous (worst) storm rainfall once with maximum intensity. The word "once" implies only one single occurrence during the whole design duration. This is tantamount to saying that the frequency, f, (probability of occurrence) f = 1/D. Hence, the frequency of 10-year (50-year) storm rainfall is 1/10 = 0.1 (1/50 = 0.02). Logic again deduces that during longer durations there are likely more intensive rainfall occurrences, and hence, the relationship between the frequency and the rainfall intensity

TABLE 1.5 Areal Distribution

Isohyets	R_i (mm)	A_i (km²)	R_iA_i	$A = \Sigma A_i$ (km²)	ΣR_iA_i	$\overline{R}$ (mm)
35−30	32.5	1103.34	35,858.55	1103.34	35,858.55	32.5
30−25	27.5	818.44	22,507.10	1921.78	58,365.65	30.4
25−20	22.5	1186.22	26,689.95	3108.00	85,055.60	27.4
20−15	17.5	924.63	16,181.03	4032.63	101,236.63	25.1
15−10	12.5	300.44	3755.50	4333.07	104,992.13	24.2
<10	7.5	36.26	271.95	4369.33	105,264.08	24.1

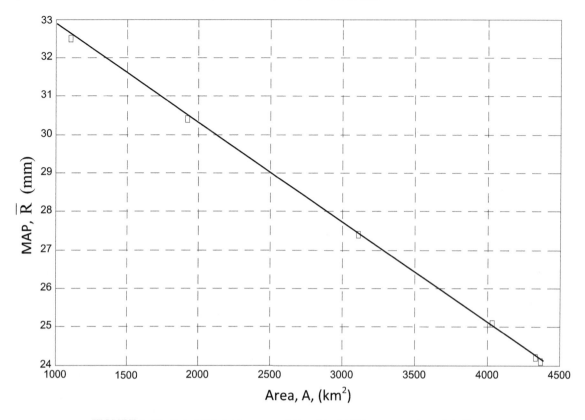

FIGURE 1.10 **Rainfall heights–area relationship.** MAP, mean areal precipitation.

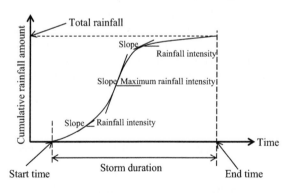

FIGURE 1.11 **Single storm cumulative rainfalls record.**

implies a nonlinear direct proportionality. In practice, the aforementioned two nonlinearity relationships yield a set of straight-lines on a double logarithmic paper. Such graphs are referred

to as the rainfall intensity-duration-frequency (IDF) curves. As an example such an IDF curve is given for Istanbul City in Figure 1.12.

The basic concern in any hydrology problem is the transformations of rainfall to runoff for surface water (runoff) and groundwater recharge. A representative example for such a transformation is shown in Figure 1.13 by considering the rainfall intensity, $I(t)$, as input and runoff discharge, $Q(t)$ and groundwater level fluctuation $H(t)$ as outputs.

In this figure there are three information sources that are necessary for hydrogeological studies. These are hyetograph, hydrograph, and well hydrograph (groundwater level fluctuations by time). Hyetograph shows the change of rainfall intensity within each storm rainfall.

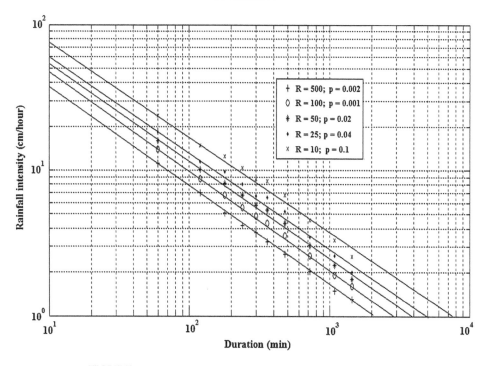

FIGURE 1.12 Istanbul City intensity-duration-frequency curves.

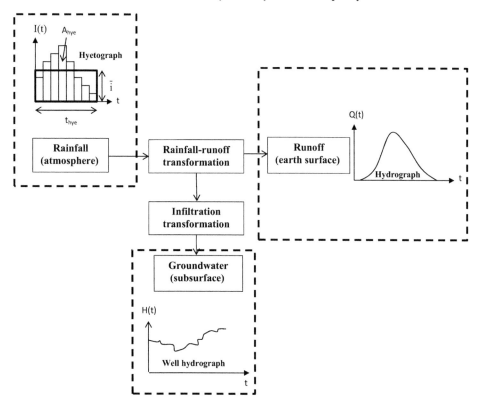

FIGURE 1.13 Rainfall transformation.

During the storm rainfall event there are many rainfall intensities in any hyetograph. In practice, the question is, "which one of these intensities to adopt for calculations?" depending on the purpose. Logically, in flood calculations the maximum intensity should be adopted, but in groundwater recharge studies an average intensity can be considered. Such an average intensity, $\bar{i}$, can be found by calculating the area, A_{hye}, below the hyetograph and then by dividing it to the hyetograph duration, t_{hye}, as,

$$\bar{i} = \frac{A_{hye}}{t_{hye}} \tag{1.1}$$

At the left hand top of Figure 1.13 the equivalent rectangle hyetograph is shown. This average intensity has two parts; the part for groundwater recharge through infiltration and the remaining part for surface runoff. In Figure 1.14, the latter part of transformation gives rise to runoff hydrograph. Rainfall-runoff transformation converts the hyetograph into hydrograph depending on the drainage basin features. Hydrograph shows the change of volume per time (discharge) by time. There are three parts in any hydrograph as the rising and recession limbs with a peak in

between. In practice, the peak discharge is taken into consideration in any design study, because it provides the most risky case among all discharges.

Logically, the total area beneath the hydrograph is equal to direct surface runoff volume at the point of hydrograph record. It is, then possible to think the spread of this volume, V_{hyd}, over the hydrograph contributing drainage area, A_{dra}, and consequently the mean runoff height, $\bar{r}$, over the drainage area can be calculated easily as,

$$\bar{r} = \frac{V_{hyd}}{A_{dra}} \tag{1.2}$$

EXAMPLE 1.2 PRECIPITATION RECORD ANALYSES

In a recording raingage total rainfall amounts are measured during a single storm as given in Table 1.6. According to these data calculate the following points (Bayazit et al., 1977).

1. Plot the total rainfall curve.
2. Plot the graph that shows the change of rainfall intensity with time (this graph is called hyetograph) during the storm.
3. Consider the rainfall amounts as the percentage of the total rainfall and the time as percentage of storm rainfall duration, and hence, plot dimensionless total rainfall curve.

Solution 1.2

1. Cumulative rainfall curve appears as R versus t relationship on the Cartesian coordinate system (Figure 1.15).
2. Hyetograph is the change of rainfall intensity $i = \Delta R/\Delta t$ with time for which the calculations are shown in Table 1.7 by making use of the data in Table 1.6 and the hyetograph is plotted in Figure 1.16.
3. In Table 1.8 the start time of rainfall 14.43 hours is taken as corresponding to zero, and R versus t change has been rearranged.

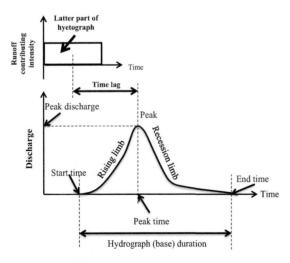

FIGURE 1.14 Direct runoff hydrograph.

TABLE 1.6 Storm Rainfall Data Record

Time, t, (hour)	Rainfall, R, (mm)	Time, t, (hour)	Rainfall, R, (mm)	Time, t, (hour)	Rainfall, R, (mm)
14.43	0.5	17.20	18.5	19.27	28.5
14.51	1.0	17.22	20.0	19.34	29.0
14.58	1.5	17.25	20.5	19.40	29.5
15.20	2.0	17.31	23.5	19.45	30.0
15.31	2.5	17.35	24.0	19.51	30.5
16.12	3.0	17.38	24.5	19.57	31.0
16.47	3.5	17.42	25.0	20.03	31.5
16.54	4.0	17.50	25.5	20.14	32.0
17.00	6	18.01	26.0	20.29	32.5
17.08	9.5	18.17	26.5	20.43	33.0
17.10	11.5	19.05	27.0	20.58	33.5
17.13	13.0	19.15	27.5	22.45	34.0
17.15	15.5	19.22	28.0		

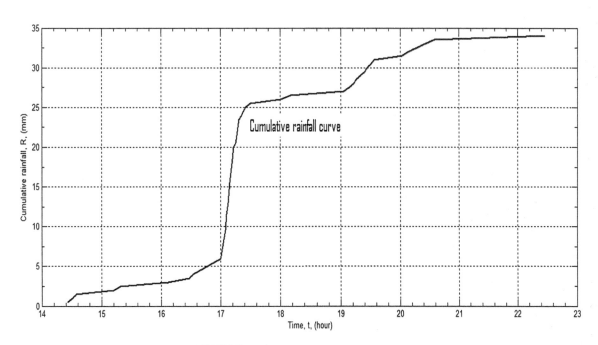

FIGURE 1.15 **Cumulative rainfall curves.**

TABLE 1.7 Rainfall Intensity

t (hour)	R (mm)	Δt (min)	ΔR (mm)	i = ΔR/Δt	t (hour)	R (mm)	Δt (min)	ΔR (mm)	i = ΔR/Δt
14.43	0.5				17.42	25.0	4	0.5	7.5
14.51	1.0	8	0.5	3.75	17.50	25.5	8	0.5	3.75
14.58	1.5	7	0.5	4.29	18.01	26.0	11	0.5	2.73
15.20	2.0	22	0.5	1.36	18.17	26.5	16	0.5	1.88
15.31	2.5	11	0.5	2.73	19.05	27.0	48	0.5	0.63
16.12	3.0	41	0.5	0.73	19.15	27.5	11	0.5	2.73
16.47	3.5	35	0.5	0.86	19.22	28.0	7	0.5	4.29
16.54	4.0	7	0.5	4.29	19.27	28.5	6	0.5	6.0
17.00	6	6	2.0	20.0	19.34	29.0	7	0.5	4.29
17.08	9.5	8	3.5	26.25	19.40	29.5	6	0.5	5.0
17.10	11.5	2	2.0	60.0	19.45	30.0	5	0.5	6.0
17.13	13.0	3	1.5	30.0	19.51	30.5	6	0.5	5.0
17.15	15.5	2	2.5	75.0	19.57	31.0	6	0.5	5.0
17.20	18.5	5	3.0	36.0	20.03	31.5	6	0.5	5.0
17.22	20.0	2	1.5	45.0	20.14	32.0	11	0.5	2.73
17.25	20.5	3	0.5	10.0	20.29	32.5	15	0.5	2.0
17.31	23.5	6	3.0	30.0	20.43	33.0	14	0.5	2.14
17.35	24.0	4	0.5	7.5	20.58	33.5	15	0.5	2.0
17.38	24.5	3	0.5	10.0	22.45	34.0	107	0.5	0.28

In the same table the rainfall height, R, is divided by the final cumulative rainfall amount (R/34) and the storm rainfall duration by 8.02 (t/8.02) for obtaining rainfall and time percentages. Finally, the plot of percentage time versus percentage cumulative rainfall yields the dimensionless cumulative rainfall curve (see Figure 1.17).

EXAMPLE 1.3 EFFECTIVE RAINFALL CALCULATION

Cumulative rainfall amounts are given in Table 1.9 for a storm rainfall event of 2 hours and 30 minutes duration. During this storm rainfall, the direct runoff amount was 350 m^3/min from an area, 200,000 m^2.

1. Interpret the given rainfall data.
2. Find the effective rainfall amount.

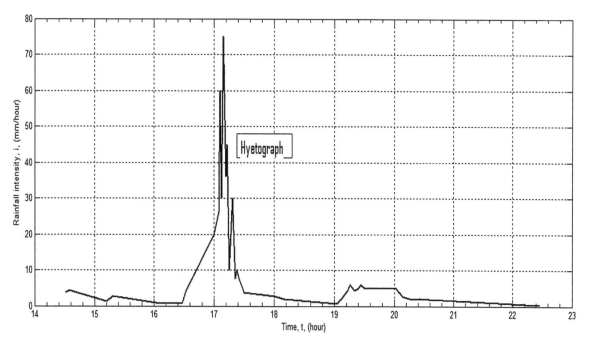

FIGURE 1.16 **Hyetograph.**

Solution 1.3

1. Interpretation through the numbers is difficult, therefore, it helps to draw the change of given data by time as in Figure 1.18

The slope of this graph at any point indicates the rainfall intensity. Herein intensity means the "speed" of rainfall as it starts rather slowly and then decreases smoothly. However, in arid and semiarid regions, rainfall occurrences start, in general, rapidly then die off slowly. Figure 1.18 has high slope at the beginning, which reduces as the time increases and at the end it becomes more or less horizontal, which means the end of storm rainfall.

Figure 1.19 indicates the change of rainfall amount at each 10-min interval. Within the first 10 minutes the rainfall intensity is the greatest and then falls off gradually, which is the characteristic rainfall pattern in arid regions.

2. The effective rainfall is the amount after the extraction of infiltration losses from the recorded precipitation amount. It is equal to the direct runoff height. For this purpose, direct runoff volume divided by the drainage area gives the direct runoff height as $350/200,000 = 0.00,175$ m/min $= 1.75$ mm/min. The area under the rainfall intensity curve starting from the peak value is scanned downward by a horizontal line. When the area scanned is equal to 1.75 mm/min then the horizontal line separates the rainfall intensity (hyetogram) into two parts. The one below the horizontal line is for infiltration and groundwater recharge.

1.4 EVAPORATION AND EVAPOTRANSPIRATION

These are losses to atmosphere from the Earth reached rainfall amounts due to mainly solar

TABLE 1.8   Percentages of Rainfall and Time

t (hour)	R (mm)	t/8.02	R/34	t (hour)	R (mm)	t/8.02	R/34
0	0.5	0	0.014706	2.59	25	0.322943	0.735294
0.08	1	0.009975	0.029412	3.07	25.5	0.382793	0.75
0.15	1.5	0.018703	0.044118	3.18	26	0.396509	0.764706
0.37	2	0.046135	0.058824	3.34	26.5	0.416459	0.779412
0.48	2.5	0.05985	0.073529	4.22	27	0.526185	0.794118
1.29	3	0.160848	0.088235	4.33	27.5	0.5399	0.808824
2.04	3.5	0.254364	0.102941	4.4	28	0.548628	0.823529
2.11	4	0.263092	0.117647	4.45	28.5	0.554863	0.838235
2.17	6	0.270574	0.176471	4.52	29	0.563591	0.852941
2.19	9.5	0.273067	0.279412	4.58	29.5	0.571072	0.867647
2.22	11.5	0.276808	0.338235	5.03	30	0.627182	0.882353
2.24	13	0.279302	0.382353	5.09	30.5	0.634663	0.897059
2.29	15.5	0.285536	0.455882	5.15	31	0.642145	0.911765
2.31	18.5	0.28803	0.544118	5.21	31.5	0.649626	0.926471
2.34	20	0.291771	0.588235	5.32	32	0.663342	0.941176
2.4	20.5	0.299252	0.602941	5.47	32.5	0.682045	0.955882
2.44	23.5	0.304239	0.691176	6.01	33	0.749377	0.970588
2.47	24	0.30798	0.705882	6.16	33.5	0.76808	0.985294
2.51	24.5	0.312968	0.720588	8.02	34	1	1

irradiation, temperature and plant transpiration effects. They are complete return water to the atmosphere without any benefit for surface or groundwater resources. Soil moisture evaporation and crop transpiration combination losses are constituents of evapotranspiration, ET. In surface runoff discharge and groundwater recharge calculations, apart from weather pattern (precipitation distribution), crop characteristics, soil type, and vegetation cover are significant input data for groundwater resources evaluation and management. ET rate from a reference surface is called the reference ET denoted as ET_O (Allen et al., 1998). It is a

necessary component of the hydrological cycle to study the evaporative demand of the atmosphere independently of crop type, crop development, and management practices. As water is abundantly available at the ET_O surface, soil factors do not affect it. ET_O expresses the evaporation power of the atmosphere at a specific location and time of the year and does not consider the crop characteristics and soil factors.

Measured or calculated ET_O values at different locations and seasons are comparable only as they refer to the ET from the same reference surface. The most significant factors affecting ET_O are climatic parameters. It is very

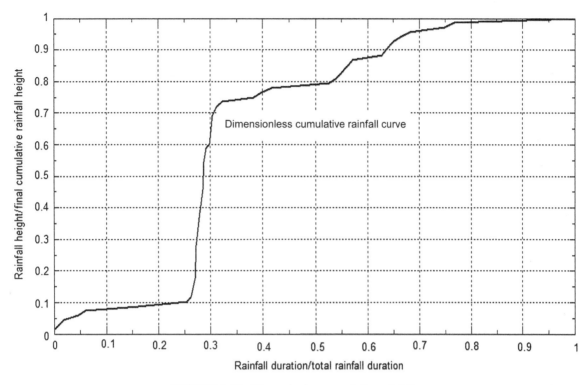

FIGURE 1.17 **Dimensionless cumulative rainfalls.**

TABLE 1.9   Rainfall Data

Time (min)	10	20	30	40	50	60	70	80	90	100	110	120	130	140	150
Accumulative rainfall (mm)	1.2	2.5	3.6	4.4	5.1	5.6	5.9	6.4	6.7	6.9	7.1	7.15	7.2	7.23	7.25

difficult to obtain accurate ET_O field measurements, and therefore, it is commonly computed from weather data, if available. A large number of empirical or semiempirical equations have been developed for assessment of ET_O from meteorological data.

1.4.1 Penman—Monteith Equation

Penman (1948) combined the energy balance with the mass transfer method and derived an equation to compute the evaporation from an open water surface based on standard climatological records of sunshine, temperature, humidity, and wind speed. Combination of crop ET methods are further developed by many researchers and extended to cropped surfaces by introducing resistance factors, which distinguish between aerodynamic and surface resistance factors. They are often combined into one parameter as the "bulk" surface resistance, which operates in series with the aerodynamic resistance. The surface resistance, r_s, describes the resistance of vapor flow through stomata openings, total leaf area, and soil surface. The aerodynamic resistance, r_a, describes the resistance from

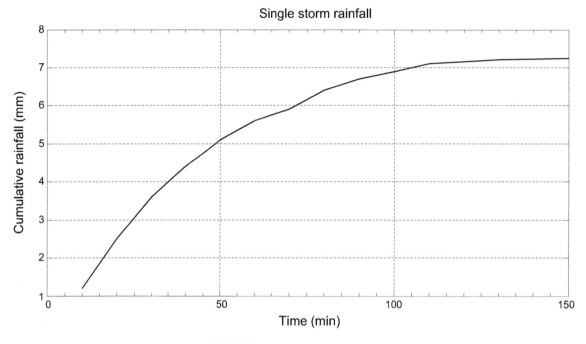

FIGURE 1.18 **Rainfall-time data.**

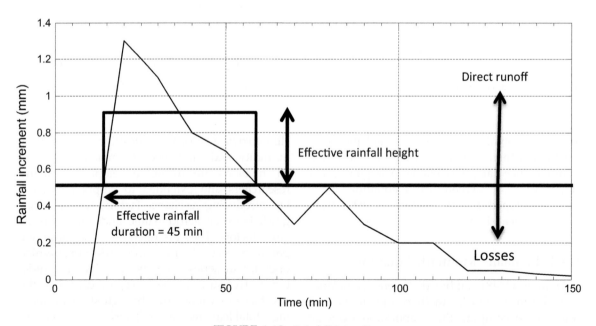

FIGURE 1.19 **Rainfall intensity.**

the vegetation upward and involves friction from air flowing over vegetative surfaces. Although the exchange process in a vegetation layer is too complex to be fully described by the two resistance factors, good correlations can be obtained between measured and calculated ET rates, especially for a uniform grass reference. The Penman–Monteith form of the combination equation is expressed as (Allen et al., 1998).

$$\text{ET}_O = \frac{0.408\,\Delta(R_n - G) + \gamma\dfrac{900}{T + 273}u_2(e_s - e_a)}{\Delta + \gamma(1 + 0.34\,u_2)}$$

(1.3)

where ET_O is the reference evapotranspiration (mm/day), R_n is the net radiation at the crop surface (MJ/m²day), G is the soil heat flux density (MJ/m²day), T is the mean daily air temperature at 2 m height (°C), u_2 is the wind speed at 2 m height (m/s), e_s is the saturation vapor pressure (kPa), e_a is the actual vapor pressure (kPa), $(e_s - e_a)$ is the saturation vapor pressure deficit (kPa), Δ is the slope of vapor pressure curve (kPa/°C), and finally, γ is the psychometrics constant (kPa/°C). In Eq. (1.3) the value of 0.408 converts the net radiation R_n expressed in MJ/m²day to equivalent evaporation expressed in mm/day. Due to the small soil heat flux compared to R_n, particularly when the surface is covered by vegetation and calculation time steps as 24 hours or longer, the estimation of G is ignored in the ET_O calculation and assumed to be zero. Allen et al. (1998) state that the soil heat fluxes beneath the grass reference surface is relatively small for that time period. Detailed calculation steps and procedures of the ET_O are given already in the FAO Irrigation and Drainage Paper No 56 (Allen et al., 1998).

1.4.2 Practical Evapotranspiration Calculations

ET is the sum of evaporation from soil moisture, surface water, and groundwater table plus transpiration from the canopy. Evaporation from the groundwater table could be ignored throughout the tropic regions because much higher evaporation appears from rainfall and flood water and in arid and semiarid regions when the water table is far from the surface. It can also be neglected throughout the dry months, when more than 80% of the land cover is dominated by crops such as rice paddy with shallow (<2 m) rooting depth (Shamsudduha et al., 2011). Three methods for the potential evapotranspiration ET_O are, modified Priestley–Taylor, Blaney–Criddle, and adjusted-pan methods.

1.4.2.1 Priestley–Taylor Method

ET_O during wet season is estimated according to the model proposed by Priestley and Taylor (1972) (evapotranspiration from saturated surface) and proved later in a study done by De Bruin (1983) to be applicable for the wet season in tropical regions. Their equation is,

$$\text{ET}_O = \alpha\frac{\Delta}{\Delta + \gamma}(R_n - G)$$

(1.4)

where $\alpha = 1.26$ (overall mean value), G is the soil heat flux density to the ground MJ/m²day, which according to Brutsaert (1982) is negligible in tropical areas (temperature stays stable during wet season.), Δ is the slope of the saturation water vapor–temperature curve at air temperature kPa/°C, γ is the psychrometric constant kPa/°C, and finally, R_n is the net radiation. De Bruin (1983) simplified Priestley–Taylor equation to be used in case when there is a lack of data in tropical regions.

$$\text{ET}_O = (0.36\,R_a - 41)\frac{n}{N} + 0.18\,R_a - 5$$

(1.5)

where n/N is the relative duration of bright sunshine and R_a is the extraterrestrial incoming shortwave radiation. This equation gives the potential evapotranspiration in W/m² during wet season and needs the relative sunshine duration, almost as the only input parameter. Among the assumptions of this expression, relative humidity and temperature are more or less constants ($T = 27$, $RH = 80\%$). R_a values can range from $200-1000$ W/m^2 depending on the time and location, which can be taken from tabulated values (Allen et al., 1998). In order to transform the evapotranspiration unit into mm/day the result in W/m^2 must be multiplied by 28.5.

1.4.2.2 Blaney–Criddle Method

Blaney and Criddle (1962) linked the percentage of hours of daylight and mean monthly temperature with the amount of water consumed by crops and presented the following reference evapotranspiration expression.

$$ET_O = p(0.48 \, \overline{T} + 8) \tag{1.6}$$

where p is the mean daily percentage of annual daytime hours and $\overline{T} = (T_{min} + T_{max})/2$ is the mean daily temperature in °C. T_{mim} (T_{max}) is calculated as the sum of all the minimum (or maximum) temperature values in a month divided by the number of days in the same month. The result of this expression is in mm/day for each month. However, this method is not very accurate and provides a rough estimate or "order of magnitude" only especially under extreme climatic conditions. In windy, dry, sunny areas, it gives underestimations up to almost 60% and in calm, humid, clouded areas overestimations are possible up to 40%. P values are taken from a convenient table based on the latitude and the month.

EXAMPLE 1.4
BLANEY–CRIDDLE
CALCULATION

A drainage basin in the northern hemisphere has latitude of 40° N. The mean maximum and mean minimum temperatures in May are 27.5 °C and 17.3 °C, respectively. Determine for the month of April the mean ET_O in mm/day using the Blaney–Criddle method. (Mean daily percentage of annual daytime hours is $p = 0.30$.)

Solution 1.4

The mean temperature is equal to the average of the maximum and minimum temperatures, $T_{mean} = (27.5 + 17.3)/2 = 22.4$ °C. Hence, the substitution of relevant values into Eq. (1.6) gives,

$$ET_O = 0.30(0.46 \times 22.4 + 8) = 5.49 \text{ mm/day}$$

1.4.2.3 Adjusted-Pan Method

Pan evaporation is estimated not to be equal to evaporation from natural water-free surfaces (Eagleman, 1967). Even if the pan and the free surface water are under the same conditions, pan evaporation is still greater than evaporation from the water body. Therefore, different conversion coefficients for pan evaporation and natural water surface evaporation can be found in literature. In order to measure practically the evaporation from free water surface, evaporation pans are used and there are different types of these instruments. Today, the most frequently used type is class A evaporation pan. In order to transit from pan evaporation readings to actual evaporation amounts, it is necessary to multiply recorded pan evaporation amounts by pan coefficient that is less than one. Logically, if measured evaporation amount from a standard pan instrument is, E_P, then the reference crop evapotranspiration can be calculated as,

$$ET_O = CE_P \tag{1.7}$$

where C is the pan coefficient, which is always less than 1. It depends on the type of pan used, the pan environment (if the pan is placed in a fallow or cropped area), and the climate especially the humidity and wind speed. For the Class A evaporation pan, C varies between 0.35 and 0.85 and on the average one can adopt it as equal to 0.70, but it changes seasonally.

The situation during dry season is more complicated (due to active groundwater abstraction for irrigation, decreased rainfall rate, soil moisture deficit). Actual evapotranspiration,

ET_A, during dry months falls far down the potential evapotranspiration. The direct relationship between ET_O and ET_A under nonpotential conditions was revised, and an inverse relationship was stated (Bouchet, 1963). In their studies, Pike (1964) and Mirza et al. (2003) applied the following modified version of Turc (1961) to calculate the ET_A by considering the precipitation value, P, as,

$$ET_A = \frac{P}{\sqrt{1 + \frac{P}{ET_O}}} \qquad (1.8)$$

De Bruin (1983) investigated previous studies regarding evapotranspiration calculation during dry months and proposed the usage of the minimum value between ET_O and $(0.9\,P + ASM)$, where ASM is the available soil moisture. However, the ET_A values are rough estimations only which can give just an impression of the actual evapotranspiration magnitude.

EXAMPLE 1.5 EVAPORATION MEASUREMENTS BY EVAPORATION PAN

Near a reservoir there is A class evaporation pan and the monthly evaporation amounts in summer months are recorded as in Table 1.10. Calculate the evaporation losses as volumes.

Solution 1.5

The pan evaporation amounts during summer months are given in Table 1.10 with corresponding pan coefficients. Hence, one can calculate for each month the ET_O height from Eq. (1.7) and the volumes, V, are obtained by multiplying these values by the reservoir surface area, A, in each month, $V = A \times ET_O$. The results are given in the last column of Table 1.10.

1.4.2.4 Meyer Formulation

Another way of evaporation calculation for free water surface is by means of empirical approaches. Meyer (1942) formulation is one of such approaches and it takes into consideration the wind speed. Meyer formulation is given as follows.

$$E = 11(e_w - e_a)\left(1 + \frac{w}{16}\right) \qquad (1.9)$$

Herein, E is the evaporation amount in mm; e_w and e_a are water and air vapor pressure in mm Hg; w is the wind speed at 8 m height over the free water surface in terms of km/hour. The value of e_w changes with air temperature. On the other hand, there is a relationship between e_w and e_a as $e_a = R_h \times e_w$, where R_h represents relative humidity amount.

TABLE 1.10 Evaporation Data

Month	E_P (mm)	Pan Coefficient, C	Reservoir Surface Area, A, (km^2)	V (m^3)
May	130	0.63	2.63	$0.63 \times 0.130 \times 2.63 \times 10^6 = 215{,}397$
June	150	0.66	2.53	$0.66 \times 0.150 \times 2.53 \times 10^6 = 250{,}470$
July	160	0.68	2.42	$0.68 \times 0.150 \times 2.42 \times 10^6 = 263{,}296$
August	145	0.70	2.35	$0.70 \times 0.145 \times 2.35 \times 10^6 = 238{,}525$
September	100	0.71	2.30	$0.63 \times 0.100 \times 2.30 \times 10^6 = 163{,}300$
Total				1,130,988

The substitution of this expression into the previous one leads to,

$$E = 11e_w(1 - R_h)\left(1 + \frac{w}{16}\right) \qquad (1.10)$$

EXAMPLE 1.6 MONTHLY EVAPORATION HEIGHT CALCULATIONS BY MEYER FORMULATION

Near a water reservoir during summer months' air temperature, T, relative humidity, R_h, saturated water pressure, e_w, and wind speed, w, are given in Table 1.11. Calculate the evaporation height over this reservoir.

Solution 1.6

In Table 1.11 average temperature for each month has been given with corresponding e_w and R_h values in addition to the wind velocity, w. The substitution of all these variables into Eq. (1.10) for each month leads to the evaporation values in the last column of Table 1.11.

TABLE 1.11 Necessary Data

Month	T (°C)	R_h (%)	e_w (mm/Hg)	W (km/hour)	E (mm)
May	12	74	10.6	11	51.2
June	16	73	13.6	8	60.6
July	17	77	14.2	8	54.0
August	18	78	15.0	9	56.7
September	13	79	11.1	8	38.5
October	9	85	8.5	11	23.7

1.4.2.5 Heat Balance

Evaporation is temperature dependent and the energy balance of water mass can also be used for its practical calculation through the following heat balance expression.

$$H_e = H_i - H_o - H_c - \Delta H \qquad (1.11)$$

where H_e is the energy used for evaporation, H_i is the input heat, H_o is the output heat, H_c is the heat loss from the lake surface due to conduction, and finally, ΔH is the necessary heat amount for evaporation from the lake surface. The relationship between H_c and H_e is given by the following expression, where R is the Bowen ratio.

$$H_c = RH_e$$

Here, R is the function of atmospheric pressure, air and water temperature, and vapor pressure difference. On the other hand, the latent heat of water, L, is related to the volume of evaporation water, E, as follows.

$$H_e = LE$$

Substitution of these last two expressions into the first energy balance equation given earlier leads to evaporation height formulation.

$$E = \frac{H_i - H_0 - \Delta H}{L(1 + R)} \qquad (1.12)$$

EXAMPLE 1.7 EVAPORATION CALCULATION FROM A LAKE BY ENERGY BALANCE EQUATION

The amount of heat energy that enters a lake of $20 \ km^2$ surface area is $600 \ cal/m^2$. 15% of this heat is reflected by the lake surface. The heat input and output amounts during one day are ignored and additionally it is assumed that during the same day the temperature of the lake has not changed. Calculate daily evaporation height in the lake. (Note: Latent heat for evaporation of water is $L = 590 \ cal/cm^2$ and the Bowen ratio is $R = 0.25$.)

Solution 1.7

The heat balance formulation given by Eq. (1.12) can be used with the given data. First, the necessary steps need to be completed are as follows.

$H_i = 600$ cal/cm^2. By taking into consideration the heat reflection percentage the output amount becomes, $H_o = 0.15 \times 600 = 90$ cal/cm^2 (because only 15% of the incipient heat is reflected). Since there is heat balance, $\Delta H = 0$. $L = 590$ cal/cm^2 and $R = 0.25$. The substitution of the relevant data into Eq. (1.12) gives,

$$E = \frac{600 - 90 - 0}{590(1 + 0.25)} = 0.69 \text{ cm}$$

Accordingly, daily water volume due to the evaporation can be found by multiplication with the lake surface area as,

$$0.0069 \times 20 \times 10^6 = 138000 \text{ m}^3$$

1.5 INFILTRATION

This is the process of water entering the soil from the Earth's surface. It is not a loss as the evapotranspiration, but feeds plant water use (root zone) and subsurface and groundwater reservoirs. Quantitatively, it is the amount of surface water that seeps into the soil from a given area in a specified time interval. Logically, infiltration rate decreases by time due to the limited saturation possibility of the soil. Initially, there is a certain amount of filtration capacity, f_i, which becomes saturated gradually and after a certain time passage (effective infiltration time) the soil is no more capable to accept infiltrating water; hence, it reaches finally the stage of saturation. The final filtration rate, f_f, becomes asymptotic to a horizontal value, which is not necessarily equal to zero. These last sentences give mental visualization of the infiltration, f (t), and change by time as a decreasing curve shown in Figure 1.20.

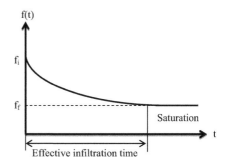

FIGURE 1.20 **Infiltration process.**

In mathematics, if a curve has intersection with one of the axis and asymptotic to any straight-line, its mathematical expression is in the form of an exponential function (Şen, 2013). This information enables one to write down the following expression,

$$f(t) = f_f + (f_i - f_f)e^{-t}$$

Such an expression provides only a single curve for given initial and final filtration rates. In order to give flexibility, it is necessary to introduce an exponential constant, k, as,

$$f(t) = f_f + (f_i - f_f)e^{-kt} \qquad (1.13)$$

where k indicates the infiltration property of the soil and it is referred to as the infiltration coefficient. This logically derived equation has been proposed initially by Horton (1940) and it is well known in the literature.

In practice, hyetograph and infiltration curve are overlapped on each other so as to find the effective precipitation amount that will cause surface flow (runoff) discharge and infiltration contribution to subsurface flow leading to the groundwater recharge possibility as in Figure 1.21.

In this figure, the hyetograph part above (or below) the infiltration curve represents effective rainfall intensity variation (effective infiltration intensity variation). The duration between the two intercepts between the hyetograph and the infiltration curve is the duration of effective rainfall.

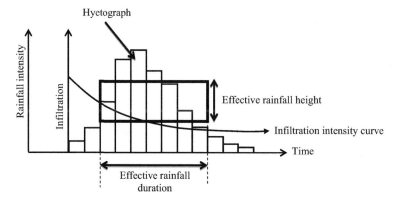

FIGURE 1.21 **Effective rainfall heights.**

In practice, since the effective rainfall intensity part of the hyetograph has irregular shape, it is converted into equivalent area rectangle with the same effective rainfall duration. The height of this rectangle is the effective rainfall height. In this manner, uniform rainfall intensity is obtained for practical uses. The average intensity of effective rainfall is the ratio of this height to effective rainfall duration. The reader should notice that, the effective rainfall duration also represents the soil saturation duration.

Another way of finding the effective rainfall height, duration, and intensity is possible; provided that hyetograph is available but not the infiltration curve. This approach is referred to as the Phi-index (Φ-index). This index is defined as the uniform infiltration height effective over the study area. Φ-index can be obtained in practice provided that runoff measurement is available at the outlet or any point within the study area. As explained in Section 1.3 after calculating the mean runoff height by Eq. (1.2), it is time to find the area in the hyetograph starting from the peak down to a horizontal line such that the area above this line is equal to the mean runoff height (Figure 1.22). Such a horizontal line can be found after little trial and error procedure by accepting practically $\pm5\%$ error limits.

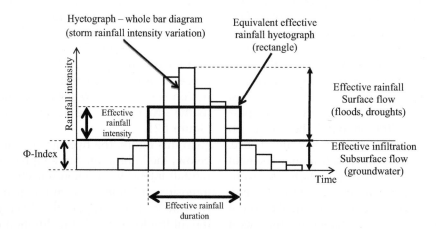

FIGURE 1.22 **Φ-index calculation.**

In this manner, the hyetograph is divided into two parts. The distance between this and the zero horizontal lines gives the Φ-index value, which is employed in the infiltration calculations for groundwater recharge studies. Such a line is shown in Figure 1.22.

EXAMPLE 1.8 STANDARD INFILTRATION CURVE DETERMINATION BY RING INFILTROMETERS

The internal diameter of a ring infiltrometer is 35 cm and the measurements with this instrument are given in Table 1.12.

1. Draw the change of infiltration capacity by time, and hence, obtain the standard infiltration curve for the soil.

2. For this soil calculate the Horton equation coefficients, f_i, f_f, and k.
3. Calculate the average infiltration amounts at 10-min and 30-min instances by using the Horton equation.

Solution 1.8

1. The change of infiltration capacity by time is calculated and shown in Table 1.13. The cross-sectional area of infiltrometer annulus is $A = 962$ cm^2. Cumulative addition of water volumes given in the third column divided by this area yield the water heights as given in the fourth column. The successive differences between these heights give infiltration height, ΔI, (in column 5). In the sixth column the division of these height differences to the corresponding time intervals yields infiltration

TABLE 1.12   Infiltration Test Results

Time (min)	0	2	5	10	20	30	60	90	150
Added water amount (cm³)	0	278	658	1173	1924	2500	3345	3875	4595

TABLE 1.13   Infiltration Capacity Changes

Time (min)	Δt (hour)	Added Water Amount (cm³)	Added Water Height (cm)	ΔI (cm)	Standard Infiltration Capacity f = ΔF/Δt (cm/hour)
1	2	3	4	5	6
0		0			
2	0.033	278	0.289	0.289	8.76
5	0.050	658	0.684	0.395	7.90
10	0.083	1173	1.219	0.535	6.45
20	0.167	1824	2.000	0.781	4.68
30	0.167	2500	2.599	0.599	3.59
60	0.500	3345	3.477	0.878	1.76
90	0.500	3875	4.028	0.551	1.10
150	1.000	4595	4.777	0.749	0.75

capacity. The standard infiltration curve is given in Figure 1.23

f, Infiltration capacity at instant t after the rainfall start.

f_i, Initial infiltration capacity.

f_f, Final infiltration capacity.

k, Infiltration coefficient.

Herein, f_i, f_f, and k are dependent on soil type and vegetation cover. As one can see from Table 1.13 and Figure 1.23, $f_f = 0.75$ cm/hour and $f_i = 8.76$ cm/hour.

2. For the k coefficient first let us consider Horton expression given in (Eq. (1.13)). By taking logarithm of both sides one can obtain,

$$Ln\left[f(t) - f_f\right] = Ln\left(f_i - f_f\right) - kt$$

According to this expression, on the axes system, $Ln[f(t) - f_c]$ versus t, the results appear around a straight-line on semilogarithmic paper. The coordinates of this straight-line are given in Table 1.14.

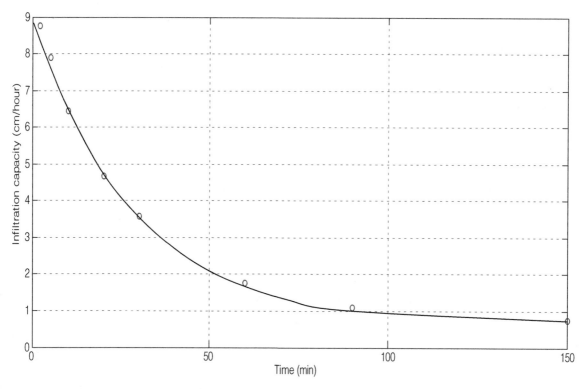

FIGURE 1.23 **Standard infiltration curves.**

TABLE 1.14 Straight-line Calculations from Semi-Logarithmic Paper

T (hour)	0.033	0.083	0.167	0.33	0.50	1.00	1.50
F (cm/hour)	8.76	7.90	0.45	4.68	3.59	1.76	1.10
$f - f_f$	8.01	7.15	5.70	3.93	2.64	1.01	0.35
$Ln\,[f(t) - f_f]$	2.08	1.97	1.74	1.37	1.04	0.01	−1.05

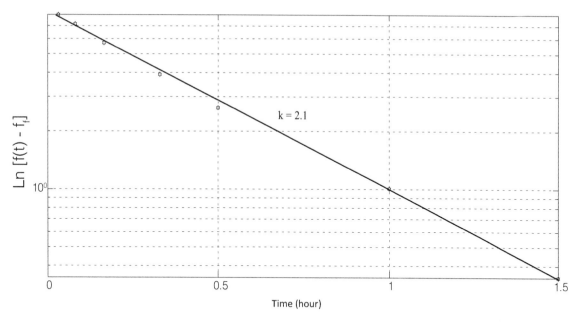

FIGURE 1.24 **Determination of k coefficient from Horton's equation.**

The plots of the first row values against the third row values' logarithms appear as a straight-line in Figure 1.24. The slope of this straight-line can be calculated as $k = 2.1$. It is necessary to notice that the time values must be substituted into the Horton equation in hours.

3. For this soil, the Horton equation becomes,

$$f = 0.75 + (8.76 - 0.75)e^{-2.1t}$$
$$= 0.75 + 8.01 \, e^{-2.1t}$$

On the other hand, the substitution of $t = 10/60$ hours and $t = 30/60$ hours into the last expression yields,

$f_{10} = 6.39$ cm/hour

$f_{30} = 3.55$ cm/hour. These values are very close to the ones in Table 1.14 column six.

EXAMPLE 1.9 INFILTRATION HEIGHT CALCULATION METHODS

Table 1.15 includes the change of infiltration capacity during a storm rainfall.

1. Draw the standard infiltration curve and calculate the coefficient k according to the Horton equation for this basin.
2. In this basin the rainfall intensities during a storm rainfall are given in Table 1.16. Hence, calculate the infiltration speed change with time.
3. Draw the curves that show the change of rainfall and runoff with time.
4. Calculate total runoff height.
5. Calculate the Φ-index.

TABLE 1.15 Infiltration Capacity Values

Time (hour)	0	2	4	6	8	10	12
Infiltration capacity (mm/hour)	5.5	3.5	1.8	1.51	1.35	1.2	1.2

TABLE 1.16   Rainfall Intensity

Time (hour)	0	1	2	3	4	5	6
Rainfall intensity (mm/hour)	2.0	2.5	4.0	6.0	5.05	2.5	1.5

Solution 1.9

1. The standard infiltration curve is plotted in Figure 1.25 from the given data. In Horton Eq. (1.13) k is the slope of the straight-line, Ln $[f(t) - f_f]$ versus t. From the given data one can find $f_f = 1.2$ mm/hour. This straight-line slope yields k coefficient. The necessary calculations are given in Table 1.17 with resulting Figure 1.25 (Note: In Figure 1.25 all the ordinates are added by 1.9 so as to make the values positive. Such an addition does not affect k calculation). Accordingly, the Horton infiltration curve for the given basin appears as follows.

$$f = 1.20 + 4.3\, e^{-0.42t}$$

2. During this rainfall the calculation procedure of the infiltration speed should be according to the following steps.
   a. After the rainfall start at time instant, t, if the rainfall intensity, i, is greater than infiltration capacity, f, (i > f) then infiltration speed is equal to infiltration capacity.
   b. If the rainfall intensity is less than the infiltration capacity (i < f) then infiltration speed is less than infiltration capacity. With this principle the change of infiltration speed change by time is given in Table 1.18, where the infiltration capacities for t = 1 hour, t = 3 hours and t = 5 hours are calculated according to Horton equation.

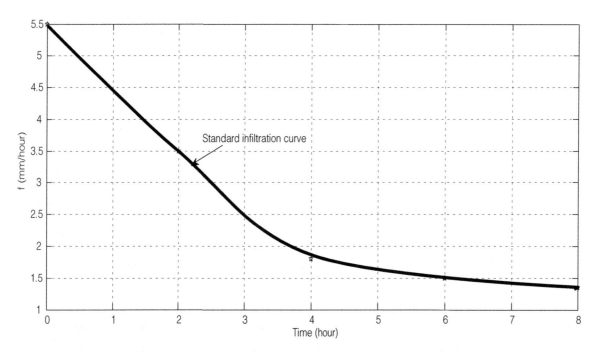

FIGURE 1.25   Standard infiltration curve based on the given data.

TABLE 1.17   Standard Infiltration Curve Calculations

t (hour)	0	2	4	6	8
f (mm/hour)	5.5	3.5	1.8	1.5	1.35
$f - f_c$ (mm/hour)	4.3	2.3	0.6	0.3	0.15
Ln $(f - f_c)$	1.46	0.83	−0.51	−1.2	−1.9

TABLE 1.18   Infiltration Speed

Time (min)	0	1	2	3	4	5	6
Rainfall intensity (mm/hour)	2.0	2.5	4.0	6.0	5.0	2.5	1.5
Infiltration capacity (mm/hour)	5.5	4.3	3.5	2.6	0.2	1.7	1.5
Infiltration speed (mm/hour)	2.0	2.5	3.5	2.6	2.0	1.7	1.5

Note: After the rainfall starts, during the first 2 hours, rainfall intensity is smaller than the infiltration capacity and the soil is less saturated. For this reason, starting from t = 2 hours, real infiltration speeds will be a little bit higher than the ones shown in this table.

TABLE 1.19   Runoff Change

Time (hour)	0	1	2	3	4	5	6
Rainfall intensity (mm/hour)	2.0	2.5	4.0	6.0	5.0	2.5	1.5
Infiltration speed (mm/hour)	2.0	2.5	3.5	2.6	2.0	1.7	1.5
Runoff intensity (mm/hour)	0	0	0.5	3.4	3.0	0	0

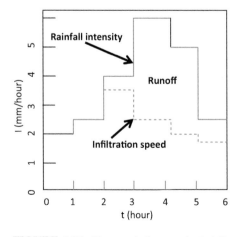

FIGURE 1.26   **Temporal change of rainfall.**

3. After the rainfall starts at time instant, t, the runoff intensity can be obtained by subtracting the infiltration speed from the rainfall intensity. According to this information, the change of runoff by time is given in Table 1.19.
   The rainfall and runoff changes with time are given in Figures 1.26 and 1.27.

4. The total area under the curve given in Figure 1.27 (or the area between the rainfall and infiltration curves in Figure 1.26) yields total runoff height, r, which is 7.7 mm.

5. In order to find Φ-index value in Figure 1.28, a horizontal line parallel to the time axis is determined in such a manner

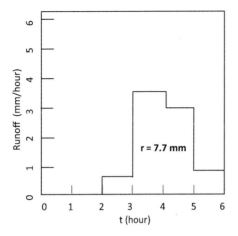

FIGURE 1.27 **Temporal variation of runoff.**

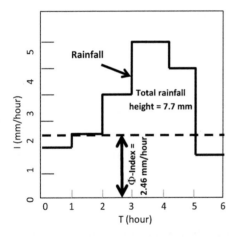

FIGURE 1.28 **Φ-Index determination.**

that the area between this line and the rainfall curve is equal to the total runoff height. The ordinate of this horizontal straight-line on the vertical axis gives

Φ-index. Such a line is determined after a succession of trial and error method. The final Φ-index value appears as 2.46 mm as in Figure 1.28.

EXAMPLE 1.10 INFILTRATION INDEX CALCULATIONS

In a basin of 400 km^2 area the height of direct runoff is 45 mm. The rainfall intensity measurements during the storm rainfall are given in Table 1.20. Calculate the Φ-index of the basin.

Solution 1.10

According to the data, the total rainfall height after the storm rainfall is $R = 150$ mm. Direct runoff height is $r = 45$ mm, and hence, the rainfall duration leads to infiltration height, F, as,

$$F = P - R = 150 - 45 = 105 \text{ mm}$$

During the rainfall ($t = 6$ hours) let us assume that the rainfall intensity is greater than infiltration capacity, and hence,

$$\Phi = \frac{F}{t} = \frac{105}{6} = 17.5 \text{ mm/hour}$$

This result indicates that during the first 1 hour duration of the rainfall there is no runoff ($i = 13$ mm/hour $< \Phi = 17.5$ mm/hour) and hence, in the Φ-index calculation, the rainfall amount along this duration will not be taken into consideration. Accordingly, the infiltration height at the end of this rainfall becomes,

$$F = (150 - 13) - 45 = 92 \text{ mm}$$

TABLE 1.20 Rainfall Intensity

Time (hour)	0	1	2	3	4	5	6
Rainfall intensity, i, (mm/hour)	13	19	46	20	33	19	

This leads to Φ-index during the remaining time duration (t = 5 hours) as,

$$\Phi = 92/5 = 18.4 \text{ mm/hour}$$

TABLE 1.21 Φ-Index Calculation

T (hour)	I (mm/hour)	Φ-Index (mm/hour)	Effective Rainfall (mm/hour)
0–1	13	18.4	–
1–2	19	18.4	0.6
2–3	46	18.4	27.6
3–4	20	18.4	1.6
4–5	33	18.4	14.6
5–6	19	18.4	0.6
Total	150	Total	45.0

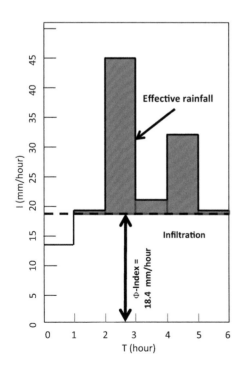

FIGURE 1.29 Φ-Index determinations by graphical method.

In Table 1.21 if Φ-index is assumed as 18.4 mm/hour then effective rainfall height can be found as 45 mm, which is equal to the direct runoff height.

The calculated Φ-index is shown in Figure 1.29.

1.6 RUNOFF

Earth reaching rainfall portion that runs over the surface is the runoff, which flows through rills, streams, creeks, wadi channels in arid regions, and rivers in humid regions. This is the most important component of the hydrological cycle, which serves humanity not only as water resource but also for transportations (through rivers in humid regions), hydroelectric power generation (dams), and environmental sustenance. Floods, droughts, sedimentation, and groundwater recharge into the rivers during dry periods are also among the most important consequences of runoff. It depends mainly on the effective rainfall but many other factors play additional active roles. Among natural factors are the geology (especially lithology), soil type and texture, vegetation cover, surface features (slope, aspect, drainage pattern, depressions, etc.), and also human activities such as land use. In recent modeling approaches and software programs these factors are required individually and then they are combined by geographical information system (GIS) in such a manner that the runoff and infiltration properties are quantified at different locations.

Since runoff is a part of the rainfall, the simplest logical and rational reasoning lead to deductions that the runoff is directly related to rainfall as a fraction, C, which is in a ratio form and it can be written linguistically as follows.

$$C = \frac{\text{Runoff}}{\text{Rainfall}} \quad (1.14)$$

By definition, it is the amount (volume) of runoff per total rainfall volume. In hydrology literature, it is referred to as the runoff coefficient. The greater is its value, the bigger is the runoff. The runoff coefficient also reflects total loss factor, L, which can be defined linguistically as,

$$L = \frac{\text{Rainfall} - \text{Runoff}}{\text{Rainfall}} \qquad (1.15)$$

Comparison of two last expressions shows that $L = 1 - C$. It is possible to call L value as loss (infiltration, evaporation, depression, transpiration, etc.) coefficient. Logically, C and L may have any value between 0 and 1, theoretically inclusive. However, in practical applications its value lies in between. For instance, most often the average value of C is taken as 0.35. There are many tables available for its adaptation depending on the hydrological soil type, topographic features, and geological as well as surface pavement practices (Maidment, 1992).

The definitions in Eqs (1.14) and (1.15) are valid for any point. Their generalization over any study area or drainage basin requires consideration of the area. Hence, by consideration of Eq. (1.14) one can write verbally as,

Runoff $=$ C $\times$ (Rainfall intensity) $\times$ Area.

Substitution of runoff (rainfall) discharge, Q, rainfall intensity, I, and area, A, into the last expression yields notationally the following formulation.

$$Q = CIA \qquad (1.16)$$

This is the most widely used formulation for runoff calculation and it is referred to in the hydrology literature as the rational method (RM). It is rational on logical bases with simplifying basic assumptions, but it does not seem physically plausible for actual flow cases over large areas. Among its assumptions, peak flow rate is produced by a constant storm rainfall intensity, which is maintained for a time equal to the period of concentration over the whole drainage basin area. This time period is defined theoretically as the time required for the surface runoff from the most remote part of the drainage basin to reach the point of interest (in general drainage basin outlet). Practically, one cannot measure it in the field, and therefore, it is calculated in an empirical manner (Kirpich, 1940; Şen, 2008). Additionally, there is a set of assumptions that the applicator should be aware of successful applications and interpretations. Otherwise, the peak discharge estimation by the classical RM may lead to unreliable conclusions.

EXAMPLE 1.11 SIMPLE RUNOFF CALCULATION BY RATIONAL METHOD

The outcrop area of a groundwater reservoir is about 5 km^2. The reservoir material is composed of medium sand and gravel. Long term rainfall studies showed that the rainfall intensity during April is 3.2 mm/s. Field work has indicated that the runoff coefficient is 0.35. What is the amount of runoff during this month?

Solution 1.11

Rational method is logical assessments of the runoff formulation and it is given by Eq. (1.16). The area, A, multiplied by the rainfall intensity, I, gives the total rainfall amount per second (discharge), Q_R, from the clouds on the surface area. Hence,

$$Q_R = AI = 5 \times 10^6 \times 3.2 \times 10^{-2}$$
$$= 160,000 \text{ m}^3/\text{s}$$

Runoff is the amount of this water that will move on the Earth's surface only. It is, therefore, necessary to know the losses (infiltration, evaporation, depression, interception) so that the remaining amount is the runoff. Logically, only a fraction of Q_R will turn into runoff; hence, it must be multiplied by a factor ($C < 1$), which depends on the geology, vegetation and land use, and surface composition of the area. It is given for the area on the average as 0.35, and therefore, the runoff discharge, Q, can be obtained from Q_R as $Q = CQ_R = CIA$.

$$Q = CQ_R = 0.35 \times 160,000 = 56,000$$
$$= 160,000 \text{ m}^3/\text{s}$$

1.7 GROUNDWATER

Groundwater is one of the most important components of the hydrologic cycle in the subsurface. Water is stored in the subsurface in pore spaces between grains, in fractures, or in solution cavities (limestone and dolomite rocks). Groundwater is found at varying depths below the Earth's surface, and wherever it is found, voids and fractures in the geological formations are saturated with water over wide areas. In the subsurface, water exists as soil moisture, vadose water, and groundwater. Provided that there is no impermeable layer above, then the upper level of this saturated zone is called water table. Above the water table is the zone of aeration and water present in this unsaturated zone is called vadose water including the zone of soil moisture (Chapter 2). Water balance studies are carried out by a number of researchers such as Shiklomanov and Rodda (2003). As stated by Singh (1992), at approximately 800 m below the water table, about four million cubic kilometers of water is present and an equal amount of water exists further down to 8 km. The most active groundwater reserves are estimated at four million cubic kilometers by L'Vovitch (1979). The volumetric amount is distributed as groundwater (95%) and the remaining 5% with

1.5% as soil moisture and 3.5% in lakes, swamps, reservoirs and river channels. It moves through the geological materials at a slower rate and residence times in the 10s, 100s, and even 1000s of years. These huge reserves of fresh groundwater are not being renewed wholly every year when compared to exploitation rate by pumping. The existing resources of fresh ground waters all over the world are estimated to last for about 550 years (Singh, 1992).

Groundwater is defined in its broadest sense as subsurface water in the saturation zone. Water that fills the openings (porous, fissure, fracture, solution cavities) of the geological formations (soil, alluvium, and rocks) is referred to as groundwater. As mentioned earlier, it originates from rainfall and subsequent infiltration process and it is the source of water for aquifers, springs, and wells (Chapter 2).

Surface water structures are not preferable in arid regions due to excessive evaporation losses except for short duration impoundment. The ratio of extensive surface area to volume of small reservoirs leads to high evaporation loss. Microstorage facilities loose, on the average, 50% of the impoundment to evaporation in arid and semiarid areas (Gleick, 1993; Sakthivadivel et al., 1997). Groundwater storage in aquifers provides water supply resources for many months and years with little or no evaporation loss. It also has the advantage that storage can be near or directly under the point of use, and immediately, available through wells and pumps on demand. The development in the tube well technology gave extra facilities for groundwater abstraction in addition to conventional large diameter hand dug wells.

1.8 CLIMATE CHANGE AND GROUND WATER

Many human activities are dependent on climate, and therefore, it is possible to classify arid and humid regions according to the context

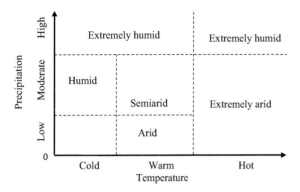

FIGURE 1.30 **Aridity and humidity (Şen, 2008).**

of interest. Meteorologically, they are classified on the basis of temperature and precipitation as in Figure 1.30.

Arid regions have dry climate with little or no rainfall, very low humidity, and high annual evaporation rates that exceed the annual rainfall with a high deficit in soil humidity, which does not give way to even dry farming practices. The scarce and sporadic rainfall is turbulent, brief, and torrential. The nature of rainfall causes an increase in runoff, erosion, and flooding, which furnish conditions that must be taken into consideration by water resources planners and operators in selecting a site for human activities.

Groundwater is the major freshwater source especially for arid and semiarid regions, but unfortunately, there has been very little attention or study on the potential climate change effects on these freshwater resources. Groundwater storages in arid and semiarid regions are replenished by floods at possible recharge outcrop areas through porous, fractured and fissured rocks, solution cavities in dolomite or limestone, and through main stream channels of Quaternary alluvium deposits. At convenient places along the main channel, engineering infrastructures such as levees, dikes, and successive small-scale dams may be constructed for groundwater recharge augmentation (Chapters 4 and 6). The groundwater recharge areas must be cared for

isolation from fine silt accumulation after each flood occurrence or at periodical intervals. Flood inundation areas are among the most significant groundwater recharge locations in arid and semiarid regions. Accordingly, their extents must be delimited by considering future climate change effects.

Understanding the relative importance of climate, vegetation, and soils in controlling groundwater recharge is critical for estimating recharge rates and for assessing the importance of these factors for aquifer vulnerability to contamination (Chapter 5). Additionally, the role of climate and vegetation in controlling groundwater recharge are also valuable in determining climate change impacts on land.

Groundwater resources and their long-term replenishments are controlled by climate conditions, and hence, present day climate changes are expected to affect their future. Increase of carbon dioxide (CO_2) and methane (CH_4) concentrations in the atmosphere has significant effects on the heat budget of the Earth's surface (Section 1.4.2.5). Such increases affect local hydrological cycle to which groundwater is one of the most significant components. Especially, shallow groundwater reservoirs, which supply much of the local societies, are most sensitive to long term climatic variations and changes. More attention must be given to the effect of climate change on shallow (unconfined) aquifers (Chapter 4). Groundwater is expected to play an even greater role for human development under changing climate conditions. As stated by IPCC (1996, p. 336; IPCC, 2007) despite the critical importance of groundwater resources in many parts of the world, there have been very few direct studies of the global warming effects on groundwater recharge. Groundwater resources are less directly and more slowly impacted by climate change as compared to surface waters. Confined and unconfined aquifers that are in contact with present day hydrological cycle will be affected by climate change more significantly due to overpumping. Since, deep (fossil groundwater storages) and especially confined aquifers are not in direct contact with the present day hydrological cycle the climate change effect on them is virtually negligible.

The following are possible climate change impacts on groundwater resources. In order to offset undesirable effects on groundwater reservoirs mitigation and adaptation activities are essential.

1. Increase in rainfall variability may decrease groundwater recharge, because more frequent heavy rainfalls may result in exceedance of soil infiltration capacity. In arid and semiarid regions the opposite case may take place.
2. An excessive or unwise exploitation of groundwater in combination with climate change may endanger future activities.
3. Projected sea-level rise due to the climate change and excessive withdrawals at coastal areas may cause salt water intrusion into fresh groundwater reservoirs (Chapter 4).
4. Higher temperatures may mean higher evaporation and plant transpiration rates, and hence, more drying up of soils. This may lead to higher soil moisture and groundwater recharge losses and soil erosion, and hence, groundwater recharge possibility decreases.
5. Overall, climate change will have an increasing impact on groundwater quantity and quality.
6. Drop in groundwater table levels due to climate change may bring extra costs in energy consumption and generation.
7. Especially in areas of shallow groundwater resources, additional water table drop due to climate change may cause desertification expansion.
8. Water saving and the use of recycled water strategies become more powerful in times of climate change.
9. Flood inundation areas are among the most significant groundwater recharge

locations in arid and semiarid regions. Accordingly, their extents should be delimited by considering future climate change effects.
10. Relatively small changes in temperature and precipitation in some areas could result in large percentage changes in runoff, increasing the likelihood and severity of droughts and/or floods.
11. Groundwater quality problems may increase where there is less flow to dilute contaminants introduced from natural and human sources.

1.9 GEOLOGY AND GROUNDWATER

The lithosphere is a continuous composition of different rock masses among which some include void spaces (intergranular fractures and solution cavities) and solids; others consist of solids only. The void space provides a place where groundwater can be stored and kept under suitable conditions for use at times of demand. Any rock mass in the lithosphere with connective voids is a permeable formation, which is also referred to as groundwater reservoir. In practical terms, reservoirs yield water easily. Many sedimentary formations, such as sandstones, lime-stones, clay, etc. have the reservoir property. Depending on the geological evolution of the void spaces, permeable reservoirs may be subdivided into three major groups as the porous, fractured, and karstic media. In this book only porous medium (intergranular voids) will be dealt with in detail.

If a rock mass does not yield water easily it is "impermeable." For instance, clay qualifies as reservoir but it is not permeable. Some permeable reservoirs do not accommodate water. Some rock masses (granite, greywacke, etc.) are without voids and they are completely impervious. The third type of geological formations from water movement point of view has semipermeable property, which allows transmission of water through voids but do not include water. Most of the fractured rocks have such a property.

The permeable reservoir rock faults allow the fluids to move from one location to another in the geological domain. Dykes have just the opposite function and they do not allow fluids to migrate to other adjacent locations. Any area characterized by faulting and fractures plays significant role on groundwater flow and distribution. When they are filled with weathered material, these structural features may act as drainage channels and aquifers that allow groundwater movement.

1.9.1 Geological Shield and Shelf

A shield is a geological term for a large area with outcrops of Precambrian rocks including crystalline, igneous, and high grade metamorphic rocks. These areas are tectonically stable with negligible earthquake risks. Shields consist of granite, granodiorite, and gneiss with very limited amount of groundwater. There are also sedimentary rocks of greenstone belts, which are normally metamorphosed green-schist, amphibolite, and granulite facies In general, due to their stability the morphology is almost flat as a result of erosion. For instance, along the Red Sea of the Arabian Peninsula, there are escarpments, which gave rise to up to almost 3000 m over MSL (Al-Sayari and Zötl, 1978).

Shield areas include replenishable groundwater in zones of fractures, weathered parts, and especially in large quantities along the Quaternary deposit filled alluvium wadis (Şen, 2008). These are dominantly in unconfined aquifer forms (Chapter 4). They provide shallow groundwater reservoirs, which can be tapped through a set of large diameter wells. The groundwater is in close connection with the present day hydrology cycle as in Figure 1.4. The groundwater resources are prone to contamination and pollution exposures, and

hence, need close attention for protection of the quality (Chapter 5). The shields are full of igneous and metamorphic-rock aquifers that can be grouped into crystalline-rock and volcanic-rock aquifers. The igneous and metamorphic rocks are permeable if fractured, and they generally yield small amounts of water. If they extend over large areas then it is possible to abstract large volumes of groundwater and, in many places, they provide the only reliable source of water supply.

Contrary to geological shield, the shelf is composed of sedimentary rock stratigraphy. They are mostly extensions of coastal plains as in the Arabian Shelf on the eastern part of the Arabian Peninsula. The sedimentary rocks are accumulated on the Precambrian basement and each layer has either groundwater potential or not.

Many sedimentary layers within the shelf areas have prolific groundwater resources, but they are at rather big depths and do not have continuous connection with the present day hydrological cycle. Therefore, the groundwater is almost of fossil type under pressure and mostly saline. Most often they discharge to the nearby seas. At some places these sedimentary rock layers are fractured and bleached, and consequently, allow groundwater exchange between adjacent layers and the sea.

1.9.2 Quaternary Deposits

These are formed as a result of weathering, erosion, and where the relief is favorable for the flow of water toward lowest elevations. Gravels as the coarsest product of erosion are moved shorter distances and are deposited in more restricted areas than are sand, clay, or mud. Quaternary deposits are widespread, especially in arid regions, where they fill the valleys of rivers, surface depressions, or fault zones.

The drainage system within the Quaternary deposits may have been altered historically by various factors such as climate changes, tectonic activity, and fluctuations in flood water discharge. The porous medium is so heterogeneous, and therefore, the rainfall has a direct relationship with the amount of groundwater stored in these deposits.

Quaternary deposits, composed of pebble, gravel, sand, and silt in wadi courses, terraces, alluvial fans, and flood plains, constitute areas with temporary surface runoffs leading to groundwater recharge. Their permeability is generally from medium to high and they conserve permanent groundwater flow in the lower courses of the wadis.

Thick Quaternary deposits of wadi sand, gravel, and silt over zones of tectonic features or due to the effects of damming areas may have concentrated surface runoff and commonly of groundwater flow. These areas are potentially favorable zones for tapping groundwater in alluvium and bedrock. Among the Quaternary deposits are alluvial fans, alluvial fills, sand dunes, sabkhahs, and basalts.

1.9.2.1 Alluvial Fans

These formations are of fluvial origin and occur where a stream leaves a steep valley and slows down as it enters a plain. The resulting cone of gravel and sand point toward upstream. The growth of alluvial fans was initiated during the Pleistocene epoch when the climate was more humid and rainfall more abundant. A good account of alluvial fans is presented by Rachocki (1981). The volume of voids in the porous medium of an alluvial fan becomes smaller by compaction, especially in zones composed of fine material (Şen, 1995).

Alluvial fans provide groundwater in coastal areas. In arid regions, they reach larger dimensions and play a more important role as potential aquifers. Groundwater in their storage is phreatic (unconfined) whereas in the lower parts it is confined. Groundwater flow in alluvial fans is replenished by percolation of surface water. Most often the water appears in the form of springs; otherwise it may continue its journey farther downstream where it emerges as surface flow.

1.9.2.2 Alluvial Fills

Fluvial gravels are widespread, especially in arid regions, where they fill the valleys of rivers, surface depressions, or fault zones. The thickness of an alluvial fill may reach several hundred meters. Such fills are initiated by the formation of alluvial fans. They are commonly coarse, interbedded with coarse and fine sands, and show a progressive decrease in size toward downstream, accompanied by a notable increase in roundness. Generally, bedding and sorting are comparatively better than in alluvial fans. The convenient features of a wadi are the flood plain, terraces, meander scrolls, swamps, and natural drainage toward the river with the result that silt-loaded flood water remains on the flood plain swamps. The drainage system on the wadi surface may have been altered by various factors such as climate changes, tectonic activity, and fluctuations in flood water discharge. Some coarse- to medium-grained pyroclastic deposits have very high porosity.

The alluvial fill of plains and valleys, consisting of gravel, sand, and clay of volcanic origin, may present important groundwater possibilities. In general, the groundwater is found in the voids of gravels. The alluvial fills make up potential groundwater reservoirs for local uses. In arid regions they are the primary locations for water-well excavation to supply the nearby villages. Because the porous medium is so heterogeneous, the rainfall has a direct relationship with the amount of groundwater stored in alluvial fills.

1.9.2.3 Sand Dunes (Eolian Sand)

Sand dunes can be regarded as one of the most isotropic and homogeneous deposits in nature. Eolian dunes include only sand and rarely gravels. From this point of view, they can be considered as a fine grained porous medium. Thick sand dune sequences of porous, permeable, cross-bedded sandstones with few impermeable barriers such as shale provide important potential reservoirs for water, gas, oil, and hydrothermal metalliferous deposits. Because of the poorly graded nature of sand dunes and their association with rapid interval drainage, no seasonal high water table is found near the Earth's surface. A geological account of sand-dune deposit evolution is presented by Anton (1991).

The geological layers underlying sand dunes may offer suitable groundwater supplies. Sand-dune materials are of uniform size and allow rapid infiltration and percolation of rainfall. Sand dunes provide huge volumes of porous reservoirs with favorable conditions for percolation and preservation of groundwater. Unfortunately, only during intense rainfalls the water infiltrates the thick cover to reach the groundwater storage. Sand dune materials have uniform size and allow rapid infiltration and percolation of rainfall. The geological layers underlying sand dunes may offer suitable groundwater supplies. These deposits act as a screen to groundwater evaporation.

1.9.2.4 Sabkhahs (Playa)

Coastal and inland saline flats appear in shallow, small, or extensive depressions with fillings of silt, clay, and muddy sand. These are referred to as sabkhah in Arabic and they are filled with brine water and at their surface white fine layers of salt appear, which is the distinguishing feature of sabkhahs.

The depth of brine water in the sabkhahs is about 1−2 m below the surface. Capillary water evaporates leaving the salt on the surface. Sabkhahs are composed of mainly silt, clay, and evaporates. They appear at the areas of water accumulation with slight or no surface flow. When water is present, it is shallow and very saline due to high evaporation rates.

In low lying areas with partial and late outlet possibilities the surface water accumulates and with strong solar irradiation power water evaporates and leaves behind saline remnants. The water table in sabkhah is at the surface from where discharge takes place. Water is always associated with sabkhahs in the form of flooding, runoff accumulation, capillary rise, and tidal fluctuation.

1.9.3 Harrats (Basalts)

Basalt and andesite are the most important extrusive rock reservoirs in which groundwater may flow freely. Originally basaltic rocks are in the form of fluid with thin flows that have considerable solution voids at the tops and bottoms of the flows. Different basalt flows frequently overlap and between them there are soil zones or alluvial materials that form permeable zones. Especially in the central parts of basalt flows developed columnar joints create possibilities for water passages and vertical water movement through the basalt layer. Among the volcanic rocks, basaltic groundwater reservoirs are the most prolific. Recent basalts cover the underlying drainage pattern, forming buried channels where the groundwater may be trapped under confining conditions. On the other hand, basalt flows block the main stream channel, resulting in a surface depression in which sedimentation of silt, clay, or volcanic ash provides an impervious layer.

The Tertiary and Quaternary age basaltic aquifers are in the forms of unconfined layers. Older basalts tend to be kindly weathered and to have their secondary openings filled by clay minerals, which may have destroyed their water transmitting capacities. The groundwater takes place due to numerous basalt flows, which are separated by palaeo-weathering surfaces and minor alluvial gravels of palaeo-drainage channels. Local aquitards exist within the massive basalt and clay-rich weathered zones. Within basaltic regions it is also possible to encounter perched aquifers (Chapter 3).

Groundwater is mostly stored especially in vesicular basalt medium store and allows groundwater movement rather easily by pathways through vesicular and fracture systems. Fracture zones are the most preferred pathways for groundwater movement. Shrinkage cracks provide more or less vertical passages that help the surface water penetrate the subsurface. In this way, either the underlying porous layers are replenished or if the underlying stratum is impervious, springs are formed at the bottom of hills. This aquifer has highly variable hydraulic parameters similar to the fractured medium and contains good-quality water, but the groundwater recharge is not sufficient for its sustainability. There are also buried channels underneath of basalt lava flows and they may include significant of groundwater.

1.10 GROUNDWATER USE IN PETROLEUM

In general, the groundwater is used for different purposes in petroleum production works as for injection and wash water, community (total residential community water), and facility and drilling (exploration and development drilling) purposes. For oil production stimulation water is injected into the oil reservoir to increase pressure. Water injection wells can be found both on- and off-shore, to increase oil recovery from an existing reservoir. Water injection helps to:

1. Support pressure of the reservoir, because water fills the emptied voids
2. Sweep or displace oil from the reservoir, and push it toward a production well

Generally, only 30% of the oil in a reservoir can be extracted, but water injection increases that percentage (known as the recovery factor) and maintains the production rate of a reservoir over a longer period.

For oil recovery water sources of injection may be coming from different origins as they are convenient and economically feasible at the production site.

1. Produced water: This is the type of water that is produced with the oil from the same reservoir. As the reservoir becomes depleted, the produced water content of the oil increases. As the volumes of water being produced are never sufficient to replace all the production volumes (oil and gas, in addition to water), additional "make-up" water must be provided. Mixing waters from different sources exacerbates the risk of scaling.
2. Seawater is obviously the most abundant and convenient source for offshore production facilities and it may also be pumped inshore for use in land fields.
3. Where available, groundwater from water bearing formations other than the oil reservoir, but in the same structure, has the advantage of purity.
4. River water will always require filtering before injection.

Unfortunately oil companies most often ignore the importance of proper water treatment. Intake water (river, sea) quality can vary tremendously which may have significant impact on the performance of the water treatment facilities. Such ignorance leads to improper and incomplete water injection, which may lead to poor water quality, clogging of the reservoir, and loss of oil production.

In any hydrocarbon reservoir oil and gas are accumulated. Porosity and permeability are the two essential features for any hydrocarbon and groundwater reservoir existence (Chapter 2). Water is compressible only to a small degree, and therefore, as the hydrocarbons are withdrawn, pressure reduction in the reservoir causes the water to expand slightly. However, if the water reservoir is large enough, this translates into a large increase in volume, which will push up on the hydrocarbons, maintaining pressure. Below the hydrocarbons may be a groundwater aquifer. An oil (petroleum) reservoir occurs in subsurface porous or fractured rock formations containing hydrocarbons (crude oil, natural gas). Hydrocarbons are fluids that are trapped by overlying lower permeability rock formations under temperature and pressure effects. The following differences exist between oil and groundwater reservoirs.

1. Oil occurs only in the sedimentary rocks (massive or fractured), but groundwater resides in all rock types.
2. Oil takes shape under pressure and temperature effect, whereas groundwater is not affected by these factors.
3. Groundwater may be at shallow or deep reservoirs, but oil is available at very deep reservoirs.
4. Shallow and to a certain extent deep groundwater reservoirs are in contact with present day hydrological cycle, which has nothing to do with oil reservoirs.
5. Oil reservoirs are under pressure and confined between impervious layers, but groundwater may be under confined, leaky, or unconfined conditions.
6. Deposition of oil takes millions of years through various stages. First as deep burial under sand or mud, pressure cooking, hydrocarbon migration from the source to the reservoir rock, and finally trapping by impermeable rock. Groundwater reservoirs have their sources from precipitation (mainly rainfall and snow).
7. Geologically there are structural oil trap, stratigraphic trap, and hydrodynamic trap. Groundwater traps are more extensive in any type of geological feature,
8. Hydrodynamic traps are caused by water pressure differences during the flow that gives rise to a tilt of the hydrocarbon—water contact.

Sandstone and limestone are by far the most common hydrocarbon reservoirs in the world, followed by dolomite and conglomerate. Siltstone, shale, or igneous rocks form hydrocarbon reservoirs only very rarely. These rocks possess also groundwater; however, groundwater may exist by itself in these formation whereas hydrocarbons do not. Hydrocarbons are specific hosts in the porous and fractured sedimentary basins of the world. They cannot occur in karstic media where groundwater is plenty.

In a hydrocarbon reservoir oil occurs above predominantly groundwater layer. These two fluids are immiscible with a transition zone between them. The grains absorb water or oil partially and do not allow their movement, and it is not possible to take out all water or oil. The amount of groundwater or oil that remains with grains is concerned with specific retention, whereas the abstractable portion is related to specific yield (Chapter 2).

References

Allen, R.G., Luis, S.P., Raes, D., Smith, M., 1998. Crop evapotranspiration — guidelines for computing crop water requirements. In: FAO Water Resources (Ed.), FAO Irrigation and Drainage Paper No. 56, p. 300.

Al-Sayari, S.S., Zötl, J.G. (Eds.), 1978. Quaternary Period in Saudi Arabia. 1: Sedimentological, Hydrogeological, Hydrochemical, Geomorphological, and Climatological Investigations in Central and Eastern Saudi Arabia. Springer Verlag, Vienna, 335 pp.

Al-Sefry, S., Şen, Z., Al-Ghamdi, S.A., Al-Ashi, W., Al-Baradi, W., 2004. Strategic Ground Water Storage of Wadi Fatimah — Makkah Region Saudi Arabia. Saudi Geological Survey, Hydrogeology Project Team (Final Report).

Al-Yamani, M.S., Şen, Z., 1997. Spatio-temporal dry and wet spell duration distributions in south-western Saudi Arabia. Theor. Appl. Climatol. 57, 165—179.

Anton, M., 1991. Ecology of coastal dune fauna. J. Arid Environ. 21, 229—243.

Bayazit, M., Avci, İ., Şen, Z., 1977. Hidroloji Uygulamalarō (Hydrology Applications). Birsen Publication Co., İstanbul, 280 pp (in Turkish).

Blaney, H.F., Criddle, W.D., 1962. Determining Consumptive Use of Irrigation Water Requirements. Tech. Bull. 1275. U. S. Dept. of Agriculture, Washington, D.C, 59 pp.

Bouchet, R.J., 1963. Evapotranspiration Réelle et Potentielle; Signification Climatique. In: General Assembly of Berkeley, vol. 62. Committee for Evaporation IAHS Publ., pp. 134—142.

Brutsaert, W.H., 1982. Evaporation into the Atmosphere: Theory, History and Applications. D. Reidel, Dordrecht, Holland.

De Bruin, H.A.R., 1983. Evapotranspiration in Humid Tropical Regions. IAHS, Hamburg, pp. 299–311.

Eagleman, J.R., 1967. Pan evaporation, potential and actual evapotranspiration. J. Appl. Meteorol. Climatol. 6 (3), 482–488.

Gleick, P. (Ed.), 1993. Water in Crisis: A Guide to the World's Fresh Water Resources. Oxford University Press, Oxford, England.

Harvey, J.G., 1982. Atmosphere and Ocean. Our Fluid Environment. The Vision Press, London, 143 pp.

Horton, R.E., 1940. An approach toward a physical interpretation of infiltration capacity. Soil Sci. Soc. Am. J. 5, 399–417.

International Panel on Climate Change (IPCC), 1996. The Science of Climate Change. http://www.ipcc.ch/ipccreports/sar/wg_I/ipcc_sar_wg_I_full_report.pdf. Entered on October 13, 2013.

International Panel on Climate Change (IPCC), 2007. Fourth Assessment Report: Climate Change. http://www.ipcc.ch/publications_and_data/publications_and_data_reports.shtml#.Ulr3iCdrPIU. Entered on October 13, 2013.

Kirpich, Z.P., 1940. Time of concentration of small agricultural watersheds. Civil Eng. 10 (6), 362. The original source for the Kirpich equation. (In PDF).

L'vovitch, M.I., 1979. World Water Resources and Their Future. American Geophysical Union, Washington.

Maidment, D. (Ed.), 1992. Handbook of Hydrology. McGraw-Hill, Inc., New York.

Meyer, A.F., 1942. Evaporation from Lakes and Reservoirs: A Study Based on Fifth Years' Weather Bureau Records. Minnesota Resources Commission, St. Paul, MN.

Mirza, M.M., Warrick, R.A., Erickson, N.J., 2003. The implications of climate change on floods of the Ganges, Brahmaputra and Meghna rivers in Bangladesh. Clim. Change 57 (3), 287–318. http://dx.doi.org/10.1023/A:1022825915791. LA – English.

Penman, H.L., 1948. Natural evaporation from open water, bare soil and grass. Proc. R. Soc. Lond. Ser. A Math. Phys. Sci. 193, 120–145.

Pike, J.G., 1964. The estimation of annual run-off from meteorological data in a tropical climate. J. Hydrol. 2 (2), 116–123. http://dx.doi.org/10.1016/0022-1694(64)90022-8.

Priestley, C.H.B., Taylor, R.J., 1972. On the assessment of surface heat flux and evaporation using large-scale parameters. Mon. Weather Rev. 100 (2), 81–92. http://dx.doi.org/10.1175/1520-0493(1972)100. <0081:OTAOSH>2.3.CO;2.

Rachocki, A., 1981. Alluvial Fans. John Wiley and Sons, New York.

Sakthivadivel, R., Fernando, N., Brewer, J., 1997. Rehabilitation Planning for Small Tanks in Cascades: A Methodology Based on Rapid Assessment. IWMI Research Report 13. International Water Management Institute, Colombo, Sri Lanka.

Shamsudduha, M., Taylor, R.G., Ahmed, K.M., Zahid, A., 2011. The impact of intensive groundwater abstraction on recharge to a shallow regional aquifer system: evidence from Bangladesh. Hydrogeol. J. 19 (4), 901–916.

Singh, N.T., 1992. Land degradation and remedial measures with reference to salinity, alkalinity, water logging and acidity. In: Deb, D.L. (Ed.), Natural Resource Management for Sustainable Agriculture and Environment. Angker Pub., New Delhi, p. 442.

Shiklomanov, I.A., Rodda, J.C. (Eds.), 2003. World Water Resources at the Beginning of the 21st Century. Cambridge University Press, Cambridge, 435 pp.

Şen, Z., 1995. Applied Hydrogeology for Scientists and Engineers. CRS, Lewis Publishers, Boca Raton, 444 pp.

Şen, Z., 1998a. Probabilistic formulation of spatio-temporal drought pattern. Theor. Appl. Climatol. 61, 197–206.

Şen, Z., 1998b. Average areal precipitation by percentage weighting polygon method. J. Hydrol. Eng. 1, 69–72.

Şen, Z., 2008. Wadi Hydrology. Taylor and Francis Group, CRC Press, Boca Raton, 347 pp.

Şen, Z., 2013. Philosophical, Logical and Scientific Perspectives in Engineering. Springer, 247 pp.

Şen, Z., 2009. Spatial Modeling Principles in Earth Sciences. Springer, New York, 351 pp.

Thiessen, A.H., 1911. Precipitation averages for large areas. Mon. Weather Rev. 39, 1082–1084.

Turc, L., 1961. Estimation of irrigation water requirements, potential evapotranspiration: a simple climatic formula evolved up to date (in French). Ann. Agronomy 12, 13–49.

Basic Porous Medium Concepts

Practical and Applied Hydrogeology
http://dx.doi.org/10.1016/B978-0-12-800075-5.00002-9

Copyright © 2015 Elsevier Inc. All rights reserved.

2.1 GEOLOGIC CONSIDERATIONS

Study of water bearing geological layers is important before any major water abstraction from the groundwater reservoirs. The proportion of solids (particles) and voids (pores, fractures, solution cavities) in any rock body defines its worth for water storage and yield capability, i.e., its water release potentiality. Voids and solids are present in any rock mass as mutually exclusive combinations, which give the rock ability to store fluids (water, gas, oil, air) in the voids. Rock masses may be potential reservoirs provided that the voids are interconnected.

The term reservoir has a broad meaning and in general, it can be defined as "any material body, which can store and release fluid." Reservoir does not necessarily mean that it can transmit water at demand levels. The proportion of interconnected pores present in the whole bulk of the rock mass further defines the permeable or impermeable nature of the reservoirs. Clay being highly porous material falls into the category of reservoirs but it is not able to yield water at significant rates, so it is an impermeable reservoir. Depending on the genesis of voids as porous, fractured, or karstic, the geological formations are regarded as fluid (groundwater, oil, and gas) reservoir. The sedimentary rocks such as sandstone, clay, and limestone are potential groundwater reservoirs. The igneous and metamorphic rocks (if not fractured) like granite and gabbro are not significant groundwater reservoirs. Depending upon the geometrical interrelationships (size, shape, orientation, interconnection, etc.) among pore spaces, the permeable reservoirs are further classified into porous, karstic, and fractured media. Porous medium can be categorized broadly into fine- and coarse-grained formations.

From volcanic rocks granite, granodiorite, diorite, and alkalic granite may have fair to good permeability related to the degree of weathering and fracturing. Groundwater recharge may take place after heavy rain in weathered zone. The main storage appears in weathered and fractured zones. Basalt and andesite occur as top-cover rocks, and therefore, their potentiality as an aquifer cannot be high, and they cause high-infiltration but low-evaporation losses. Basalt thickness may reach up to about 60–80 m and the underlying volcanic rocks may include water in fractures and weathered zones. The groundwater quality is expected to be good.

Lithology describes the physical makeup, including the mineral composition, grain size, and grain packing of the sediments. Stratigraphy informs about the geometrical and age relations between various lenses, beds, and sedimentary formations. Structural features, such as cleavages, fractures, folds, and faults, are the geometrical properties of geologic systems produced by deformation after deposition or crystallization. In unconsolidated sedimentary deposits, lithology and stratigraphy give clues about the most important geologic controls. Recognition of depositional environments of geologic systems is also of utmost importance. Direction and magnitude of the depositing agency include anisotropy and heterogeneity in the deposited geologic formations.

2.2 GRAIN SIZE

Grain (particle) size refers to the diameter of individual sediment grains or lithified particles in clastic rocks. The term may also be applied to other granular materials that can range from very small colloidal particles, through clay, silt, sand, and gravel, to boulders. ISO 14688-1 establishes the basic principles for the identification and classification of soils on the basis of those material and mass characteristics most commonly used for soils for engineering purposes (Table 2.1).

Grain-size distribution analysis is among the early interests of geologists and hydrogeologists

TABLE 2.1 Grain-Size Classification (ISO 14688-1:2002 at scribd.com)

Name			Size Range (mm)
Very coarse soil		Large boulder, LBo	>630
		Boulder, Bo	200–630
		Cobble, Co	63–200
Coarse soil	Gravel	Coarse gravel, CGr	20–63
		Medium gravel, MGr	6.3–20
		Fine gravel, FGr	2.0–6.3
	Sand	Coarse sand, CSa	0.63–2.0
		Medium sand, MSa	0.2–0.63
		Fine sand, FSa	0.063–0.2
Fine soil	Silt	Coarse silt, CSi	0.02–0.063
		Medium silt, MSi	0.0063–0.02
		Fine silt, FSi	0.002–0.0063
	Clay, Cl		<0.002

for fluid flow assessments (Krumbain, 1934). Such analysis is useful in distinguishing among different depositional environments. Interconnected voids give rise to hydraulic conductivity, which is the measure of fluid flow ease through porous material. Certain relationships are expected to exist between hydraulic conductivity and statistical parameters that describe the grain-size distribution of the depositional medium. Intuitively, the greater the grains of the same size the more will be the voids, and hence, more possibility of water storage. Mixtures of different grain sizes at various proportions result in different void/solid ratios. From groundwater point of view, increase in interconnected void space is convenient, but undesirable for foundation-engineering purposes.

Gradation of grain sizes is achieved by means of mechanical analysis, which is conducted through a nest of standard sieves with the coarsest on the top and the finest at the bottom covered with a lid on the top and a pan at the bottom as receiver. In order to perform the test, a representative sample of porous medium, about 150–500 g, is taken by quartering oven-dried weight poured into the top sieve. The whole nest is shacked for about 5 min and the total percentage passing each sieve is then calculated and the data are plotted on a semilogarithmic paper (Figure 2.1).

The logarithmic horizontal axis is for the grain sizes and along the same axis three categories of grains are shown as fine and course gravel; sand as coarse, medium or fine classifications; and finally, fine grains consisting of silt and clay classes. The main benefits from a grain-size distribution curve can be summarized as follows:

1. To know the proportions of different sizes of particles in the porous material.
2. To understand the proportions of porous media material; whether coarse, medium, or fine grained.

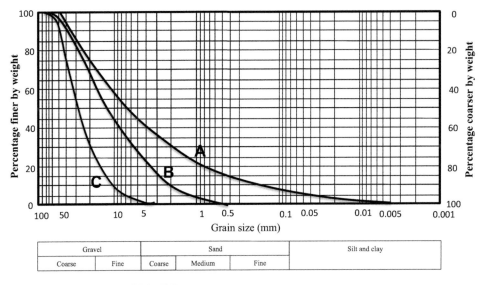

FIGURE 2.1 **Grain-size distribution curves.**

3. To deduce some information about the geological history of the porous material. In this, the very fine material is probably due to wind-deposited sand dunes; the very graded mixture of boulders, sand and mud would be identified as a glacial till, etc.
4. To determine some design quantities such as average grain size, effective diameter, uniformity coefficient (Driscoll, 1986).
5. To learn much about the properties of the medium and to estimate some of the parameters such as the hydraulic conductivity (Krumbain, 1934; Al-Yamani and Şen, 1993).

In Figure 2.1, three different representative grain-size distribution curves A, B, and C are plotted for comparison purposes. Curve A has the widest domain of grain-size distribution, whereas C has the smallest range. Although A has all grain-size types fine, medium and coarse, C has only two types, coarse and fine gravel. Curve B is composed of gravel and sand only. The degree of sorting for sediments is measured by the uniformity coefficient, U_c, which is defined as the ratio of the grain size that is 60% finer by weight, D_{60}, to the grain size that is 10% finer by weight, D_{10}, which is also known as the effective particle size.

$$U_c = \frac{D_{60}}{D_{10}} \qquad (2.1)$$

In general, any sample with $U_c < 4$ ($U_c > 6$) is well (poorly) sorted. The smallest uniformity coefficient is with curve C, the biggest is with A, and B has the uniformity coefficient in between.

From groundwater point of view, fine-grained aquifer materials allow for the validity of Darcy's (1856) law (Section 2.8), and hence, theoretical derivation of the groundwater movement equations become simpler with a set of configurational assumptions as will be mentioned in the next two chapters.

2.3 RESERVOIR CLASSIFICATIONS

Quantification of groundwater problems needs for further classification of the reservoirs. On the basis of transmission and storage properties, geological formations are classified generally into three hydrogeological units as aquifers, aquitards, and aquicludes. However, in groundwater literature, the words aquifer and aquitard are more in common use.

2.3.1 Aquifer

There are at least three separate definitions that are useful for practicing hydrogeologists (Şen, 1995). First, an aquifer is a saturated geological unit that can transmit water easily in significant amounts under normal conditions. In this statement, "saturation," "easiness," and "significance" are among the basic water-related descriptions of any aquifer. The second definition is that an aquifer is a water-bearing geological formation or stratum that yields significant water quantity for economic extractions through wells. This definition appeals to groundwater engineers, who are not concerned in detail with geology but rather with hydraulics, management, and economic aspects of water. The third definition is that an aquifer is a zone of rock through which groundwater moves. This last statement attracts nonspecialists, who are concerned with the movement of groundwater only.

Rock formations that serve as aquifers are gravel, sand and sandstone, alluvium, cavernous limestone, fissured marble, fractured granite, weathered gneiss and schist, heavily shattered quartzite, vesicular basalt, and jointed slate. Two basic properties of aquifers are the storage and transmission of water within the reservoir. Depending on the lack of either or both of these properties, a geologic formation is considered as aquitard or aquiclude.

2.3.2 Aquitard

An aquitard is any geological formation of a rather semipervious nature that transmits water at slower rates than an aquifer. Freeze and Cherry (1979) describe an aquitard as the less-permeable beds in a stratigraphic sequence. These beds may be permeable enough to transmit water in quantities that are significant in the study of regional groundwater flow, but their permeability is not sufficient to allow the completion of production wells. In an interlayered sand-silt sequence, the sand may be considered as aquifer, whereas in the silt-formation aquitard clays, shale, and silty clays are the stratigraphic units, which can be considered as aquitards.

2.3.3 Aquiclude

Aquiclude as a saturated geological formation, although capable of absorbing water slowly, will not transmit it fast enough to yield significant supply for a well (Şen, 1995). In other words, it can store water but cannot transmit it easily. Clay lenses and shale layers are main examples. However, Freeze and Cherry (1979) say that an aquiclude is a saturated geologic unit that is incapable of transmitting significant quantities of water under ordinary hydraulic gradients.

2.4 HYDRAULIC TERMINOLOGIES

For the appropriate description and qualitative and quantitative understanding of aquifer potentiality, there is a need to define a set of terminology in the groundwater context. They provide basic information about the aquifer behavior and help in the development of qualitative demarcations and analytical models, which are used ultimately to evaluate and assess the available potential of an aquifer.

2.4.1 Hydraulic Head (Piezometric Level) and Gradient

In practice, hydraulic head is defined as the height to which water can rise in a piezometer or a well relative to a datum. In order to see the regional flow behavior of water in geological formations, water levels are measured at a set of points (wells) with respect to some datum

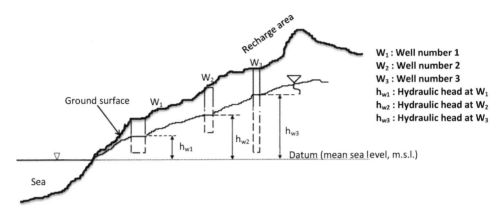

FIGURE 2.2 Hydraulic head.

(usually m.s.l.). The height of the water column shown in Figure 2.2 from the datum is called hydraulic head (water table or piezometric level).

Groundwater in the geological formations remains in a continuous dynamic state depending upon the regional hydraulic head positions. The hydraulic heads measured at two points in the subsurface geological formations can be connected by a straight line. If this line is horizontal with respect to the reference datum, then the groundwater is in the static (steady state) condition, otherwise flow will occur from high toward low hydraulic head positions. It can be said that a gradient exists between the two points and it is called hydraulic gradient. Numerically, hydraulic gradient, i, is defined as the ratio of difference in the hydraulic heads (h_{w1} and h_{w2}) at two well locations (say, W_1 and W_2) to the horizontal distance, Δl, between them (Figure 2.3). It is a dimensionless quantity defined as,

$$i = \frac{h_{w2} - h_{w1}}{\Delta l} = \frac{\Delta h}{\Delta l} \approx \frac{dh}{dl} \qquad (2.2)$$

where Δh is the piezometric head difference along Δl distance. The groundwater movement reflects the aquifer parameters through hydraulic gradient, which is defined as the piezometric-level difference over unit horizontal distance. The greater the hydraulic gradient the greater is the groundwater velocity.

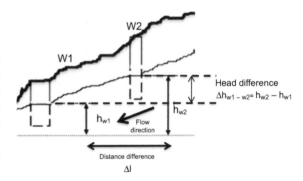

FIGURE 2.3 Hydraulic gradient.

The geological structure affects piezometric surface and hydraulic gradient. If the geological structure is discontinuous including faults, fractures, joints, dikes, solutions cavities, etc. (but not anticlines and synclines) or continuous with facial changes, then its impacts are reflected on the piezometric surface. In the discontinuity locations such as main fractures, the gradient change has discontinuities. It is worthy to consider that the hydraulic gradient is also dependent on the aquifer material composition apart from recharge and discharge events. This is equivalent to saying that the groundwater velocity is a function of not only the hydraulic gradient, but also the material properties among which hydraulic conductivity is the most

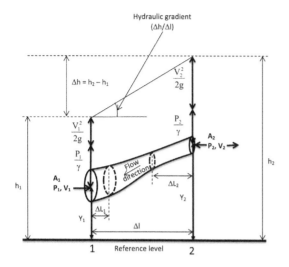

FIGURE 2.4 **Fluid (water, oil, gas) flow.**

significant factor for groundwater movement (see Section 2.7.3).

For theoretical derivation of the hydraulic head, consider Figure 2.4, which indicates an imagination of the water flow tube with two points, at Δl apart, each with area (A_1 and A_2), pressure (p_1 and p_2), velocity (v_1 and v_2), geodetic elevations (y_1 and y_2), and movement lengths (ΔL_1 and ΔL_2).

The works, W_1 and W_2, done by fluid pressure forces at cross-sections 1 and 2 can be written from physical point of view as,

$$W_1 = p_1 A_1 \Delta L_1$$

and because of the opposite direction,

$$W_2 = -p_2 A_2 \Delta L_2$$

respectively. For incompressible fluids, the fluid volumes at two cross-sections are equal to each other, i.e., $A_1 \Delta L_1 = A_2 \Delta L_2$. Accordingly, the work by the gravity, W_g, can be written as,

$$W_g = -mg(y_2 - y_1)$$

where m is the mass. The net work, W_n, on the fluid volume is the summation of all these works.

$$W_n = p_1 A_1 \Delta L_1 - p_2 A_2 \Delta L_2 - mg(y_2 - y_1)$$

Up to this point, the velocities are not taken into consideration. Each velocity gives rise to kinetic energy at any cross-section, and because of the energy conservation principle in physics, the difference between the kinetic energies must be equal to the total net work done on the fluid volume.

$$\frac{1}{2}mV_1^2 - \frac{1}{2}mV_2^2 = p_1 A_1 \Delta L_1 - p_2 A_2 \Delta L_2 \\ - mg(y_2 - y_1)$$

In general, mass is equal to specific density, ρ, multiplied by water volume, $m = \rho A_1 \Delta L_1$ or $m = \rho A_2 \Delta L_2$, the substitutions of which into the previous expression give after the necessary arrangements,

$$p_1 + \frac{1}{2}\rho V_1^2 + \rho g y_1 = p_2 + \frac{1}{2}\rho V_2^2 + \rho g y_2$$

So far in the derivations, friction or any other loss between the cross-sections is not taken into consideration. This last expression is valid between two points within the flow domain, and finally, the hydraulic head, h, can be written as,

$$h = \frac{P}{\gamma} + \frac{V^2}{2g} + y_1 = \text{constant} \qquad (2.3)$$

This final expression is known as the Bernoulli's equation in the hydraulics context. The hydraulic head is also known as the piezometric level.

After the aforementioned physical explanations, it is obvious that the water flow is possible only under the hydraulic head difference, $h_2 - h_1$. The same difference may be between very close or far-away points. It is, therefore, necessary to find a dimensionless quantity for the driving factor of the movement. It is the hydraulic gradient, i, as defined in Eq. (2.2).

EXAMPLE 2.1 HYDRAULIC HEAD AND GRADIENT CALCULATIONS

The positions of four wells (W_1, W_2, W_3, and W_4) in a confined aquifer are given in Figure 2.5. Their location elevations from m.s.l. are 235.5, 225.0, 243.3, and 218.7 m, respectively, and the distances from the well head to the piezometric levels are 15.2, 5.4, 8.9, and 3.2 m, respectively. Hence,

1. Calculate the hydraulic head elevations in each well.
2. What are the hydraulic gradients between the wells?

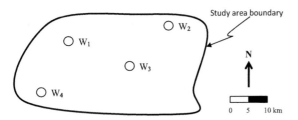

FIGURE 2.5 Well locations.

Solution 2.1

1. The hydraulic head elevations must be referred to the same reference level, which is m.s.l. For this purpose, the water-level distances must be subtracted from the well-head elevations. Hence, the hydraulic heads in the wells are,

$$h_{W_1} = 235.5 - 15.2 = 220.3 \text{ m}$$
$$h_{W_2} = 225.0 - 5.4 = 219.6 \text{ m}$$
$$h_{W_3} = 243.3 - 8.3 = 235.0 \text{ m}$$
$$h_{W_4} = 218.7 - 3.2 = 215.5 \text{ m}$$

2. In order to calculate hydraulic gradients between the wells, it is necessary to know the distances between each pair of wells. This can be achieved by taking into consideration the scale bar. The distances, d, between the wells

are calculated from Figure 2.5 and the results are given in km in the following matrix.

	W_2	W_3	W_4
W_1	25.0	18.2	14.3
W_2		14.7	33.4
W_3			11.2

After knowing the distances, the hydraulic gradients can be calculated as the piezometric head per distance as follows:

$$i_{W_1-W_2} = \frac{h_{W_1} - h_{W_2}}{d_{W_1-W_2}} = \frac{220.3 - 219.6}{25.0 \times 10^6}$$
$$= +0.028 \times 10^{-6}$$

$$i_{W_1-W_3} = \frac{h_{W_1} - h_{W_3}}{d_{W_1-W_3}} = \frac{220.3 - 235.0}{18.2 \times 10^6}$$
$$= -0.81 \times 10^{-6}$$

$$i_{W_1-W_4} = \frac{h_{W_1} - h_{W_4}}{d_{W_1-W_4}} = \frac{220.3 - 215.5}{14.3 \times 10^6}$$
$$= +0.34 \times 10^{-6}$$

$$i_{W_2-W_3} = \frac{h_{W_2} - h_{W_3}}{d_{W_2-W_3}} = \frac{219.6 - 235.0}{14.7 \times 10^6}$$
$$= -1.05 \times 10^{-6}$$

$$i_{W_2-W_4} = \frac{h_{W_2} - h_{W_4}}{d_{W_2-W_4}} = \frac{219.6 - 215.5}{33.4 \times 10^6}$$
$$= +0.123 \times 10^{-6}$$

$$i_{W_3-W_4} = \frac{h_{W_3} - h_{W_4}}{d_{W_3-W_4}} = \frac{235.0 - 215.5}{11.2 \times 10^6}$$
$$= +1.74 \times 10^{-6}$$

Some of the hydraulic gradients have negative sign. For instance, the hydraulic gradient between wells W_1 and W_3 is negative. This means that although in the calculations W_3 hydraulic head is subtracted from W_1 implying that the groundwater flow is from W_1 to W_3, but the negative sign says just the opposite that the groundwater flow is from W_3 to W_1. The other negative signs have similar interpretations.

2.5 EQUIPOTENTIAL LINES

From groundwater point of view, any line along which there is no groundwater flow (constant hydraulic head, i.e., constant energy) is an equipotential line. These are the lines along which the hydraulic head remains the same; hence along them the hydraulic gradient is equal to zero. Since groundwater velocities are small, the kinetic energy components according to Eq. (2.3) being the square of this velocity ($V^2/2g$) become even smaller. In practical applications, groundwater is assumed to have only potential (position) energy, because kinetic energy component is negligible.

The water molecule prefers the minimum work path which is perpendicular to the equipotential lines in a homogeneous and isotropic medium (Section 2.9). It is obvious that another set of lines for water movement can be drawn on the same map showing flow lines, which are also referred to as the streamlines. None of these lines can have corners, they have smooth traces only.

2.5.1 Flow Nets

Equipotential and flow lines (streamlines) form a flow net, where the flow lines are imaginary impervious barriers because there is no cross-flow through them. Such nets are the basis of groundwater equipotential (piezometric) or water table contour maps (Figure 2.6).

Water table is a collection of points where the absolute pressure equals the atmospheric pressure and the relative pressure, i.e., the pressure due to water only (gage pressure) equals zero. The location of the water table is depicted by the water level in an open well or piezometer. In a regional context, water table is a surface drawn through hydraulic heads in an area. Water table map includes equal hydraulic headline contours.

Sometimes, groundwater in geological formations is under pressure. Geologically, the aquifer is sandwiched between two impervious layers and the water is under pressure (Chapter 3). Hydrogeologically, the aquifer is not in touch directly with the Earth's surface. If any such an aquifer is tapped by drilling a hole (well), as the drilling gets below the upper impervious layer, water rises to the piezometric-level (hydraulic head) position in the well. A surface drawn through a set of piezometric levels is piezometric (potentiometric) surface, which is imaginary everywhere except at well locations. Interpretation of these maps provides useful information to identify areal variations in aquifer

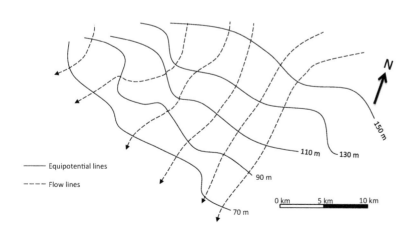

FIGURE 2.6 Flow net.

behavior. The following information can be deduced from a flow net (Şen, 1995):

1. The water table separates the saturated zone from the capillary water zone. Its fluctuations help calculate recharge and discharge locations and time within the aquifer (Chapter 4).

2. The slope of piezometric surface (water table) is related to the rate of transmission and/or hydraulic conductivity of the aquifer. The slope varies directly with the velocity, and inversely with the ease of movement which is hydraulic conductivity (permeability) as will be explained later in Section 2.7.3.

3. The piezometric-level intersection with the Earth's surface delimits the surfaces subject to influent and effluent seepages as well as springs. Permanently wet lands, such as swamps, indicate the intersection of the piezometric level with the Earth's surface.

4. Geologically, water table separates the belt of weathering, oxidation, rock decomposition, and section from the underlying belt of mineral precipitation and rock cementation.

5. Hydraulic heads can be calculated for any desired point within the flow net domain by interpolation.

6. Hydraulic gradients can be known between any two successive contour lines and change of hydraulic gradient between any two points can be calculated.

7. Closed equipotential lines indicate either recharge (domes) or discharge zones (depressions).

8. Convergence of equipotential lines indicates difficulty (less-permeable zones) in the water movement within the media, whereas divergence implies easy groundwater movement.

9. Discontinuity in the equipotential lines corresponds to geological discontinuities such as faults, dikes, fractures, etc.

10. Existence of two or more equal-value equipotential lines indicates stagnant water zones.

11. Divergence and then reconvergence of flow lines show the zones of possible impervious lenses.

12. Parallel equipotential lines imply parallel flow lines, and hence, parallel (one-dimensional) flow in the case of homogeneous and isotropic media.

13. Highest points on the flow net map imply a groundwater divide line similar to water divide lines in hydrology, but in groundwater context these divide lines are not stable, they change by time.

14. Refraction of flow lines implies the change in lithology, facies, and similar features.

15. The groundwater direction at any point is tangent (perpendicular) to the flow line (equipotential line) at the same point provided the groundwater reservoir material is homogeneous and isotropic.

16. The aquifer area between any two adjacent streamlines is referred to as the stream tube and the tube walls can be considered as impervious boundaries since there is no cross flow through these boundaries.

Groundwater flow toward a vertical well in a homogeneous and isotropic aquifer will have flow lines radially with equipotential lines as perpendicular circles (Figure 2.7).

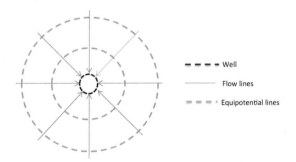

FIGURE 2.7 **Radial flow equipotential and flow lines (streamlines).**

EXAMPLE 2.2 PIEZOMETRIC SURFACE MAP

During a field trip, a well inventory has been prepared with the locations and the static piezometric levels by considering six wells as shown in Table 2.2 and Figure 2.8. Prepare the piezometric surface map for the aquifer.

Solution 2.2

Based on the piezometric levels given in Table 2.2, Figure 2.9 is prepared with piezometric levels at each well location.

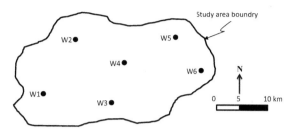

FIGURE 2.8 **Well locations.**

The following steps lead to the groundwater piezometric map:

1. Connect the neighboring well locations by straight lines (Figure 2.9).
2. Obtain triangularization through all the possible straight-line connections.
3. Decide about the equal piezometric line intervals. Herein, 10 m is considered for application.
4. Make linear interpolation between a pair of neighboring wells to find equipotential lines. The first glance gives the impression that the maximum equipotential line will be 300.0 m and the minimum 260.0 m with in-between equipotential lines of 270.0, 280.0, and 290.0 m.
5. After the completion of interpolations, connect equal piezometric points along smooth curves as in Figure 2.10.

TABLE 2.2   Well Locations and Piezometric Levels

Well Number	Well Location (m)		Piezometric Level (m)
	Easting	Northing	
1	50.2	80.3	312.4
2	55.3	120.9	302.5
3	68.9	74.3	298.7
4	71.5	105.2	286.4
5	92.6	118.7	263.9
6	99.1	100.6	252.6

FIGURE 2.9 **Piezometric level.**

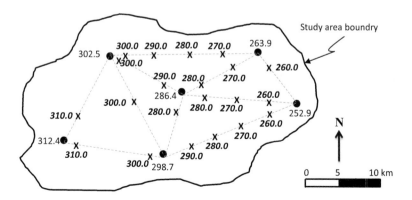

FIGURE 2.10 **Equipotential lines.**

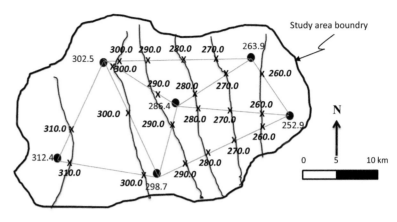

2.6 AQUIFER TYPES

Aquifers can be classified as unconfined, confined, leaky, and perched types according to their geological composition, prevailing hydrological conditions, and water pressures at particular time and place.

2.6.1 Unconfined

The general configuration of such an aquifer is presented in Figure 2.11. The following three features are important for an aquifer to be considered as unconfined:

1. Geologically, there are two layers as permeable and underlying impermeable.

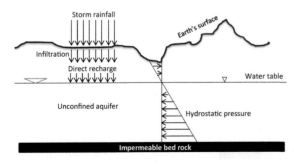

FIGURE 2.11 **Unconfined aquifer.**

2. Meteorologically, the water table is in contact with air, and hence, the water is under atmospheric pressure.
3. Hydrologically, there is direct groundwater recharge through shallow infiltration and percolation (deep infiltration).

The water table receives recharge directly from the surface and geologically there is no upper confining layer. They are shallow aquifers in the sense close to Earth's surface. They occur mainly in the Quaternary deposits.

2.6.2 Confined Aquifer

It has at least three layers, upper and lower confining aquitards and in between a sandwiched permeable geologic unit that is fully saturated with water under pressure (Figure 2.12).

In the well, the water level is above the upper confining layer. If the water in the well does not reach to the Earth's surface then artesian conditions exist. Rise of water above the Earth's surface implies flowing artesian well. The main geological features for a confined aquifer are synclines, monoclines, depressions, troughs, grabens, and tectonically fractured zones. Majority of large confined aquifers appear in the sedimentary layers such as sandstone and dolomite. Confined aquifer gets recharge from nearby

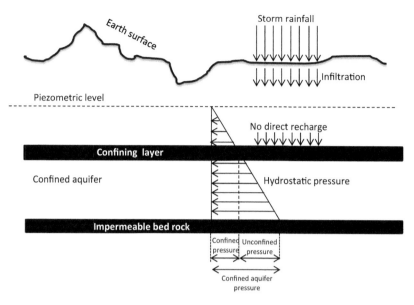

FIGURE 2.12 Confined aquifer.

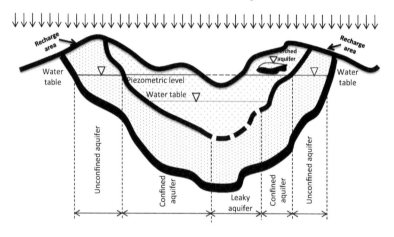

FIGURE 2.13 Composite aquifer.

unconfined aquifers through outcrops (see Figure 2.13).

2.6.3 Leaky Aquifer

The geologic configuration is similar to that of confined aquifer with upper or lower semipermeable confining layer. The semipermeable layer transmits water from one aquifer to the other depending upon the vertical hydraulic head difference (Figure 2.14). In this figure, the groundwater flow takes vertically from the upper unconfined aquifer to the lower semiconfined aquifer.

These aquifers are abundant in multilayer sedimentary formations and there are groundwater exchanges between adjacent aquifers through aquitards.

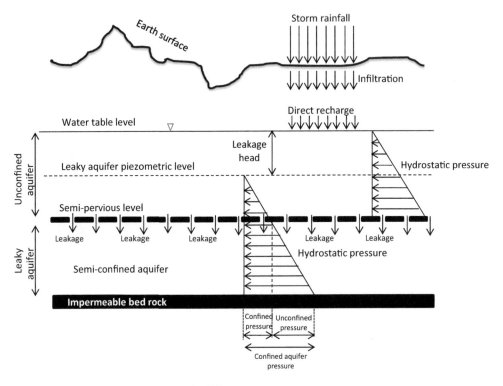

FIGURE 2.14   **Leaky aquifer.**

2.6.4 Perched Aquifer

Occurrence of clay lenses in alluvial formations is widely spread in Quaternary deposit aquifers. Such lenses, even of a few centimeters' thickness, can produce water table conditions in the unsaturated zone of the main unconfined aquifer and they are called perched aquifers (Figure 2.15). They are formed temporally at the time of infiltration after each storm rainfall. Wells drilled in a perched aquifer cannot sustain water supply for long time durations.

2.6.5 Composite Aquifer

In nature, different aquifer types may be next to each other in the form of a composite aquifer. Different aquifer types may have groundwater exchanges. Discharge dynamism is triggered by groundwater recharge through outcrop areas after each storm rainfall or water abstraction through well pumping. Figure 2.13 is a representative cross-section of composite aquifer case.

2.7 AQUIFER PARAMETERS

In practice, the most essential work is aquifer parameter determination. The following parameters are necessary for any groundwater exploration, quantity and management studies:

1. Porosity,
2. Specific yield and retention,
3. Hydraulic conductivity (coefficient of permeability),
4. Transmissibility (transmissivity),
5. Storage coefficient and specific storage.

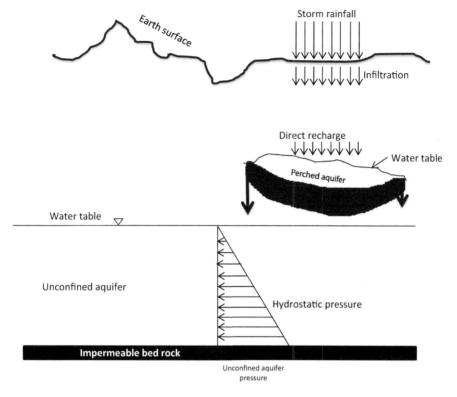

FIGURE 2.15 **Perched aquifer.**

2.7.1 Porosity

Porosity, n, is the capacity of a rock mass to accommodate fluid. It is defined as the ratio of void volume, V_V, to the total volume, V_T, of the rock body as,

$$n = \frac{V_V}{V_T} \qquad (2.4)$$

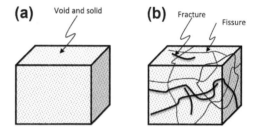

FIGURE 2.16 **Porosity: (a) primary, (b) secondary.**

In practical terms, it indicates void volume percentage. Porosity has primary and secondary types depending upon environmental deposition. Its formation during the time of diagenesis is called primary porosity. Any change in this porosity due to post processes such as faulting, fracturing, folding and jointing, and solution activity generates secondary porosity (Figure 2.16).

Effective porosity is that portion of the total void space of a porous material that is capable of transmitting fluid and it is almost similar to the specific yield. Porosity is the primary rock property that is used in all fluid (groundwater, oil, and gas) reservoir evaluations and assessments. It helps to evaluate various reservoir properties such as,

1. Available fluid volumetric calculation in the reservoirs,
2. Fluid saturation degree calculation,
3. Geologic characterization of the reservoir.

Porosity values of different rock materials are shown in Table 2.3 as a modification from Freeze and Cherry (1979) and Şen (1995).

2.7.2 Specific Yield and Retention

Specific yield is the amount of water that can be extracted under the gravitation force, and specific retention cannot be separated from the grain surfaces because the fluid (water, oil) is attached to grain surfaces due to adhesion forces. The specific yield, S_y, is the storage term used directly for unconfined aquifers. It is defined as the drainable water volume, V_d, from storage per unit surface area of the aquifer per unit decline in the water table, which is shown diagrammatically in Figure 2.17. This definition implies that it is a dimensionless quantity (m^3/m^2m).

The porosity is the combination of specific yield and specific retention, S_r, and they complement each other,

$$n = S_y + S_r \qquad (2.5)$$

Meinzer (1923) has defined specific yield as "the specific yield of a rock or soil, with respect to water, is the ratio of the volume of water which, after being saturated, it will yield by gravity to, its own volume."

It is a part of porosity as given by Eq. (2.5) and treated as a storage term, independent of time that in theory accounts for the instantaneous release of water from storage, but in nature the release of water is not instantaneous. Rather, the release can take an exceptionally long time, especially for fine-grained sediments. King (1899) determined S_y to be 0.20 for fine sand,

TABLE 2.3   Approximate Porosity Values

Sediment Classification	Porosity (%)
Well-sorted sand and gravel	25–50
Mixed sand and gravel	20–35
Mixed sand and gravel	20–35
Glacial till	10–20
Silt	35–50
Clay	40–70
Limous	35–50
Argiles	45–50
Marn	47–50
Limestone	0.5–17
Karst limestone	5–50
Calcaireoolitique	3–20
Marble	0.1–0.2
Schist	1–10
Dolomite	3–5
Granite	0.02–15
Gypsum	3–4
Basalt	0.1–12
Fractured basalt	5–15
Sandstone	5–30
Shale	0–10
Fractured crystalline rock	0–10
Dense crystalline rock	0–5

however, it took two-and-a-half years of drainage to obtain that value. The limitations of this definition are noted by Meinzer (1923), who also points out that it does not account for temperature and chemical effects. These problems have no doubt contributed to the wide range of values that are reported in the literature. Since in coarser sediments the groundwater

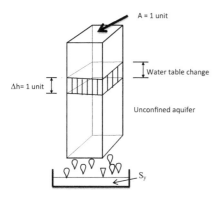

FIGURE 2.17 **Specific-yield definitions.**

of water that a soil can hold against gravity. This is the specific retention, S_r, capacity. It is also essential for plant-water-use calculations. The term "field capacity" was introduced by Veihmeyer and Hendrickson (1931) as "the amount of water held in the soil after the excess gravitational water has drained away and after the rate of downward movement of water has materially decreased." In the light of the previous sentences, one can conclude that field capacity is equivalent to specific retention. Its definition is slightly more satisfying than that presented for S_r. Specific retention is the complementary part of specific yield within the porosity and is defined as,

$$S_r = \frac{V_r}{V_T} \qquad (2.7)$$

where V_r is retained water volume. Depending on the material type, the percentages of S_y and S_r vary within n. The major role in the ratio of such contribution is played by the particle surface area. The smaller the average grain size the larger is the surface area of the medium. Larger surface areas attract more water, and accordingly, their specific retention values are greater. As a general rule, the finer the grain sizes the greater is the specific retention. It is tantamount to saying that there exists a reverse proportionality between the grain size and specific retention. A similar proportionality exists between porosity and the grain size. In the light of these proportionalities and Eq. (2.5), one can conclude that the specific yield is relatively small for very coarse and fine materials and attains its maximum value somewhere in between. Empirical studies have resulted in a graph similar to the one shown in Figure 2.18.

The maximum specific yield corresponds to the range of sand and this is one of the reasons why most prolific aquifers in the world are found in sedimentary formations. Some representative porosity, specific yield, and specific retention values are given in Table 2.4.

drainage is faster than the finer sediments, S_y is less dependent on time and temperature. The fact that S_y is not constant but varies as a function of depth of the water table is described from the perspective of soil physics by Childs (1960). That analysis takes into account the moisture-retention curve and is briefly described here.

Classical porosity is concerned with stagnant water content. Specific yield is useful to define part of porosity referring only to the movable (abstractable) water in the rock. It is also known as effective porosity in engineering.

$$S_y = \frac{V_w}{V_T} \qquad (2.6)$$

where V_w is the volume of drainable water due to gravitation. Usually $0.01 < Sy < 0.3$, as mentioned by Freeze and Cherry (1979). High S_y values are indicative that the water is released from the storage of unconfined aquifers by dewatering of the pores, while in confined aquifers the release of water is due to secondary causes such as expansion of water and compaction of aquifer. The same yield from unconfined aquifers can be obtained with less head changes over less-extensive areas when compared to confined aquifers.

Many researchers (soil scientists, hydrologists, water engineers) need to define the amount

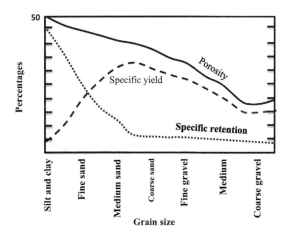

FIGURE 2.18 **Porosity, specific yield, and retention relations.**

TABLE 2.4   Representative Porosity, Specific Yield, and Specific Retention Values

Grain Size	Material	Porosity (%)	Specific Yield (%)	Specific Retention (%)
Very fine	Clay	50	2	48
Fine	Sand	25	22	3
Course	Gravel	20	19	1

The specific yield percentages are given for unconfined aquifers in unconsolidated sedimentary formations as in Table 2.5.

In hydrogeological studies, the specific yield concept is more important than porosity, since the latter does not give any clue of water abstraction from aquifers. For instance, in igneous rocks, porosity is not more than 2% but they yield all the available water in their voids up to approximately 100%. On the contrary, clays have the maximum porosity but they yield less than 5% of stored water.

If a granular material with an average porosity of 30% exists below the water table, its

TABLE 2.5   Sediment Specific Yield (Johnson, 1967)

Material	Minimum	Average	Maximum
Clay	0	0.02	0.05
Sand clay	0.03	0.07	0.12
Silt	0.03	0.18	0.19
Fine sand	0.10	0.21	0.28
Medium sand	0.15	0.26	0.32
Coarse sand	0.20	0.27	0.35
Gravelly sand	0.20	0.25	0.35
Fine gravel	0.21	0.25	0.35
Medium gravel	0.13	0.23	0.26
Coarse gravel	0.12	0.22	0.26

voids are saturated containing 30% water per unit volume. If the water table is lowered, the soil will eventually give up about two-thirds of its water, equivalent to 20% of the total volume. The remaining 10% water by volume is held by surface tension. In hydrological terminology, one says that the specific yield is 0.2 and specific retention is 0.1.

EXAMPLE 2.3 POROSITY CALCULATIONS

A loose soil sample of 45 cm^3 is collected from the field. It is poured into a graduated cylindrical cup and then filled with water. It is determined that 25.2 cm^3 of water is in the voids. What is the porosity of this soil?

Solution 2.3

The definition of porosity by Eq. (2.4) leads after the substitution of the relevant numerical parameters into the following expression as,

$$n = \frac{V_T - V_v}{V_T} = \frac{45.0 - 25.2}{45.0} = 0.44$$

EXAMPLE 2.4 MIXTURE POROSITY

If 0.5 m^3 of sand with porosity 0.23 is mixed with 1.1 m^3 of gravel of 0.32 porosity, what will be the porosity of the mixture?

Solution 2.4

Pore volume of sand (gravel) is $0.5 \times 0.23 = 0.115$ m^3 ($1.1 \times 0.32 = 0.352$ m^3). Porosity can be calculated as the weighted average of individual porosities. Hence, one can obtain,

$$n = \frac{0.115 + 0.352}{0.5 + 1.1} = 0.2918$$

2.7.2.1 *Specific Yield Determination Methods*

Specific yield plays key role in many hydrogeological and engineering works, and therefore, its accurate determination is important. For this purpose, there are different field and laboratory works . Specific yield is very significant for unconfined aquifers and groundwater recharge calculations.

2.7.2.1.1 UNCONFINED AQUIFER TEST METHODS

Specific yield, S_y, and transmissibility (Section 2.7.4), T, can be obtained from properly performed aquifer tests over a period of hours or days in the field for unconfined aquifers. Time-drawdown data preferably from observation wells are matched against theoretical type curves developed by Boulton (1963), Neuman (1972), and Moench (1994, 1995). Time versus drawdown is plotted on a log–log scale, and type curves at the same scale are overlaid on the plot. The curves are shifted (keeping coordinate axes parallel) until most of the data points lie on a curve. An arbitrary match point is then selected and the values of drawdown and time at that point are used to calculate T and S_y (Chapter 4). Aquifer tests provide in situ measurements of S_y and other aquifer properties that are integrated over fairly large areas, but the methods are not without some drawbacks.

2.7.2.1.2 VOLUME-BALANCE METHOD

The volume-balance method helps to determine the specific yield through a water budget of the cone of depression (Section 2.10). This approach has been in practical use since 1917 (Clark, 1917; Wenzel, 1942). Nwankwor et al. (1984) and Şen (1995) have used it to analyze results from aquifer tests. Nwankwor et al. (1984) have noticed from field experiments in Canada that the results show a trend of increasing S_y with increasing time; the longer they ran the aquifer test, the larger the calculated value of S_y became. They attributed the trend to delayed drainage from the unsaturated zone and questioned the appropriateness of Neuman's (1972) assumption of instantaneous drainage from the unsaturated zone. An analysis by Neuman (1987) showed that these results could be explained by flow from outside the assumed radial extent of the cone of depression.

Moench (1994) showed that type-curve estimates can agree with those of volume-balance methods when composite drawdown plots from more than one well are used and when partial penetration is taken into account.

2.7.2.1.3 WATER-BUDGET METHODS

Another way to estimate specific yield can be achieved by the application of a simple water budget for a basin, which can be written as follows,

$$(P + Q_{in}) - (ET + Q_{out}) = \pm \Delta S$$

where P is precipitation plus irrigation; ET is the sum of bare-soil and open-water evaporation and plant transpiration; Q_{in} and Q_{out} are surface and subsurface water inflow and outflow, respectively; and finally, ΔS is change in water storage. The change in storage may be due to three different components as in the surface reservoirs, ΔS_{sr} (including water bodies as well as ice and snow packs); the unsaturated zone, ΔS_{uz}; and the saturated zone, ΔS_{sz} and one can write that,

$$\Delta S = \Delta S_{sr} + \Delta S_{uz} + \Delta S_{sz}$$

This expression is suggested by Walton (1970) in conjunction with Eq. (2.6) for estimating S_y during periods of water level rise in winter months when ET is usually small and the soil is nearly saturated, so that the change in water storage in the unsaturated zone is also small. Consideration of this situation gives rise to the water balance equation as,

$$S_y = (P + Q_{in} - Q_{out} - ET - \Delta S_{sr} - \Delta_{uz})$$
$$\times /\Delta h/\Delta t$$

Gerhart (1986) and Hall and Risser (1993) used this method for winter periods with the assumption that ET, ΔS_{uz}, and net subsurface flow were zero. If surface and subsurface flows are easily determined, then this method yields the best results. In this method, implicit assumptions are that no difference exists in base flow prior to and during the rise in groundwater levels. Rasmussen and Andreasen (1959) have also used the water-budget approach to estimate S_y but they called the term gravity yield.

Pan lysimeters are installed in general at depths of 1–2 m beneath undisturbed soil columns. They are designed to capture all downward-moving water. The rate of water percolation measured by the lysimeters is assumed to be equal to the recharge rate. After measuring water levels in wells near the lysimeters, S_y can be calculated from Eq. (2.6).

2.7.2.1.4 FURTHER THOUGHTS ON SPECIFIC YIELD

The preceding discussions indicate that considerable uncertainty remains as to what value of S_y to use in a practical study because any values will be rather different from each other. In fact, the estimation of specific yield cannot be achieved without errors and uncertainties. Since Hazen (1893) and King (1899) very little progress has been made in addressing the limitations of S_y. Laboratory S_y values are generally greater than those obtained from field tests. Laboratory measurements of S_r are largely dependent on the amount of time allowed for drainage as well as the length of the test column. Meyboom (1967) was well aware of this discrepancy between laboratory and field estimates of S_y. He multiplied laboratory-determined values of S_y by 0.5 to arrive at a field value that he then used with Eq. (2.6) to estimate recharge in a prairie setting.

2.7.3 Hydraulic Conductivity

Hydraulic conductivity is one of the parameters, which tells about the transmission properties of an aquifer. It depends upon the specific yield (effective porosity) of the aquifer, which means the degree of interconnection of the pores. It can be defined as the volume of water per unit time passing through per unit cross-sectional area of the aquifer under the effect of unit hydraulic gradient as shown in Figure 2.19. Its

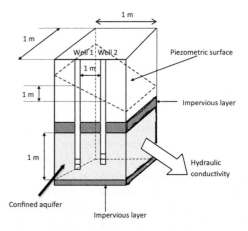

FIGURE 2.19 Definition of hydraulic conductivity.

dimension is $[L^3/T/L^2]$ but generally written as L/T, which is equivalent to the velocity dimension in physics.

Under the light of this definition, one can write notationally the hydraulic conductivity, K, as,

$$K = \frac{Q}{Ai} \qquad (2.8)$$

where Q is the discharge, A is the cross-sectional area, and i is the hydraulic gradient. Hydraulic conductivity is a property that describes the ease with which water can move through the interconnected void spaces. Table 2.6 presents the hydraulic conductivity classification by taking into consideration both numerical and linguistic specifications. Values are for typical fresh groundwater conditions—using standard values of viscosity and specific gravity for water at 20 °C and 1 atm (Bear, 1979).

Additionally, the average hydraulic conductivity values for unconsolidated and consolidated rocks are given by Domenico and Schwartz (1990) in Table 2.7.

Domenico and Schwartz (1990) also presented hydraulic conductivity values for a set of crystalline rocks as in Table 2.8

TABLE 2.7 Unconsolidated and Consolidated Aquifer Materials

Material	Hydraulic Conductivity (m/s)
UNCONSOLIDATED SEDIMENTARY MATERIALS	
Gravel	$3 \times 10^{-4} – 3 \times 10^{-2}$
Coarse sand	$9 \times 10^{-7} – 6 \times 10^{-3}$
Medium sand	$9 \times 10^{-7} – 5 \times 10^{-4}$
Fine sand	$2 \times 10^{-7} – 2 \times 10^{-4}$
Silt, loess	$1 \times 10^{-9} – 2 \times 10^{-5}$
Till	$1 \times 10^{-12} – 2 \times 10^{-6}$
Clay	$1 \times 10^{-11} – 4.7 \times 10^{-9}$
Unweathered marine clay	$8 \times 10^{-13} – 2 \times 10^{-9}$
SEDIMENTARY ROCKS	
Karst and reef limestone	$1 \times 10^{-6} – 2 \times 10^{-2}$
Limestone, dolomite	$1 \times 10^{-9} – 6 \times 10^{-6}$
Sandstone	$3 \times 10^{-10} – 6 \times 10^{-6}$
Siltstone	$1 \times 10^{-11} – 1.4 \times 10^{-8}$
Salt	$1 \times 10^{-12} – 1 \times 10^{-10}$
Anhydrite	$4 \times 10^{-13} – 2 \times 10^{-8}$
Shale	$1 \times 10^{-13} – 2 \times 10^{-9}$

TABLE 2.6 Hydraulic Conductivity Classifications

K (cm/s)	10^2	10^1	10^0	10^{-1}	10^{-2}	10^{-3}	10^{-4}	10^{-5}	10^{-6}	10^{-7}	10^{-8}	10^{-9}	10^{-10}
Relative permeability	Pervious				Semi-pervious				Impervious				
Aquifer	Good				Poor						None		
Unconsolidated	Well		Well sorted		Very fine sand, silt								
Sand and gravel	Sorted gravel		Sand or sand and gravel		Loess, loam								
Unconsolidated Clay and organic					Peat		Layered clay				Fat/unweathered clay		
Consolidated rock	Highly fractured				Oil reservoir rock		Fresh sandstone				Fresh limestone, dolomite		Fresh granite

TABLE 2.8   Crystalline Rock Values

Crystalline Rocks	
Permeable basalt	$4 \times 10^{-7}–2 \times 10^{-2}$
Fractured igneous and metamorphic rock	$8 \times 10^{-9}–3 \times 10^{-4}$
Weathered granite	$3.3 \times 10^{-6}–5.2 \times 10^{-5}$
Weathered gabbro	$5.5 \times 10^{-7}–3.8 \times 10^{-6}$
Basalt	$2 \times 10^{-11}–4.2 \times 10^{-7}$
Unfractured igneous and metamorphic rock	$3 \times 10^{-14}–2 \times 10^{-10}$

Another view about the hydraulic conductivity values of the unconsolidated and hard rocks are given by de Marsily (1986) as in Table 2.9.

Permeability can be defined as the capacity of a rock for transmitting a fluid (water, oil, gas)

TABLE 2.9   Hydraulic Conductivity for Unconsolidated and Hard Rocks

Medium	K (m/day)
UNCONSOLIDATED DEPOSITS	
Clay	1–5
Fine sand	5–20
Medium sand	$20–10^2$
Coarse sand	$10^2–10^3$
Gravel	$5–10^2$
Gravel and gravel mixes	$10^3–10^{-1}$
HARD ROCKS	
Chalk (very variable according to fissures if not soft)	30.0
Sandstone	3.1
Granite, weathered	1.4
Limestone	0.94
Dolomite	0.001

and it is also a measure of the relative ease with which a porous medium can transmit a liquid. The main factors that affect the permeability are:

1. Textural features such as voids, grain size, distribution and shapes, and gravel packing,
2. Clay type, distribution, and amount,
3. Secondary porosity,
4. Reactive fluids,
5. Turbulent (high-velocity) flow,
6. Overburden pressure.

According to experimental evidences hydraulic conductivity, K, is directly proportional to the grain size, d, in a nonlinear manner. The nonlinearity is expressed as the square of the grain size, d^2 (Hazen, 1893; Krumbain, 1934; Al-Yamani and Şen, 1993).

$$K = c\,d^2 \qquad (2.9)$$

where c is a coefficient that reflects the characteristic of the rock formation. The dimension of permeability is L^2, which shows its relation to an area, which is the throat area of the pores, because water flows through the pores. It is possible to represent permeability variation with grain size linguistically as in Figure 2.20.

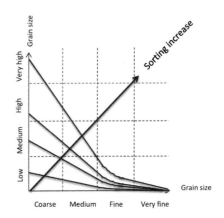

FIGURE 2.20   **Gravel-sorting effect on the permeability.**

EXAMPLE 2.5 HYDRAULIC CONDUCTIVITY CALCULATION BY TRACER

Three wells, W_1, W_2, and W_3, are at various horizontal distances from each other as in Figure 2.21. The hydraulic heads (piezometric levels) in these wells are 212, 202 and 198 m, respectively. The horizontal distance between W_1 and W_2 (W_1 and W_3) is 65 m (136 m). A tracer like potassium permanganate is dropped at well number W_1 and after 48 and 25 min it reached W_2 and W_3, respectively. So what can one say about the hydraulic conductivity of the aquifer? Make necessary interpretations.

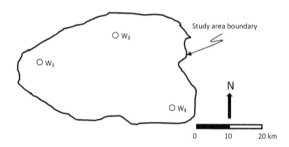

FIGURE 2.21 **Well locations.**

Solution 2.5

The groundwater flow is from W_1 toward W_2 and W_3, because the hydraulic head at W_1 is the highest. It is first necessary to calculate the hydraulic gradient between well pairs (W_1, W_2) and (W_1, W_2). The first pair has the hydraulic gradient as $(212 - 202)/65 = 0.154$ and the other $(212 - 198)/136 = 0.103$. If the aquifer is homogeneous and isotropic with uniform thickness, then the hydraulic conductivity between W_1 and W_2 is more than between W_1 and W_3, hence, under the given conditions groundwater flows faster from W_1 to W_2 than from W_1 to W_3.

2.7.3.1 *Hydraulic Fracturing*

Among completion techniques that facilitate oil or groundwater production by augmenting hydraulic conductivity is the hydraulic fracturing method. It helps augment the openings in the low-permeability storage medium and especially shale (high porosity and low permeability) in oil production. Hydraulic fracturing process involves pumping fluid (commonly water) through perforations in the wellbore that are located at target-formation depths. Pressure exerted by the fluid creates small cracks or fractures in the rocks that enable groundwater (or hydrocarbon) to flow toward the wellbore. It helps the propagation of fractures in a rock layer caused by the presence of pressurized fluid. This process is used to facilitate the easy movement of groundwater (or hydrocarbon) through the aquifer by enlarging the fracture media. This procedure leads to the augmentation of the following fracture-medium properties.

1. The number of fractures,
2. Extension (length) of the fractures,
3. Fracture aperture,
4. Decrease of fracture-surface roughness resistance against the fluid (water, hydrocarbon) flow.

It is often shortened as "fracking" or "hydrocracking" and it is known colloquially as a "frac job" due to fracture generation from a wellbore drilled into reservoir rock formations. In general, there are two types, namely, low-volume hydraulic fracturing for stimulation of high-permeability reservoirs and high-volume hydraulic fracturing used in tight rocks such as shale.

A hydraulic fracture is formed after pumping the fluid into the wellbore at a rate sufficient to increase pressure down hole to exceed the fracture resistance of the rock. The rock cracks and the fracture fluid continue farther into the rock, extending the crack still farther, and so on.

2.7.4 Transmissivity

Various definitions of transmissivity as it stands in the groundwater hydraulics literature fall into one of the following categories:

1. The rate of flow under unit hydraulic gradient through a cross-section of unit width over the whole saturated thickness of the aquifer (Bear, 1979; Kruseman and de Ridder, 1990),
2. The product of the thickness of the aquifer and the average value of the hydraulic conductivity (Hantush, 1964; Bouwer, 1978; Freeze and Cherry, 1979),
3. The ratio at which water of prevailing density and viscosity is transmitted through a unit width of an aquifer or confining bed under a unit hydraulic gradient. It is a function of the properties of the liquid, the porous media, and the thickness of the porous media (Fetter, 1980; Todd, 1980).

It is another very important transmission property of an aquifer, which is different from hydraulic conductivity in that it includes the whole saturation thickness, m, of the aquifer while K is defined for unit saturation thickness only. One can write transmissivity, T, in terms

TABLE 2.10   Aquifer Potentiality

Transmissivity (m²/day)	Potentiality Description
$T < 5$	Negligible
$5 < T < 50$	Weak
$50 < T < 500$	Moderate
$T > 500$	High

of aquifer thickness and hydraulic conductivity by considering Eq. (2.8) as follows.

$$T = Km = \frac{Q}{Wi} \qquad (2.10)$$

where W is the aquifer width. In words, T can be defined as volume of water per unit time passing from per-unit width under unit hydraulic gradient through the whole saturation thickness (Figure 2.22).

Potentiality implies extraction possibilities of groundwater from the aquifers. In groundwater movements instead of hydraulic conductivity transmissivity must be adopted for objective decisions. Logically, high transmissivity values imply high potentiality. Generally accepted

FIGURE 2.22   **Aquifer transmissivity.**

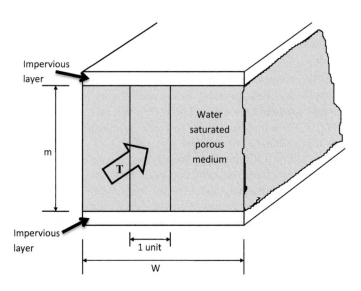

numerical boundary values and their verbal explanations are given in Table 2.10 (De Wiest, 1965).

EXAMPLE 2.6 AQUIFER POTENTIALITY (PRODUCTIVITY) CALCULATION

An aquifer has 723 m thickness and the hydraulic conductivity is calculated as 2×10^{-6} m/s. What is the aquifer potentiality?

Solution 2.6

Aquifer potentiality is related to its transmissivity value and also the unit of this parameter. Potentiality classification can be decided according to the m^2/day value of transmissivity. By definition, $T = Km$, where m is the saturation thickness of the aquifer and K is the hydraulic conductivity. Substitution of the given data into this definition yields $T = 723 \times 2 \times 10^{-6} = 1444 \times 10^{-6}$ m^2/s. This must be converted to m^2/day, which is $T = 1444 \times 10^{-6} \times 60 \times 60 \times 24 = 124.76$ m^2/day. Since this value falls between 50 m^2/day and 500 m^2/day, according to Table 2.9, the aquifer is moderately potential. The potentiality originates not from hydraulic conductivity but from aquifer thickness.

Transmissivity definition in Eq. (2.10) assumes that hydraulic conductivity remains constant along the saturation thickness. However, if it changes vertically with thickness, $K(z)$, then the transmissivity will be the summation along the thickness as,

$$T = \int_{bottom}^{top} K(z)dz \qquad (2.11)$$

After some pondering upon the definitions of K and T, it becomes evident that hydraulic conductivity is defined for only unit width and unit saturation thickness while transmissivity takes into consideration unit width and the whole saturation thickness. This leads to define another transmission parameter of aquifer, which takes into account the whole saturation thickness, and width of the aquifer together. It is named here as "Gross Transmissivity" (G_T). For this purpose, from Eq. (2.10) one can define it by considering that $A = mW$ as,

$$G_T = KmW = \frac{Q}{i} \qquad (2.12)$$

Furthermore, G_T can be defined as the volume of water per unit time under unit hydraulic gradient passing through the whole cross-sectional area of the aquifer. The dimension of G_T is L^3/T, which has the same dimension as the discharge.

EXAMPLE 2.7 HETEROGENEOUS MEDIA TRANSMISSIVITY CALCULATION

If the hydraulic conductivity variation in an aquifer is given as in Figure 2.23, then calculate its transmissivity along the same depth.

Solution 2.7

For transmissivity calculation, water level is visualized to increase steadily from 0 to 80 m. Hence, according to Eq. (2.13), the transmissivity values will be added on each other. The transmissivity calculations are achieved along the following four steps.

1. If the aquifer is saturated up to 30 m, then the transmissivity is $T_{30} = 30 \times 3 = 90$ m^2/day.
2. If saturation reaches to 40 m, then additional transmissivity will come from the second layer as $10 \times 6 = 60$ m^2/day. Hence, total transmissivity is $T_{40} = T_{30} + 60 = 90 + 60 = 150$ m^2/day.
3. Likewise, in the case of 60 m saturation, additional transmissivity is $20 \times 1 = 20$ m^2/day.

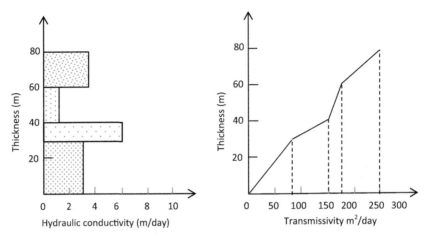

FIGURE 2.23 **Vertical hydraulic conductivity variations.**

Addition to the previous step leads to $T_{60} = T_{40} + 20 = 150 + 20 = 170 \text{ m}^2/\text{day}$.

4. Finally additional 20 m rise in the saturation level leads to additional transmissivity of $20 \times 4 = 80 \text{ m}^2/\text{day}$. The total transmissivity becomes $T_{80} = T_{60} + 80 = 170 + 80 = 250 \text{ m}^2/\text{day}$.

In real field studies, however, most aquifers consist of a combination of different geological layers of varying hydraulic conductivities with different thicknesses. The transmissivity must be worked out by calculating the contribution of each layer.

2.7.5 Specific Capacity and Productivity Index Comparison

Specific capacity is defined as the ratio of discharge to the drawdown in a well. It is discharge per unit drawdown and usually expressed in (l/min)/m (Logan, 1964). It provides a measure for well behavior and productivity. Wells are productive provided that small drawdowns are coupled with high discharges. The bigger the specific capacity the better is the

well. Table 2.11 gives limits for well classification from the specific capacity point of view.

The specific capacity value of a well is not constant but dependent on time. It is a function of aquifer parameters, hydraulic conductivity, transmissivity, and the storage coefficient. Logically, for the same discharge it is directly proportional to the transmissivity but indirectly related to storativity.

In a confined aquifer for radius of influence, $R_o = 3000 \text{ m}$ and $r_w = 0.20 \text{ cm}$, the specific capacity is given as follows (Şen, 1995):

$$\frac{Q}{s_w} = \frac{T}{1.6} \tag{2.13}$$

TABLE 2.11 Specific Capacity Well Classification (Şen, 1995)

Specific Capacity (l/min/m)	Well Productivity Classification
>300	High
300–30	Moderate
30–3	Low
3–0.3	Very low
<0.3	Negligible

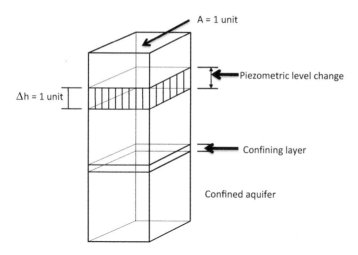

FIGURE 2.24 **Storativity definition.**

In unconfined aquifers, the radius of influence is practically smaller, and hence, by adopting $R_o = 300$ m and $r_w = 0.20$ cm, the same equation gives,

$$\frac{Q}{s_w} = \frac{T}{1.2} \qquad (2.14)$$

It should be born in mind that these are only rough approximations which may include up to 40% error.

A high specific capacity indicates an efficient well design and construction. Comparison of two or more wells from the productivity point of view is possible provided that the steady state or quasi-steady- state drawdowns are considered. Such a comparison is valid only when the well diameters are equal to each other.

2.7.5.1 Storage Coefficient (Storativity) and Specific Storage

The amount of water taken into or released from groundwater reservoir under unit change in the piezometric level and unit horizontal area is the definition of storage coefficient, which is also called as storativity. Any fall (rise) in the piezometric level implies the release (intake) of water from the groundwater reservoir storage. The word storativity (storage coefficient), S,

is used especially for the storage properties of confined aquifers and interchangeably with specific yield for unconfined aquifers. It is defined as the volume of water that an aquifer releases from storage per unit surface area of the aquifer per unit decline in the hydraulic head normal to that surface (Figure 2.24).

Şen (1995) defined the storage coefficient, S, as the ratio of withdrawn water volume, V_W, from the storage to the total volume of dewatered zone, V_D.

$$S = \frac{V_W}{V_D} \qquad (2.15)$$

All the time, $V_D > V_w$, because the water is available in the voids only, whereas V_D includes the void and solid volumes and consequently $S < 1$ always. Theoretically, it cannot be equal or even close to one and its maximum value may be equal to the specific yield. Practically, it is much less than 0.5. The most prominent characteristic of a confined aquifer is that as the water is withdrawn the aquifer remains fully saturated. The water from the storage is released in two ways. First, by the expansion of water as the pressure is lowered and second by the specific yield (as will be defined below) of the aquifer material. The usual range of confined aquifer storage coefficient, S_c, is 10^{-3} to 10^{-5}; roughly 40% of which result from the expansion of the water and 60% from the compression of the medium as stated by Bear (1979).

Storage coefficient depends upon the whole saturation thickness of the aquifer and the division of storativity by saturation thickness, m, is called specific storage, S_S. Storativity involves no dimensions, while S_S has the dimension of inverse length (L^{-1}).

$$S_S = \frac{S}{m} \qquad (2.16)$$

Physically, S_S is described as the volume of water that a unit volume of aquifer takes into storage or releases from storage under a unit decline in hydraulic head.

Practical studies have indicated possible ranges of storage and specific storage coefficients and specific yield from experiments all over the world as follows (Rushton, 2003),

$$10^{-6} < S_s < 10^{-4}\, m^{-1}$$
$$10^{-5} < S_c < 3 \times 10^{-3}$$
$$10^{-3} < S_y < 2.5 \times 10^{-1}$$

The specific storage of various aquifer materials are given by Younger (1993) as in Table 2.12.

Furthermore, specific values of the storage coefficients are given in Table 2.13 for aquifer material type classification.

2.7.5.2 Storage Coefficient and Compressibility

The ability of an aquifer to store water is one of the most important hydraulic properties. The

TABLE 2.13   Storage Properties

Lithology Types	Specific Storage (m^{-1})
Clay	9.81×10^{-3}
Silt, fine sand	9.82×10^{-4}
Medium sand, fine	9.87×10^{-5}
Coarse sand, medium gravel, highly fissured	1.05×10^{-5}
Coarse gravel, moderately fissured rock	1.63×10^{-6}
Rock without fissures	7.46×10^{-7}

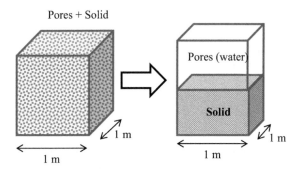

FIGURE 2.25   **Bulk density.**

groundwater in rest is under the influence of gravity and hydrostatic pressure. The former is the combined weight of solid and water above a horizontal area at any depth in the aquifer. As shown in Figure 2.25, the bulk weight is the sum of weights of solids and water.

On the other hand, bulk weight can be expressed as $\gamma_b V_T$ where γ_b is the bulk density

TABLE 2.12   Specific Storage by Assumption of Equivalent 15% Porosity (Younger, 1993)

Storage Type	Aquifer Type	Minimum	Maximum
Specific storage	Confined (pressure)	$10^{-6}\, m^{-1}$	$10^{-4}\, m^{-1}$
Storage coefficient	Confined (pressure)	10^{-5}	10^{-3}
	Unconfined (drainage)	10^{-3}	2.5×10^{-1}

and V_T is the total volume of aquifer. The bulk weight per unit area can be expressed as,

$$G = \gamma_b g h \qquad (2.17)$$

where h is the depth of saturation zone above unit area. In Eq. (2.17), the only variable is h, and consequently, its increase gives rise to an increase in total gravity force. Deep sediment is compacted under the force of gravity. The maximum compaction occurs when the voids are not filled with water. On the other hand, the hydrostatic pressure, p, of water at the same depth is,

$$p = \gamma_w g h \qquad (2.18)$$

in which γ_w is the water density. The direction of this pressure is upward provided that the bottom of representative volume is horizontal. The effective pressure, p', on the bottom is,

$$p' = (\gamma_b - \gamma_w) g h \qquad (2.19)$$

Since γ_b and γ_w are constants, any small increment in the pressure is directly related to the small increments in h as,

$$dp' = (\gamma_b - \gamma_w) g d h \qquad (2.20)$$

Furthermore, $\gamma_b > \gamma_w$ and the gravity force are always bigger than the hydrostatic pressure. The hydrostatic pressure at a point is equal at every direction with the same magnitude. It means that upward pressure works as a lifting force opposite to the gravity force whereas downward it is combined with the gravity force, and hence, these two resultant forces try to compact the formation material. The effective pressure acts as a compression stress on the solids and on the water in the voids. As a result, solid and water are compressed to a certain extent. Although the compressibility of water is smaller compared to the solids in large areas, it adds up to a significant volume. Equation (2.20) indicates that change in effective pressure (dp′) is related directly to change in the hydraulic head (dh). Table 2.14 gives the range of material compressibility in an aquifer.

TABLE 2.14   Earth Material Compressibility Ranges

Material	Compressibility Range (m²/Newton)
Clay	$10^{-6}-10^{-8}$
Sand	$10^{-7}-10^{-9}$
Gravel	$10^{-8}-10^{-10}$
Jointed rock	$10^{-8}-10^{-10}$
Competent rock	$10^{-9}-10^{-11}$
Water	$4.4 \times 10^{-10}-4.6 \times 10^{-10}$

One can conclude that the storage ability of a reservoir depends on its effective pressure. If there were no pressure, the porosity helps evaluate the available water volume in a certain reservoir; and the specific yield is for finding the volume of abstractable water. Such a situation is valid for unconfined aquifers with rather shallow saturation thickness. The specific yield concept becomes invalid in the reservoirs where the groundwater is stored under pressure; instead compressibility takes place due to the aquifer material and water compressibility.

The most distinctive characteristic of confined aquifers is that as the water is withdrawn the aquifer remains fully saturated. The overburden is supported partly by the solid grains and partly by the pressure of the water. Removal of water from the aquifer occurs at the cost of pressure drop, and therefore, more overburden must be taken by the solid grains, which results in slight compression leading to consolidation of the layer. Meanwhile, a slight expansion of water takes place. In unconfined aquifers, the volume of water derived from expansion of the water and compression is negligible. Expansion of water is relatively smaller than compression of the aquifer, and accordingly, the storage coefficient is smaller than unconfined aquifers. Field experiences have shown that in unconfined aquifers, S varies, generally, between 0.3 and 0.1, whereas in confined aquifers it has a range from 10^{-5} up to 10^{-4}. For most practical purposes, the storage coefficient of an unconfined aquifer is taken as equal to the specific yield, i.e., $S \cong S_y$. For an unconfined unit, the storativity must be calculated by the formula,

$$S = S_y + \Delta h S_s \qquad (2.21)$$

where h is the thickness of the saturated zone. In reality, the value of S_y is several orders of magnitude greater than $\Delta h S_s$ for an unconfined aquifer. For a fine-grained unit, S_y may be very small, approaching the same order of magnitude as $\Delta h S_s$.

Depending on piezometer (water table) fall, Δh, released water volume, V_w, can be calculated by the use of the storage coefficient, S, and the plan area, A, of the aquifer as follows:

$$V_w = SA\Delta h \qquad (2.22)$$

EXAMPLE 2.8 SPECIFIC YIELD AND STORAGE COEFFICIENTS

A confined aquifer has 346 km^2 areal extent with saturation and confining layer thicknesses as 125 m and 12 m, respectively. Various field tests indicated that the average storativity coefficient is 1.9×10^{-4} and the aquifer material has a specific yield of 0.18. The initial piezometric level is at 73 m above the confining layer. If there is 128 m drop in the piezometric level, then what amount of water is taken from the aquifer? Discuss the aquifer situation, in general.

Solution 2.8

A relevant figure for the exposition of the case is given in Figure 2.26 where all the relevant values are shown.

The aquifer is confined initially, but after a long period of withdrawal the initial piezometric level falls to 128 m, which causes this aquifer to behave as an unconfined aquifer. The unconfined aquifer starts after $73 + 12 = 85 \text{ m}$ fall from the initial piezometric level. This means that until 85 m drop from the initial piezometric level the aquifer is under confined conditions and for the remaining $128 - 85 = 43 \text{ m}$ the unconfined aquifer case prevails. First, the water volume, V_{co}, which comes from the confining condition, can be calculated according to Eq. (2.22) as,

$$V_{co} = 1.9 \times 10^{-4} \times 346 \times 10^6 \times 85$$
$$= 558.8 \times 10^6 \text{ m}^3$$

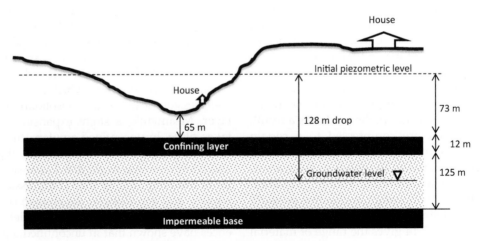

FIGURE 2.26 Aquifer composition.

The hydraulic head drop within the unconfined condition is 43 m and the use of specific yield value in the same equation yields the unconfined aquifer water volume, V_{un}, as,

$$V_{un} = 0.18 \times 346 \times 10^6 \times 43 = 2678.0 \times 10^6 m^3$$

Drilling any well in the valley gives rise to flowing well condition and the house may be affected due to groundwater inundation.

EXAMPLE 2.9 GROUNDWATER RESERVOIR VOLUME CALCULATIONS

In a confined aquifer, water-bearing layer has 10 m thickness and its area is about 100,000 m². The piezometric level is at 17 m above the bottom impermeable layer. The thickness of the overlying impermeable layer is 2 m. The hydrogeological aquifer parameters are hydraulic conductivity, 3 m/s; storage coefficient, 2×10^{-3}; specific yield, 0.01; and porosity, 0.35. Hence, calculate aquifer reservoir volume in case of mining (theoretically abstraction of even the last drop of water)

1. Confined condition,
2. Unconfined condition.

Solution 2.9

First, the aquifer configuration can be drawn even as a rough sketch as in Figure 2.27.

It is obvious that the water pressure, pw, on the lower boundary of the confining layer is only 7 m of the water column (this is the hydraulic head), which causes to the compressibility of water in the groundwater reservoir. In groundwater terminology, this is the drop in the piezometric level, Δh.

1. The confined aquifer case water volume, V_{co}, will be calculated by considering the storage coefficient, S. The water volume can be calculated from Eq. (2.22) as,
   $$V_{co} = 2 \times 10^{-3} \times 100,000 \times 7 = 1400 \text{ m}^3$$
2. In order to find the water volume, V_{wu}, under the unconfined condition, specific yield comes to play a role in the calculations, and hence, the amount of available water again from Eq. (2.22) by considering the whole saturation thickness as $\Delta h = 10$ m one can calculate,

$$V_U = 0.35 \times 100,000 \times 10 = 35,000 \text{ m}^3$$

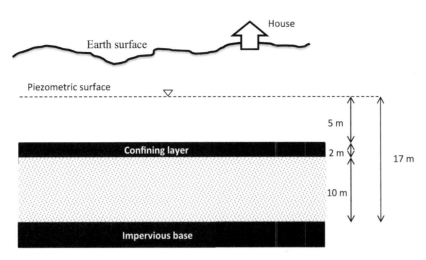

FIGURE 2.27 Aquifer configuration.

This example shows that the amount of water under the confining conditions is comparatively very small with respect to the unconfined condition. For aquifer mining, the initial (static) piezometric level should drop to the level of the bottom confining level. Such a drop can happen at two stages. Hence, total abstractable water volume, V_T, coming from the aquifer has two parts, namely, abstractable water under the confining, V_{wc}, and unconfined, V_{wu}, conditions. In general,

$$V_T = V_{co} + V_{un}$$

Substitution of the valid numerical values leads to,

$$V_T = 1,400 + 35,000 = 36,900 \text{ m}^3$$

EXAMPLE 2.10 WATER TABLE DROP AND ABSTRACTABLE WATER CALCULATION

An unconfined aquifer has 15 m saturation zone and the static water table is 8 m below the Earth's surface. During 15 days, the groundwater table dropped by 2.5 m. Previous field tests indicated that the aquifer storage coefficient varies between 0.03 and 0.05. How much maximum (minimum) water can be abstractable from this aquifer, considering its areal extent as 5 km²?

Solution 2.10

The given quantities of the problem are shown in Figure 2.28 with aquifer configuration.

The maximum (minimum) implies the consideration of the storage coefficient at its lowest (highest) values. The storage coefficient is the main factor that plays a role in water availability. Abstractable water can be calculated first by calculating the total aquifer volume, V_T. This is equal to area multiplied by the water drop, which becomes $5 \times 10^6 \times 2.5 = 12.5 \times 10^6 \text{ m}^3$. This volume consists of voids and grains (solids). Since the amount of water that can be drawn from 1 m² of the aquifer area under 1 m water drop is equal to the storage coefficient (Section 2.7.5.1), then the total abstractable water can be calculated by multiplying the above volume by the storage coefficient. Finally, the maximum (minimum) abstractable water volume from the aquifer is $0.05 \times 12.5 \times 10^6 = 0.625 \times 10^6 \text{ m}^3$ $(0.03 \times 12.5 \times 10^6 = 0.375 \times 10^6 \text{ m}^3)$.

EXAMPLE 2.11 HYDRAULIC HEAD DROP AND ABSTRACTABLE WATER

An extensive (500 km²) sandstone aquifer of thickness 500 m is overlain by 10 m shale layer and the static piezometric level is 622 m above

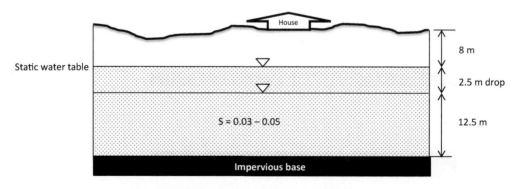

FIGURE 2.28 Groundwater description.

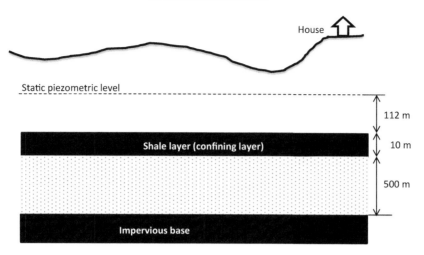

FIGURE 2.29 **Aquifer layout.**

the upper bedrock surface. The aquifer material has a storage coefficient equal to 3.7×10^{-5}. If 1.2×10^6 m^3 water should be withdrawn from this aquifer by a set of wells, then what would be the amount of piezometric level fall? Will the aquifer remain in the confining state?

Solution 2.11

Again, the drawing of the given configuration is helpful for imagination of the given quantities and configuration in Figure 2.29.

If the drop in the static piezometric level is indicated as unknown by Δh, then similar to the previous example, the total volume of the drop space from Eq. (2.22) is,

$$V_T = 3.7 \times 10^{-6} \times 500 \times 10^6 \times \Delta h$$

This volume must be equal to the volume withdrawn, and therefore, one can write that

$$3.7 \times 10^{-5} \times 500 \times 10^6 \times \Delta h = 1.2 \times 10^6$$

from which one can obtain $\Delta h = 64.86$ m. Since this is less than the confining pressure level of 122 m, the aquifer remains under the confining condition. This means that 1.2×10^6 m^3 of water is withdrawn due to the water compressibility only.

EXAMPLE 2.12 WATER WITHDRAWAL VOLUME CALCULATION FROM ADJACENT AQUIFERS

In a geological syncline, unconfined and confined aquifers exist as neighbors. The average perpendicular distance between the lower and upper impervious layers is 55 m. The aquifer material has confined and unconfined storage coefficients as 1.3×10^{-6} and 0.032, respectively, with a porosity of 0.45 and the specific retention equal to 0.235. Initially, the piezometric level is 250 m from the upper boundary of the bottom impervious layer. The areal extent of the confined aquifer is 52 km^2, whereas on both sides the total unconfined aquifer horizontal area is 23 km^2. After a 45 m fall in the piezometric level, the total unconfined aquifer area increases to 26.6 km^2 and the confined aquifer retreat ends up to 42.6 km^2. Calculate the amount of withdrawn water from this combined aquifer.

Solution 2.12

As usual, a rough figure is useful for visual inspection and interpretation of various aspects (Figure 2.30).

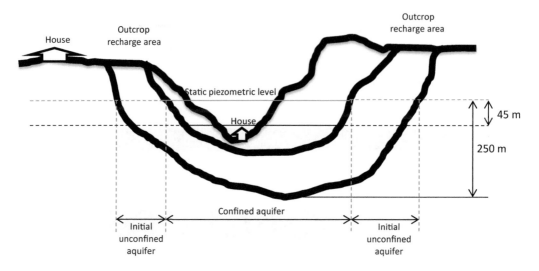

FIGURE 2.30 **Aquifer composition.**

The calculation will be achieved according to Eq. (2.22) and for its application the unconfined and confined aquifers' areal extents are necessary. Since before and after 45 m drop in the piezometric (water table) level the aquifer areas are known, one can depend on the mean areas in the calculations. Hence, the unconfined aquifer mean area, A_u, during the water level fall is, $A_u = (23.0 + 26.6)/2 = 24.8$ km^2. Similar calculation yields the confined aquifer mean area as, $A_c = (45.0 + 42.6)/2 = 43.8$ km^2. Hence, the substitution of relevant quantities into Eq. (2.22) for confined aquifer leads to water withdrawal volume,

$$V_{TU} = 45 \times 43.8 \times 10^6 \times 1.3 \times 10^{-6}$$
$$= 2562.3 \text{ m}^3$$

For the unconfined case, the specific yield is necessary and its value from Eq. (2.5) is $S_y = 0.450 - 0.235 = 0.215$. Further consideration of Eq. (2.22) with the suitable values gives the unconfined aquifer water withdrawal volume as,

$$V_{TU} = 45 \times 24.8 \times 10^6 \times 0.215$$
$$= 239.94 \times 10^6 \text{ m}^3$$

Finally, the total groundwater withdrawal is almost negligible from the confined aquifer, because the ratio of confined aquifer to unconfined aquifer withdrawal is 0.00107 percent.

EXAMPLE 2.13 SPECIFIC RETENTION AND STORATIVITY

In an unconfined aquifer there is about 6.7 m drop in the water table. The aquifer area is 8 km^2. The aquifer material is composed of sand with porosity 0.37 and specific retention 0.10.

1. Calculate the specific retention.
2. Calculate the volume change in the aquifer storage.

Solution 2.13

1. The porosity is composed of specific yield and specific retention summation as in Eq. (2.5). The necessary arrangement in this equation gives,

$$S_y = n - S_r = 0.37 - 0.10 = 0.27$$

2. The change in water volume is given by Eq. (2.22), and the substitution of convenient numerical values into this equation gives,

$$V_w = 0.27 \times 8 \times 10^6 \times 6.7 = 1.45 \times 10^7 \ m^3$$

2.7.6 Groundwater Velocity

Physically, velocity is defined as the ratio of distance to time. In the case of groundwater, it is almost impossible to trace actual flow path length, because water moves along irregular paths. Groundwater velocity is defined in three different ways. First, the real velocity, V_r, is defined as the ratio of flow-line path length to time as shown in Figure 2.31. The real velocity can be expressed as the ratio of the flow line path length, L, to time t.

$$V_r = \frac{L}{t} \tag{2.23}$$

On the other hand, the actual velocity, V_a, is defined as the ratio of straight-line distance, D, between two sections to the time taken by the water molecules to cover this distance as shown in Figure 2.32. It is possible to measure actual velocity on the field by using tracers.

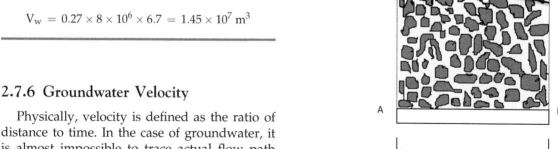

FIGURE 2.32 **Actual groundwater velocities.**

$$V_a = \frac{D}{t} \tag{2.24}$$

The third is referred to as "filtration velocity" or "specific discharge" or "Darcy velocity," V_f. It is defined as the volume of water passing through the unit cross-sectional area, A, per unit time, which is by definition the discharge, Q, (see Figure 2.33).

$$V_f = \frac{Q}{A} = \frac{Q}{Wm} \tag{2.25}$$

In the last definition, water is assumed to cross the whole aquifer section as if there were

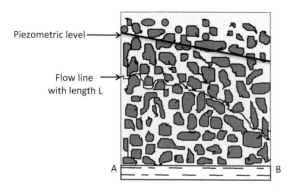

FIGURE 2.31 **Real groundwater velocities.**

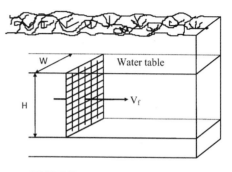

FIGURE 2.33 **Filtration velocity.**

TABLE 2.15   Critical Specific Discharge Values (m/day)

Porosity (%)	Grain Size (mm)					
	0.1	0.2	0.5	1	2	4
25	39.0	19.6	7.8	3.9	1.9	0.4
30	30.5	15.2	6.0	3.0	1.5	0.3
35	24.3	12.1	4.8	2.4	1.2	0.24
0	19.6	9.8	3.9	1.9	1.0	0.21

no grains. The following relation can relate filter velocity to real velocity through porosity.

$$V_f = nV_r \qquad (2.26)$$

Flow through a medium of solids and voids (grains, fractures, solution cavities) allows passage of a certain volume of water, V_w, during a certain time interval, Δt, and hence, the discharge, Q, is defined as the volume of water per time.

$$Q = \frac{V_w}{\Delta t}$$

Since the volume is area, A, multiplied by the length of water movement path, L, ($V_w = AL$), then the discharge can be rewritten as,

$$Q = \frac{AL}{\Delta t} \qquad (2.27)$$

Physically, $L/\Delta t$ is the filter velocity, v_f, of the water flow, and hence, the discharge can be written also as,

$$Q = Av_f \qquad (2.28)$$

This is the most frequently used discharge formulation in practical applications. The filter velocity is the one which is used in groundwater problem solutions and it is within the Darcy's law.

Depending on the porosity and the grain size, critical specific discharge values are given in Table 2.15 (Şen, 1995).

2.8 DARCY'S LAW

Darcy (1856) suggested his law after a series of experimental studies. He performed experiments to deduce the nature of flow laws in saturated porous media. His simple experimental setup is shown in Figure 2.34 in which he used sand as a medium and water as a fluid.

The water flow is possible under the hydraulic head, Δh. Herein, rational and logical rules will be used for the Darcy's law. Logically,

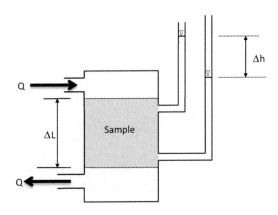

FIGURE 2.34   Darcy apparatus setup.

groundwater filter velocity, V_f, is directly proportional to the hydraulic gradient, $i = \Delta h/\Delta L$. Hence, in general, one can write the proportionality relationship as,

$$V_f \, \alpha \, \frac{\Delta h}{\Delta l}$$

Herein, α shows the proportionality sign. In order to convert this proportionality into an equation form, it is necessary to import a constant, K, which yields,

$$V_f = K\frac{\Delta h}{\Delta l} \qquad (2.29)$$

This is the logical derivation of Darcy's law. The hydraulic gradient is dimensionless and the proportionality constant has the same unit as the velocity. It is equivalent to the groundwater velocity per unit hydraulic gradient. K should reflect the medium feature for fluid flow. In general, the more is the interconnected void percentage in the flow cross-section, the bigger will be the K value. This again logically implies that the coarser the grains in a porous medium the bigger is the K value. Logic cannot tell numerical values, and therefore, determination of K value requires field or laboratory experiments. In groundwater literature, K is referred to as the permeability of hydraulic conductivity (Section 2.7.3). If one considers the product of Eq. (2.29) by unit cross-sectional area, then it is the amount of discharge that passes from the unit area. This is named as the specific discharge, q. Darcy succeeded in formulating an empirical relationship among different variables as volumetric flow rate through a homogeneous and isotropic media, perpendicular to the unit cross-sectional area, which is directly proportional to the hydraulic gradient.

$$q = \frac{Q}{A} = K\frac{\Delta h}{\Delta l} \qquad (2.30)$$

in which Q is the discharge through the cross-sectional area, A, perpendicular to the flow direction, K is the proportionality constant known as the hydraulic conductivity of the medium, and $\frac{\Delta h}{\Delta l}$ is the hydraulic gradient.

EXAMPLE 2.14 CONSTANT HEAD PERMEAMETER

As shown in Figure 2.35, a constant head permeameter, similar to Darcy's experiment in Figure 2.33, has a soil sample of 30 cm length with the cross-sectional area equal to 100 cm². As 0.1 cm³/s discharge goes through this soil sample, the hydraulic head falls 9 mm. Calculate the hydraulic conductivity of the soil.

Solution 2.14

Permeameters are used for hydraulic conductivity measurements in the laboratory. The given quantities with usual notations are as follows.

The length of soil sample	: L = 30 cm
The cross-sectional area	: A = 100 cm²
Head fall	: Δh = 9 mm
Discharge	: Q = 0.1 cm³/s

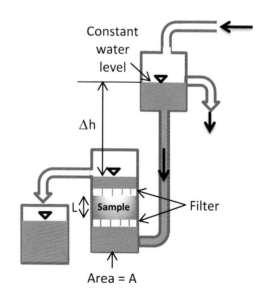

FIGURE 2.35 Constant head permeameter.

The velocity of the flow is $V = Q/A = 0.1/100 = 0.0001$ cm/s.

The slope of the piezometer line is $i = \Delta h/L = 0.9/30 = 0.03$.

The substitution of these values into basic Darcy's law (Eq. (2.29)) yields the hydraulic conductivity of the soil as,

$$K = V/i = 0.0001/0.03 = 3.3 \times 10^{-2} \text{ cm/s}$$

EXAMPLE 2.15 VARIABLE HEAD PERMEAMETER

Figure 2.36 indicates a variable head permeameter where a soil sample of 50 cm length is located with its cross-sectional diameter of 20 cm. During the test in the vertical 2 mm diameter pipe the hydraulic head falls from 100 to 70 cm during 19 s. Calculate the hydraulic conductivity of the soil.

Solution 2.15

The hydraulic conductivity measurements of low conductivity soils are difficult to measure.

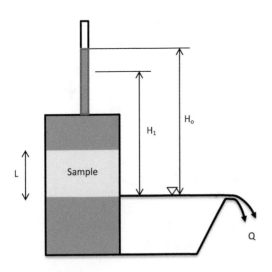

FIGURE 2.36 Falling head permeameter.

Therefore, it is preferable to measure hydraulic conductivity of such soils in falling head permeameters. The data for the problem are as follows.

a   : The cross-sectional area of the pipe of water entrance,

A   : The cross-sectional area of the soil sample,

L   : The length of soil sample,

H_o : Initial water height in the permeameter (hydraulic load),

H_1 : The second water height in the permeameter.

During the time interval, dt, the head fall is dH and the water volume decrease in the pipe is equal to adH/dt. The amount of water through the sample can be calculated from Darcy's law (Eq. (2.30)) as AKdh/dl. These two amounts' equality gives,

$$\frac{AKH}{L} = -a\frac{dH}{dt}$$

The sign on the right-hand side is negative because of the head fall. It is possible to write this expression as follows,

$$-a\frac{dH}{dH} = \frac{KAdt}{L}$$

Its integration leads to,

$$-a\int_{H_O}^{H_1} \frac{dH}{H} = K\frac{A}{L}\int_0^{t_1} dt$$

Finally

$$K = \frac{a}{A}\frac{L}{t_1}\text{Ln}\left(\frac{H_o}{H_1}\right)$$

The substitution of the numerical values into this last expression yields,

$$K = \frac{\pi}{100\pi} \times \frac{50}{18} \times \text{Ln}\left(\frac{100}{70}\right) = 0.01 \text{ cm/s}$$

EXAMPLE 2.16 HYDRAULIC CONDUCTIVITY MEASUREMENT WITH TRACER

The distance between two wells is 50 m. In well A the static water level is 125 m, whereas in well B it is 124.5 m. A radioactive tracer, given into the well A, reaches well B after 4 hours. The sample porosity of the soil is 14%. Hence,

1. Find the hydraulic gradient in the aquifer.
2. In which direction does the groundwater flow? What is the groundwater real velocity? Calculate the filtration velocity.
3. Calculate the hydraulic conductivity of the aquifer.
4. The thickness of this aquifer is 20 m, hence accordingly calculate the transmissivity coefficient of this aquifer.

Solution 2.16

1. The slope of the groundwater surface is the hydraulic gradient as,

$$\Delta I = \frac{dH}{dL} = \frac{125.0 - 124.5}{50} = 0.01$$

2. The flow direction is from A to B, because the groundwater level in well A is higher than in well B. The real velocity of the flow can be found from the physical definition of the velocity as,

$$V_s = \frac{L}{t} = \frac{50 \times 100}{4 \times 3600} = 0.35 \text{ cm/s}$$

Hence, the filtration velocity can be calculated by use of the porosity value as,

$$V_f = n \times V_s = 0.14 \times 0.35 = 0.049 \text{ cm/s}$$

3. By use of Darcy's law, the hydraulic conductivity of the same aquifer can be calculated as,

$$K = \frac{V_f}{i} = \frac{0.049}{0.01} = 4.9 \text{ cm/s}$$

4. The thickness of the aquifer is m = 20 m and accordingly the aquifer transmissivity can be calculated as,

$$T = mK = 20 \times 100 \times 4.9 = 9800 \text{ cm}^2/\text{s}$$

2.8.1 Non-Darcian Flow

Darcy's law is valid for laminar flow, which appears at very low velocities where water molecules move along almost smooth and linear paths parallel to the pore's solid boundaries. Increase in the velocity of water causes more energy losses and steeper hydraulic gradients and water moves along irregular paths with random movements and in this case Darcy's law is not valid. The flow regime is expressed through the Reynolds number, Re, which is given as,

$$\text{Re} = \frac{\rho \, vD}{\mu} \tag{2.31}$$

where v is the fluid (groundwater) velocity; D is the dimension of the water conduit; ρ is the density of fluid; μ is the viscosity of fluid (g/sm-s); and finally, μ/ρ is the kinematic viscosity of the fluid. The following criteria is valid for deciding whether the flow is laminar (abide with Darcy's law) or turbulent.

- If Re = <2100 the flow is laminar.
- If Re is in between 2100 and 4000, it is intermediate.
- If Re > 4000 the flow is turbulent.

In the vicinity of a pumping well, the flow often enters into turbulent domain because of high hydraulic gradients, vertical flows due to partially penetrating wells, etc. These cases happen near the well vicinity causing extra well losses (Chapter 3). The important thing about Darcy's law is that it is a linear flow law and should be used strictly in that context. Laboratory experiments by many investigators and field tests have shown already that in cases of high hydraulic conductivities and/or steep hydraulic gradients it is necessary to care for nonlinear laws to describe the flow. Such flows are also related to coarse-grained porous medium, where the groundwater can move easily at high velocities causing nonlinear (turbulent) flow as in Figure 2.37 (Basak, 1977; Şen, 1989).

In general, deviations from Darcy's law are expected due to two groups of factors. In the first group, there are factors that are related to flow properties such as specific discharge and hydraulic gradients, whereas in the second group there are characteristics of the aquifer material such as effective porosity and hydraulic conductivity in the coarse-grained aquifers.

There is now a large body of experimental and theoretical evidence to show that groundwater flow occurs through porous media under the influence of gradients other than that of the hydraulic gradient (Freeze and Cherry, 1979; Gurr et al., 1952; Philip and de Vries, 1957; Hoekstra, 1966; Harlan, 1972).

Chemical gradients can also cause the flow of water (as well as the movement of chemical constituents through the water) from regions of higher salinity to lower salinity, even in the absence of other gradients including the hydraulic gradient. The role of chemical gradients can be important, but their direct influence on the movement of chemical constituents is of major importance in the analysis of groundwater contamination. Figure 2.38 is a very simple experimental design, which indicates clearly that changes in hydraulic conductivity with distance influence the hydraulic gradient although initial and final piezometric levels remain constant.

It is necessary to have groundwater flow models with changing hydraulic conductivity effects so as to decide whether conventional Darcian groundwater flow actually occurs or a modified Darcian groundwater flow takes place due to various effects. One of such effects is the hydraulic conductivity gradient (Şen, 2013a). Consideration of each gradient leads to a more general flow law than the classical Darcy equation, which can be written as (Freeze and Cherry, 1979)

$$v(r) = c_1 \frac{dh(r)}{dr} + c_2 \frac{dT(r)}{dr} + c_3 \frac{dc(r)}{dr} \qquad (2.32)$$

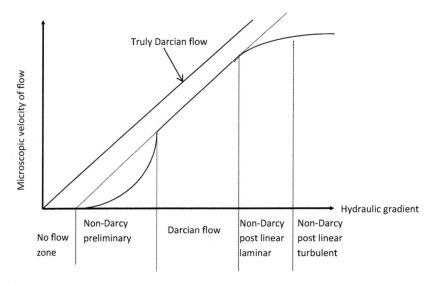

FIGURE 2.37 **Different flow zones.**

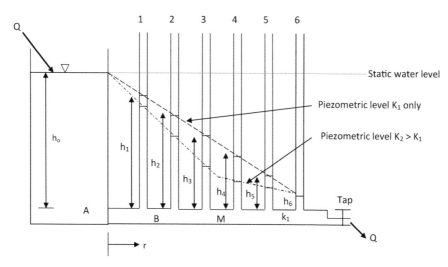

FIGURE 2.38 **Experiment to illustrate the effect of permeability change.**

where $h(r)$ is hydraulic head, $T(r)$ is temperature, and $c(r)$ is chemical concentration; r is the radial distance; c_1, c_2, and c_3 are constants of proportionality. If $dc(r)/dr = 0$, then the fluid flow occurs in response to both a hydraulic head gradient and a temperature gradient;

$$v(r) = c_1 \frac{dh(r)}{dr} + c_2 \frac{dT(r)}{dr} \qquad (2.33)$$

Without temperature difference, Eq. (2.33) becomes identical to the classical Darcy's law when $c_1 = k$.

In the classical analytical studies, the groundwater flow velocity is assumed to change linearly through a constant hydraulic conductivity with hydraulic gradient only (Eq. (2.29)). In nature, hydraulic conductivity gradient (change) also leads to changes in the groundwater velocity coupled with the hydraulic gradient. In the classical aquifer tests, hydraulic conductivity around the wells is assumed as homogeneous, isotropic, and hence, constant. Is it not possible that it may change with distance from the well? Physically, hydraulic conductivity, and hence,

transmissivity, is not constant but varies with distance (Şen and Wagdani, 2008).

2.8.1.1 Non-Darcian Flow Transmissivity Definition

In groundwater literature, the concept of transmissivity is taken as granted universally valid under whether the flow regime is linear (Darcian) or nonlinear (non-Darcian). Nonlinearity in the flow at the pumping wells is attributed solely to "well losses" as will be explained in Chapter 3. Transmissivity definitions in the literature are valid only for Darcian groundwater flows. Therefore, erroneous results are obtained in groundwater resources evaluation, if the groundwater flow is non-Darcian, which is exemplified by polynomial (Forchheimer) or power (Escande) laws.

In the classical definition of transmissivity, the aquifer material is assumed to be homogeneous and isotropic; the flow is Darcian and two dimensional. However, Bear (1979) stated that in three-dimensional flow through porous media the transmissivity concept is meaningless. In nature, there are many occasions where Darcy's law is not valid strictly and therefore, the classical transmissivity concept will not be valid in such situations. A comprehensive literature review of nonlinear flow regime laws is presented by Şen (1989, 1995). These laws are categorized into two major groups as "polynomial" or "power" laws. The most widely used polynomial law is due to Forchheimer (1901) who expressed the hydraulic gradient in terms of groundwater velocity, v, as,

$$i = av + bv^2 \qquad (2.34)$$

in which a and b are constants depending on the aquifer material composition within the aquifer and i is the hydraulic conductivity. Note that for low v, $av \gg bv^2$, hence $i = av$ which is equivalent to Darcy's law provided that $K = 1/a$ and $v = Ki$. However, for high v, $bv^2 \gg av$, and hence, the flow is completely turbulent with power flow law as $i = bv^2$. In the case of Escande (1953) power law, the exponent 2 is replaced by n ($v = K'i^n$, K' is a constant). Generally, in Eq. (2.34), "a" can be referred to as laminar flow coefficient whereas "b" is the turbulence factor. Values of a and b for an extensive list of different types of porous media are given by Basak (1977). If v is made the subject, from Eq. (2.34), the only physically plausible result emerges as,

$$v = \frac{-a + \sqrt{a^2 + 4bi}}{2b} \qquad (2.35)$$

This leads to an indeterminate form $0/0$ for $b \to 0$. In order to get rid of this indeterminate form according to L'Hospital's rule by taking individually the derivatives of numerator and denominator, one can obtain after the necessary algebra $v = i/K = Ki$. After having proved that the classical transmissivity definition is not valid in non-Darcian groundwater flow cases, the practical questions are:

1. How to decide that a non-Darcian groundwater flow occurs in the aquifer? Practical answers are already presented to a certain extent by Şen (1989, 1990).

2. How to calculate the discharge that crosses through a given cross-section?

In a non-Darcian flow aquifer test analysis, transmissivity is defined in a similar way to the Darcian flow but its direct use in groundwater resources assessments is quite invalid. However, the non-Darcian transmissivity calculation plays an interim role in giving rise to some non-Darcian flow parameters such as a and b in the polynomial and K' and n in the power-law cases. For instance, Şen (1988) defined the Forchheimer flow transmissivity, T_F, as,

$$T_F = \frac{m}{a} \qquad (2.36)$$

and for the power law as (Şen, 1989)

$$T_P = m(K')^{1/n} \qquad (2.37)$$

respectively. None of these definitions corresponds to unit hydraulic gradient. However, the substitution of Eq. (2.35) into Eq. (2.28) leads to Forchheimer discharge, Q_F, expression as,

$$Q_F = \frac{\sqrt{a^2 + 4bi} - a}{2b} \qquad (2.38)$$

EXAMPLE 2.17 NON-DARCIAN TRANSMISSIVITY CALCULATION

It is rather difficult to find nonlinear flow aquifer test data in the literature (Şen, 1989). The first of such data are presented by Dudgeon et al. (1973) who treated the data with the numerical solution of the groundwater movement equation based on the Forchheimer flow law (Eq. (2.34)). The basis for using the polynomial solution in the first example is the fractured medium with large blocks, which lead to turbulent flow. They calculated the relevant aquifer parameter estimates as the transmissivity from Forchheimer flow $T_F = 2.85 \text{ ft}^2/\text{min}$ and $b = 6.5 \text{ (min/ft)}^2$. The aquifer thickness is $m = 10 \text{ ft}$ and the use of Eq. (2.36) leads to $a = 0.285 \text{ ft/min}$. With the assumption of Darcian flow in the aquifer, the discharge that passes from a cross-section, say, of width 5 ft under a hydraulic gradient 0.1 becomes from Eq. (2.30) for Darcian flow case $Q_D = 1.425 \, ft^3/min$. Under the light of the previously mentioned facts, this is an unreliable result since the flow regime is non-Darcian and therefore, the discharge should be calculated from Eq. (2.38) which gives after the substitution of relevant quantities, $Q_F = 5.20 \, ft^3/min$. Hence, the Darcian flow transmissivity concept underestimates the actual discharge by almost 73% relative error. This result indicates a very significant error in discharge calculations if non-Darcian flow is ignored and the flow is treated as if it was Darcian.

On the other hand, the equivalent transmissivity T_E under the assumption of Darcian flow can be obtained by the linear fit to actual non-Darcian data, which is different for T_F and T_P as defined previously through Eqns (2.36) and (2.37), respectively. The equivalent transmissivity for the first example is $T_E = mK = 1.176 \times 10^1 \, ft^2/min$. The estimated discharge, from Eq. (2.30), becomes $5.88 \, ft^3/min$. The actual discharge was calculated in the paper to be $5.20 \, ft^3/min$ and it overestimates the Darcian discharge by 13% relative error.

Another example is due to Şen (1989) who has presented type-curve solutions for the power type (Escande) of groundwater flow law. After performing the necessary type-curve matching procedure to the data, the relevant aquifer parameters are calculated as $T_P = 200 \, m^2/day$ with $n = 1.1$. If the conventional transmissivity definition is considered, then the amount of groundwater flow through a cross-section of 6 m width under a hydraulic gradient of 0.01 becomes, from Eq. (2.30), $Q_D = 12 \, m^3/day$. However, the non-Darcian flow consideration invalidates the transmissivity usage in such calculations and rather Eq. (2.37) should be adopted. Consequently, after the substitution of relevant quantities into this expression and after necessary calculations one can find that $Q_P = 7.57 \, m^3/day$. This indicates that in the case

of the Darcian law, 59% overestimation occurs in the groundwater discharge.

2.9 HETEROGENEITY AND ANISOTROPY

As already mentioned, transmission properties of the aquifers are related to hydraulic conductivity, K, of the geological materials. It may show variations at different points in a geologic formation or along different directions at the same point. The former is called heterogeneity and the latter is referred to as anisotropy. The presence of heterogeneities and anisotropies in geological materials is an indication of the scale of variation in terms of time and space, and geological processes are responsible for the fabrication of these materials. Researchers, who are inclined very much to classical techniques, take heterogeneity and anisotropy as malicious conspiracies of nature to maximize their interpretive and analytical difficulties as mentioned by Freeze and Cherry (1979). A reservoir medium is regarded as an anisotropic domain if its basic hydraulic properties are dependent on direction.

An isotropic medium where permeability is equal in all directions in an aquifer is a theoretical assumption, which is used for the simplification of governing groundwater equation analytical solutions in porous media. This assumption is considered valid both in large-scale hydrology (Dullien, 1992) and comparatively in small-scale flow through permeable sediments (Huettel et al., 1998). On the contrary, natural sediments are all anisotropic in their simplest modes, where vertical hydraulic conductivity is different from horizontal conductivity.

Heterogeneity is the property of the aquifer medium where hydraulic conductivity is different in one place from that measured in another. It is observed through many field studies that natural sediments have both anisotropy and spatial heterogeneity which affect both the pattern and rates of porous medium flow (Dullien, 1992; Dagan, 1982; Freeze and Cherry, 1979). It has been stated by Vogel and Roth (2003) that a change of scale through several orders of magnitude in length is necessary to achieve mapping of permeability at basin scale.

In order to better understand and solve groundwater flow problems, there is the need for a scientific methodology which should be able not only to locate the hidden variability of the geological systems, but also to join them in a model to obtain quantitative solutions. Heterogeneous and anisotropic nature of the groundwater systems is discussed by Şen (1995), Kruseman and de Ridder (1990), Schad and Teutsch (1994), and Knochenmus and Robinson (1996). In nature, isotropy (anisotropy) and homogeneity (heterogeneity) occur in pair-wise combinations as shown in Figure 2.39.

The general definition of transmissivity has been given already by Eq. (2.11). Let the thicknesses and hydraulic conductivities be m_1, m_2, m_3, ..., m_n and k_1, k_2, k_3, ...,k_n, respectively, where n is the number of different layers in a multiple aquifer such as in Figure 2.40.

First, the horizontal flow from left to the right is considered with total discharge, Q. Since no water is gained or lost in passing through the various layers, the principle of continuity leads to,

$$Q_h = Q_1 + Q_2 + \dots + Q_n$$

For unit width (W = 1) of the aquifer cross-section, the individual discharges can be written by considering Darcy's law as,

$$Q_1 = k_1 m_1 i_1$$
$$Q_2 = k_2 m_2 i_2$$
$$\vdots \quad \vdots$$
$$Q_n = k_n m_n i_n$$

Horizontal flow in each layer has the same hydraulic gradient, and hence, the substitution of these equations into the previous one gives,

$$Q_h = (k_1 m_1 + k_2 m_2 + \dots + k_n m_n)i_h \quad (2.39)$$

Substitution of this equation into the specific discharge expression (Eq. (2.30)) yields,

$$q_h = \frac{k_1 m_1 + k_2 m_2 + \dots\dots k_n m_n}{m_1 + m_2 + \dots\dots m_n} i_h$$

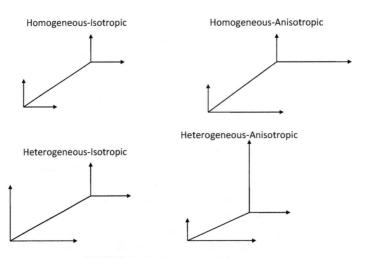

FIGURE 2.39 **Isotropy and homogeneity.**

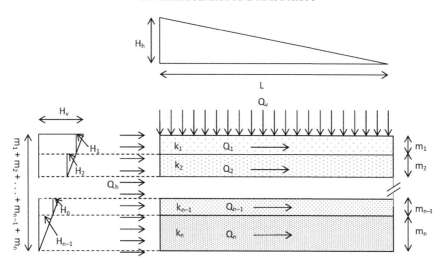

FIGURE 2.40 Multilayer aquifers (heterogeneity).

which means that the horizontal hydraulic conductivity, k_h, for the section considered is,

$$k_h = \frac{k_1 m_1 + k_2 m_2 + \ldots\ldots k_n m_n}{m_1 + m_2 + \ldots\ldots m_n} \quad (2.40)$$

This last expression implies that horizontal hydraulic conductivity is the weighted average of the individual hydraulic conductivities with layer thicknesses being the weights. It is interesting to notice that if the layers have the same thicknesses, then Eq. (2.40) takes the form as,

$$k_h = \frac{k_1 + k_2 + \ldots\ldots k_n}{n} \quad (2.41)$$

where n is the number of layers. This last expression implies that only in the case of equal layer thicknesses, the average hydraulic conductivity is equivalent to the arithmetic average of hydraulic conductivities.

As a second case, let us consider vertical flow for which the overall hydraulic gradient is equal to the summation of the individual hydraulic heads divided by the total thickness (see Figure 2.40).

$$i_v = \frac{H_v}{m} = \frac{H_1 + H_2 + \ldots\ldots + H_n}{m_1 + m_2 + \ldots\ldots + m_n} \quad (2.42)$$

Each layer allows passage of the same total vertical discharge Q_v, and therefore, the specific discharge in each layer, q_v, is the same. The application of Darcy's law in each layer gives the individual head losses as,

$$H_1 = \frac{m_1}{k_1} q_v$$

$$H_2 = \frac{m_2}{k_2} q_v$$

$$\vdots \quad \vdots \quad\quad (2.43)$$

$$H_n = \frac{m_n}{k_n} q_v$$

Substitution of Eq. (2.43) into Eq. (2.42) yields,

$$q_v = \frac{m_1 + m_2 + \ldots\ldots m_n}{\frac{m_1}{k_1} + \frac{m_2}{k_2} + \ldots\ldots \frac{m_n}{k_n}}$$

which is tantamount to saying that the vertical hydraulic conductivity, k_V, is,

$$k_V = \frac{m_1 + m_2 + \ldots\ldots + m_n}{\frac{m_1}{k_1} + \frac{m_2}{k_2} + \ldots\ldots + \frac{m_n}{k_n}} \qquad (2.44)$$

Most frequently, multiple layers occur in the sedimentary rocks. In fact, they are often anisotropic with respect to hydraulic conductivity because they contain grains which are not spherical but elongated in one direction. During deposition, these grains settle with their longest axes more or less horizontally and this usually causes the horizontal hydraulic conductivity to be greater than vertical conductivity in a single layer. However, when many layers are considered then the bulk hydraulic conductivity of sediments is usually much greater than the vertical counterpart. In order to prove this point from the previous equations by considering two layers, only the ratio k_h/k_V can be written as,

$$\frac{k_H}{k_V} = \frac{m_1^2 + \left(\frac{k_1}{K_2} + \frac{k_2}{K_1}\right)m_1m_2 + m_2^2}{m_1^2 + 2m_1m_2 + m_1^2} \qquad (2.45)$$

The hydraulic conductivity ratio terms in the denominator are always greater than 2, since if x is any number then $x + (1/x) > 2$. Consequently, always $k_H > k_V$. It is also true for alluvial deposits, which are usually constituted by alternating layers or lenses of sand and gravel on occasional clays.

EXAMPLE 2.18: HETEROGENEITY ASSESSMENT

In an area five field tests are carried out each with transmissivity and storativity coefficients as in Table 2.16. According to ±5% relative error definition, the aquifer might be classified into different sets.

1. Is this aquifer homogeneous from storativity point of view?
2. Is it homogeneous from transmissivity point of view?

3. Which group of wells is homogeneous from storativity side?
4. Which group of wells is homogeneous from transmissivity side?
5. Which group of wells is homogeneous from both parameter sides?

Solution 2.18

For the solution, it is necessary to consider the maximum and minimum parameter values as in Table 2.17.

The relative error percentages are defined herein as the absolute value of difference between the maximum and minimum divided by the maximum and the result multiplied by 100. The result is the relative error percentage as in the last column of Table 2.17.

1. Since the relative error is 85%, the aquifer does not have regional storativity homogeneity.
2. Since the relative error is 60%, the aquifer does not have regional transmissivity homogeneity.
3. For this purpose, the following matrix of relative error percentages is prepared for the storage coefficients.

TABLE 2.16   Well Data

Well Number	W1	W2	W3	W4	W5
Storativity ($\times 10^{-3}$)	1.3	2.4	2.75	8.72	2.3
Transmissivity (m^2/day)	136	129	147	70.1	59.8

TABLE 2.17   Relative Errors

Well Number	Max.	Min.	Relative Error (%)
Storativity ($\times 10^{-3}$)	8.72	1.3	85.0
Transmissivity (m^2/day)	147	59.8	60.0

	W2	W3	W4	W5
W1	45.8	52.7	85.1	43.5
W2		12.7	72.5	4.2
W3			68.6	15.8
W4				73.6

There is only one case less than 5%, which is between W_2 and W_5. So, one can conclude that at 5% relative error levels of W_2 and W_5 have similarities.

4. Similar relative error matrix is prepared by taking into consideration the transmissivity pairs.

	W2	W3	W4	W5
W1	5.1	7.5	48.6	56.0
W2		12.2	43.3	50.9
W3			52.3	59.3
W4				14.7

From transmissivity point of view at 10% level wells W_1, W_2, and W_3 make a homogeneous group.

2.10 WATER BUDGET

This term is used commonly in practical water-related sciences but its physical synonyms are continuity equation, mass balance, or mass conservation principle. Herein, the water budget will be considered from theoretical and practical application points of view. The former helps to derive analytical formulations in groundwater sciences whereas the latter is for practical uses only. However, both of them have the same scientific philosophical, logical, and linguistic information.

2.10.1 Theoretical Formulations

In the following sequel, two different theoretical derivations are presented based not on complicated mathematical principles but first on rational and logical physical principles linguistically and then arithmetically leading to mathematical formulations.

2.10.1.1 Confined Condition

This has some different physical mechanisms than the unconfined (free surface) case (Chapter 4). The water is considered under pressure, and hence, the compressibility of water is taken into consideration. Water balance (equilibrium) is concerned with a certain representative volume of water (control volume) whether it is in balance (equilibrium) or not (nonequilibrium). A certain representative control volume, as in Figure 2.41, provides a basis for further deductions.

The quantity, Q, (water, oil, gas, or any concentration of material) is in the state of balance if there are no source, sink, and compressibility effects on and within the volume. Nonequilibrium implies that the total input, total output, and the change in the control volume storage summations are equal to zero. However, in

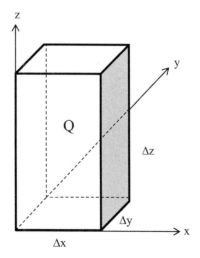

FIGURE 2.41 **Volume of water.**

equilibrium state total input is equal to total output without any change in the storage. In nature, there are always changes, which are characterized with respect to space dimensions (x, y, and z as geometric directions) and/or time dimension, t. Apart from balance status (equilibrium), there are three types of changes,

1. Spatially with respect to geometrical dimensions only,
2. Temporally with respect to time only,
3. Spatio-temporal variation with simultaneous geometrical and time considerations.

In fluid dynamics, balance status is referred to as steady state and the others are all in unsteady state regimes. Let us explain these cases on the basis of rational and logical thinking. In order to visualize the steady state case, one can imagine that there are quantity changes along each spatial direction. The condition that the quantity remains the same means that there is no quantity change, and therefore, the summation of these changes must be equal to zero. Linguistically, one can write down the expression as,

(change in x-direction) + (change in y-direction) + (change in z-direction) = zero.

This expression can be translated into mathematical terms by considering the changes along each direction as dQ/dx, dQ/dy, and dQ/dz, which render previous linguistic expression into the mathematical form as,

$$\frac{dQ}{dx} + \frac{dQ}{dy} + \frac{dQ}{dz} = 0 \qquad (2.46)$$

In case of radial flow, there is only one dimension (radial direction, r), and hence, this last expression takes the following form.

$$\frac{dQ}{dr} = 0 \qquad (2.47)$$

There are implied assumptions behind these formulations. The following points are worth to ponder about.

1. Since the same Q is considered along each direction, the medium is considered as isotropic.
2. The control-volume dimensions are kept constant, because the representative volume is within the fluid domain entirely without any moveable boundary.
3. Instead of partial variations, ordinary variation is considered. Partial variations are expressed by ∂ sign as mathematical convenience.

After all these points Eq. (2.46) can be written in the form of partial variations along each orthogonal axis as follows,

$$\frac{\partial Q_x}{\partial x} + \frac{\partial Q_y}{\partial y} + \frac{\partial Q_z}{\partial z} = 0 \qquad (2.48)$$

where Q_x, Q_y, and Q_z express anisotropy of the medium and/or fluid.

In case of unsteady state flow, time variability also plays role in the overall changes, and therefore, another time-change term must be added to Eq. (2.48). The only possible change with time, t, is the hydraulic head, h(x, y, z, t), and hence, in an infinitesimally small time duration this partial change can be expressed mathematically as $\partial h(x, y, z, t)/\partial t$. The collective change becomes,

$$\frac{\partial Q_x}{\partial x} + \frac{\partial Q_y}{\partial y} + \frac{\partial Q_z}{\partial z} + c \frac{\partial h(x, y, z, t)}{\partial t} = 0 \quad (2.49)$$

Geometric and time changes have different units and in order to convert time changes into geometry domain units, the time change is multiplied by a converter constant, c.

2.10.1.2 Unconfined Condition

It has phreatic surface (water table), which is defined as the upper boundary of the saturation layer through which the aquifer receives direct recharge as in Figure 2.42. In this figure, there are two independent systems. The first one is the Cartesian coordinate system and the other

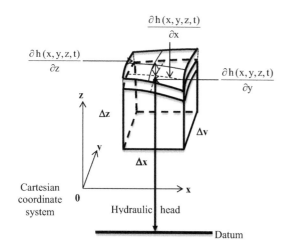

FIGURE 2.42 **Phreatic groundwater levels.**

is the hydraulic head system with respect to a datum.

Whatever is the object for movement within the groundwater material there will not be movement without hydraulic gradient. In general, there are three geometric gradients as $\partial(x, y, z, t)/\partial x$, $\partial(x, y, z, t)/\partial y$, and $\partial(x, y, z, t)/\partial z$; and the time gradient is $\partial(x, y, z, t)/\partial t$. In case of phreatic groundwater, the only time-wise change takes place vertically with addition (infiltration, recharge) or abstraction (withdrawal) of water. This causes vertical rise or fall in the phreatic surface, which is represented by $\partial(x, y, z, t)/\partial z$. Two factors physically play a role in groundwater movement.

1. Hydraulic gradient changes in x, y, and z directions. No hydraulic gradient means no causative effect on possible groundwater movement.
2. Even though there may be causative hydraulic gradients, if the groundwater medium does not allow movement, there will not be any groundwater movement.

If the groundwater medium's ability to allow water movement along individual directions are m_x, m_y, and m_z, then the movement takes place

logically when above mentioned two factors are in functional operation. Let us consider these directional groundwater movements along the following steps.

1. In order to have movement along x direction, both causative factor (hydraulic gradient, $((x, y, z, t)/\partial x)$, and medium groundwater movement ability, m_x,) must occur simultaneously, which implies logical ANDing operation (Şen, 2013b). Hence, the overall movement capacity can be expressed arithmetically only by multiplication operation, which yields $m_x\partial(x, y, z, t)/\partial x$ term.
2. Similar arguments for y direction yield that the movement is possible only with the effective role of the two factors along this direction leading to $m_y\partial(x, y, z, t)/\partial y$ term.
3. The movement along the vertical Cartesian coordinate direction due to the phreatic surface can take place either as a result of addition (subtraction) of water with time leading to vertical time-wise change, i.e., $\partial(x, y, z, t)/\partial t$ or the ability of vertical movement allowance of the medium, m_z.

After all these explanations, the water budget (balance) for the water column in Figure 2.27 can be expressed as follows,

$$m_x\frac{\partial h(x, y, z, t)}{\partial x} + m_y\frac{\partial h(x, y, z, t)}{\partial y}$$
$$+ m_z\frac{\partial h(x, y, z, t)}{\partial z} + \alpha\frac{\partial h(x, y, z, t)}{\partial t} = 0$$

(2.50)

In this expression, α is a logical unit conversion factor, which appears as a result of dimensional analysis. In the derivation of this expression, the compressibility of water has not been taken into consideration. This is the correct formulation derived by rational and logical reasoning.

However, in many books including Bear (1972), Batu (1998), and Delleur (2007) the same mistake is being repeated, which has been explained by some mathematical derivations (Chen et al., 2009).

2.10.2 Practical Formulations

The representative geometry in Figure 2.41 is considered as perfect prism, which is not available in practical studies. In practice, most often one is concerned with irregular geometries such as drainage area, A, as in Figure 1.5, and therefore, the principle of water balance must be considered for such irregular areas. For balance, one can logically state in a rational manner that the total input, I, must be equal to total output, O, during a certain time interval. This sentence can be translated into mathematical form as,

$$\sum_{i=1}^{n} I_i = \sum_{i=1}^{n} O_i \qquad (2.51)$$

where $I_i(O_i)$ implies i-th input (output). All the quantities must be considered within certain time duration such as hour, week, month, year, etc. On the other hand, in case of imbalance (nonequilibrium), the difference between total input and total output must not be equal to zero. The difference can be either positive or negative. The former (latter) implies increasing (decreasing) change in the drainage basin storage, ΔS, and hence, the valid mathematical formulation can be written as,

$$\sum_{i=1}^{n} I_i - \sum_{i=1}^{n} O_i = \Delta S \qquad (2.52)$$

If $\Delta S > 0$ ($\Delta S < 0$), then there is an increase (decrease) in the storage. Water budget is the same in any part of the world, but its components have different significance in arid, semi-arid, and humid regions. In its most explicit form verbally, the water balance equation says.

Input − Output = change in storage.

The components in the application of water balance equation depend on the study purpose.

For instance, in cases of the surface flow (runoff), the evaporation and infiltration components gain distinctive roles. If evaporation rates are higher than the rainfall amounts, which is the case in arid regions (Chapter 1) it is not possible to apply the water balance equation for long periods confidently.

EXAMPLE 2.19 WATER BALANCE IN A DAM RESERVOIR

In July, 1970, the water surface area of a surface reservoir was 400 km^2. During this month, the storage volume of the reservoir dropped by $150 \times 10^6 \text{ m}^3$. Near to the dam, the evaporation height was measured by evaporation pan during the same month as 20 cm. The evaporation pan coefficient is 0.75. In the same month, the average water outlet from the dam was $130 \text{ m}^3/\text{s}$. According to these data, calculate the input discharge to this dam.

Solution 1.19

The entrance water to this dam is equal to what the river brings as input. It is assumed that during July there is no precipitation in the area. The output volumes from the dam are runoff, V_r, and evaporation, V_e. It is possible to calculate the output runoff as,

$$V_r = 130 \times 31 \times 86400 = 348 \times 10^6 \text{ m}^3$$

In order to calculate the evaporation amount from the reservoir, it is necessary first to multiply the evaporation volume, V_e, measurement in the evaporation pan by the pan coefficient, and hence, the evaporation loss from the reservoir becomes,

$$V_e = 0.75 \times 0.2 \times 400 \times 10^6 = 60 \times 10 \text{ m}^3$$

On the other hand, the change in the storage is given as decrease, $\Delta S = -150 \times 10^6 \text{ m}^3$.

By the application of water balance expression given by Eq. (2.52), one can calculate the input discharge volume, V_i as,

$$V_i - (348 + 60) \times 10^6 = -150 \times 10 \times 10^6 \text{ m}^3$$

and hence,

$$V_i = 258 \times 10^6 \text{ m}^3$$

It is now possible to calculate monthly discharge of the river as follows.

$$Q = 258 \times 10^6/(31 \times 86400) = 96 \text{ m}^3$$

EXAMPLE 2.20 WATER BALANCE IN ON ISLAND

On an island in the delta of a river, there are agricultural activities, where the annual evapo-transpiration height is 180 cm. During one year, precipitation height is 87 cm, but in the same year during 4 dry months additional water supply volume of 0.44×10^6 m^3 is brought to this island by pipes and during the remaining 8 months on the average daily 0.28×10^6 m^3 of water is given back to the delta. The increase in the soil moisture during the same year is 4 cm. According to these data, calculate the leakage water volume that comes from the groundwater to the island.

Solution 1.20

Let us first calculate the elements of water balance equation for one year. Water volume that comes from the precipitation is,

$$V_{pr} = 0.87 \times 50 \times 10^6 = 435 \times 10^6 \text{ m}^3$$

Water transfer volume through the pipes is,

$$V_{pi} = 122 \times 0.44 \times 10^6 = 53.7 \times 10^6 \text{ m}^3$$

Evapotranspiration volume loss is,

$$V_{ev} = 1.80 \times 50 \times 10^6 = 90 \times 10^6 \text{ m}^3$$

Return water volume to the delta is,

$$V_{re} = 243 \times 0.28 \times 10^6 = 68 \times 10^6 \text{ m}^3$$

Increase in the soil moisture is,

$$\Delta S = +0.04 \times 50 \times 10^6 = 2 \times 10^6 \text{ m}^3$$

By substituting these values into the water balance expression in Eq. (2.52), one can then calculate the groundwater leakage volume, V_L, to the island as,

$$43.5 \times 10^6 + 53.7 \times 10^6 + V_L$$
$$- \left(90 \times 10^6 + 68 \times 10^6\right) = +2 \times 10^6$$

Hence, one can find that

$$V_L = 62.8 \times 10^6 \text{ m}^3$$

EXAMPLE 2.21 WATER BALANCE IN A LAKE

The surface area of a lake is 40 km^2 and the average inflow during June is 0.56 m^3/s with the average outflow of 0.48 m^2/s. Average monthly precipitation and evaporation amounts are 45 mm and 105 mm, respectively. The infiltration height from the bottom of the lake during the same month is 25 mm. Calculate the volume change in this lake during the month.

Solution 2.21

In any hydrological system, if the total inflow and outflow are I_T and O_T, respectively, then the change in the lake storage, ΔS, can be calculated by Eq. (2.52).

In this problem, the inflows are runoff, I_{ro}, and precipitation, I_{Pr}, which have the volumes as,

$$I_{ro} = 0.56 \times 30 \times 86400 = 1.45 \times 10^6 \text{ m}^3$$
$$I_{Pr} = 0.045 \times 40 \times 10^6 = 1.80 \times 10^6 \text{ m}^3$$

On the other hand, O_T has also three components as outflow runoff volume, O_{ro}, evaporation, O_{ev}, and infiltration, O_{in}, volumes as losses, which can be calculated as,

$$O_{ro} = 0.48 \times 30 \times 86400 = 1.25 \times 10^6 \text{ m}^3$$
$$O_{ev} = 0.105 \times 40 \times 10^6 = 4.2 \times 10^6 \text{ m}^3$$
$$O_{in} = 0.025 \times 40 \times 10^6 = 1.00 \times 10^6 \text{ m}^3$$

Substitution of the inflow and outflow amounts into Eq. (2.52) yields the change in the lake volume as,

$$\Delta S = (1.45 \times 10^6 + 1.8 \times 10^6) - (1.25 \times 10^6$$
$$+ 4.2 \times 10^6 + 1.00 \times 10^6)$$
$$= -3.20 \times 10^6 \text{ m}^3$$

According to this result, during June the volume of the lake has decreased by $3.2 \times 10^6 \text{ m}^3$. This corresponds to lake-water-level drop as follows.

$$3.2 \times 10^6 / 40 \times 10^6 = 0.08 \text{ m}.$$

2.11 GROUNDWATER ELEMENTS

Water balance equation is a very convenient way for groundwater loss or gain estimations. It can be applied through a water year, which starts from October and ends in September. Seasonally, most often the water year is divided into two, namely, dry and wet spells. For instance, in many countries the wet season is considered from October to March and dry season from April to September. In general, as groundwater resources are concerned then the comparison of aquifer loss or gain becomes important.

2.11.1 Total Groundwater Input Elements

Among the most significant elements of groundwater inputs into an aquifer are the following factors:

1. **Rainfall recharge (Q_{RRE}):** This can be calculated by considering precipitation, P,

potential evaporation, ET, and infiltration as (Richard and Allen, 1998),

$$Q_{RRE} = 0.8(P - \phi \log ET)^{0.5} \qquad (2.53)$$

2. **Lateral subsurface inflow (Q_{LSI}):** Its calculation necessitates the lateral cross-sectional area perpendicular to groundwater flow direction. In practical applications, the cross-sectional profile is obtained by geophysical prospecting and then the irregular cross-section can be converted into an equivalent rectangle with saturation depth, m, and width, W. From the piezometric (water table) map, it is possible to calculate hydraulic gradient, i, at the cross-section location. The volume of lateral inflow per time can then be calculated by consideration of Darcy's law as,

$$Q_{LSI} = mWKi \qquad (2.54)$$

where K is the hydraulic conductivity at the cross-section location.

3. **Recharge from river beds (Q_{RRB}):** Its calculation requires completion of surface water hydrological studies. Direct or indirect recharges (Q_W and Q_D) to the aquifer system can be obtained for wet and dry periods. The total average of annual stream bed infiltration can be calculated as follows,

$$Q_{RRB} = Q_W + Q_D \qquad (2.55)$$

4. **Recharge due to irrigation returns (Q_{RIR}):** First, the total amount of water withdrawals from wells, Q_W, qanats, Q_Q, and springs, $Q_{S,}$ must be calculated. Hence, the recharge due to irrigation return can be calculated as,

$$Q_{RIR} = C_{RIR}(Q_W + Q_Q + Q_S) \qquad (2.56)$$

where C_{RIR} is a percentage. Based on Richard and Allen (1998) and considering soil characteristics, cultivation type, and ground-level condition about 35% of irrigation water can be considered as recharge from irrigation return.

5. **Sewage infiltration (Q_{SIN}):** It is, first, necessary to know total water consumption including domestic, Q_D, and industrial, Q_I, uses. Again, a certain percentage of these two components' summation gives the sewage infiltration as,

$$Q_{SIN} = C_{SIN}(Q_D + Q_I) \qquad (2.57)$$

where C_{SIN} is a certain percentage depending on the wastewater dump to aquifer prior or after treatment. In practical studies, it may be zero or in cases of actual wastewater return, its value may reach up to 0.65–0.75.

After all the aforementioned components, it is now possible to write the total input as,

$$T_I = Q_{RR} + Q_{LSI} + Q_{RRB} + Q_{RIR} + Q_{SI} \quad (2.58)$$

2.11.2 Total Groundwater Output Elements

The total amount out an aquifer consists of irrigation water uses, discharge from springs and qantas, lateral subsurface outflow, evaporation from groundwater table, and domestic and industrial water uses. Their explanations are presented in the following steps.

1. **Output irrigation water uses (Q_{IWU}):** Irrigation starting and ending months must be determined for the study area. In general, it is from April to October. If there is n number of plant types in the area, each one must be determined with their total irrigation area, A_i, and crop water use, Q_i. Hence, irrigation water use can be calculated as,

$$Q_{IWU} = \sum_{i=1}^{n} A_i Q_i \qquad (2.59)$$

If all water use is to be covered by the groundwater resource, then with Q_{IWU} value, one can assess the number of existing wells, if any,

and additional well number, if necessary. Each well should pump discharge in such a quantity that there will not be any interference among the adjacent wells.

2. **Contribution to surface water (Q_{CSW}):** There are different surface water resources such as dams, Q_D, springs, Q_S, lakes, Q_L, and qantas, Q_Q. Hence, the total surface water contribution can be calculated according to the following equation,

$$Q_{CSW} = Q_D + Q_S + Q_L + Q_Q \qquad (2.60)$$

3. **Subsurface outflow (Q_{SSO}):** This amount can be considered during the wet and dry seasons each by considering Darcy's law similar to lateral subsurface flow input as explained above.

4. **Evaporation from groundwater table (Q_{EGT}):** This happens only in cases of shallow unconfined aquifer situation, especially when the water table is less than 2 m from the Earth's surface and the unsaturated material is rather course medium. Depending on the groundwater table depth pan evaporation, E_P, measurements are multiplied with a certain percentage as,

$$Q_{EGT} = C_{EGT} E_P \qquad (2.61)$$

5. **Domestic and industrial water uses (Q_{DIW}):** If there are factories or industrial areas based on the number of population, one can estimate the domestic water use. It is necessary to estimate the population at time t, $N(t)$, and the consumption rate per person per day, Q_{PPD}, and then the amount of total domestic use, Q_{TDU}, can be calculated from the following equation.

$$Q_{TDU} = N(t)Q_{PPD} \qquad (2.62)$$

In many projects, the amount per capita per day is taken between 160 l/s and 110 l/s. However, in cases of water scarcity and drought periods, it can fall down to 70 l/day or 50 l/day.

Additionally, total industrial water use, Q_{TIU}, can be obtained from the receipts, and hence,

$$Q_{DIW} = N(t)Q_{PPD} + Q_{TIU} \qquad (2.63)$$

After all the output components' explanations given above, the total output, T_O, from an aquifer can be written as,

$$T_O = Q_{IWU} + Q_{CSW} + Q_{SSO} + Q_{EGT} + Q_{DIW} \qquad (2.64)$$

Depending on the comparison of the total input and output values, the aquifer either stores ($T_I > T_O$) or losses ($T_I < T_O$) water. However, if these two totals are equal to each other then there is a balance, which may preserve the safe and/or sustainable yield.

References

Al-Yamani, M.S., Şen, Z., 1993. Determination of hydraulic conductivity from grain-size distribution curves. Groundwater 31, 551–555.

Basak, P., 1977. Non-Darcian flow and its implication to seepage problems. Irrig. Drain. Div. Am. Soc. Civ. Eng. 103 (IR4), 459–473.

Batu, V., 1998. Aquifer Hydraulics – A Comprehensive Guide to Hydrogeological Data Analysis. Wiley Interscience, New York.

Bear, L., 1972. Dynamics of Fluids in Porous Media. American Elsevier Publication Co., New York.

Bear, J., 1979. Hydraulics of Groundwater. McGraw-Hill Book Co., New York, 567 pp.

Boulton, N.S., 1963. Analysis of data from non-equilibrium pumping tests allowing for delayed yield from storage. Proc. Inst. Civ. Eng. 6, 469–482.

Bouwer, H., 1978. Groundwater Hydrology. McGraw-Hill Koyakusha, Ltd., 480 pp.

Chen, C., Kuang, X., Jiao, J.J., 2009. Methods to derive the differential equation of the free surface boundary. Groundwater 48 (3), 229–232.

Childs, E.C., 1960. The non-steady state of the water table in drained land. J. Geophys. Res. 5, 780–782.

Clark, W.O., 1917. Groundwater for irrigation in the Morgan hill area, California. U.S. Geol. Surv. Water-Supply Paper 400, 84–86.

Dagan, G., 1982. Analysis of flow through heterogeneous random aquifers 2. Unsteady flow in confined aquifers. Water Resour. Res. 18 (5), 1571–1585.

Darcy, H., 1856. Les fountaines publique de la ville de Dijon. Victor Dalmont, Paris, France.

de Marsily, G., 1986. Quantitative Hydrogeology—Groundwater Hydrology for Engineers. Academic Press, Orlando, Florida, 440.

De Wiest, R.J.M., 1965. Geohydrology. John Wiley, New York.

Delleur, J.W. (Ed.), 2007. The Handbook of Groundwater Engineering, second ed. CRC Press, Boca Raton, Florida.

Domenico, P.A., Schwartz, F.W., 1990. Physical and Chemical Hydrogeology. John Wiley & Sons, New York, 824 p.

Driscoll, F.G., 1986. Groundwater and Wells. Johnson Division, 1089 page.

Dudgeon, C.R., Kuyakorn, P.S., Swan, W.N.C., 1973. Hydraulics of flow near wells in unconsolidated sediments. University of New South Wales, Australia.

Dullien, F.A.L., 1992. Porous Media: Fluid Transport and Pore Structure, second ed. Academic Press.

Escande, L., 1953. Experiments concerning the filtration of water through rock mass. Proc. Minn. Int. Hydraul. Congr.

Fetter, C.W., 1980. Applied Hydrogeology. Charles E. Merrill Publishing Co., 488 pp.

Forchheimer, P.H., 1901. Wasserbewegung durch Boden. Zitschrifft Ver. Dtsch. Ing. (49), 1736–1749 and No.50: 1781–1788.

Freeze, R.A., Cherry, J.A., 1979. Groundwater. Prentice-Hall, Inc, Englewood Cliffs, New Jersey, 604 pp.

Gerhart, J.M., 1986. Groundwater recharge and its effect on nitrate concentrations beneath a manured field site in Pennsylvania. Groundwater 24, 483–489.

Gurr, C.G., Idarshall, T.J., Hutton, J.T., 1952. Movement of water in soil due to a temperature gradient. Soil Sci. 74, 335–345.

Hall, D.W., Risser, D.W., 1993. Effects of agricultural nutrient management on nitrogen fate and transport in Lancaster county, Pennsylvania. Water Resour. Bull. 29, 55–76.

Hantush, M.S., 1964. Hydraulics of wells. In: Chow, V.T. (Ed.), Advances of Hydrosciences, vol. 1. Academic Press, New York and London, pp. 281–432.

Harlan, R.L., 1972. Ground conditioning and the groundwater response to surface freezing. In: The Role of Snow and Ice in Hydrology. Proceedings of Banff Symposium. International Association of Hydrological Science Publication 107, vol. 1, pp. 326–341.

Hazen, A., 1893. Some Physical Properties of Sand and Gravel. Massachusetts State Board of Health, 24th Annual Report.

Hoekstra, P., 1966. Moisture movement in soils under temperature gradients with the cold side temperature below freezing. Water Resour. Res. 2, 241–250.

Huettel, M., Ziebis, W., Luther, G.W., 1998. Advective transport affecting metal and nutrient distributions and interfacial fluxes in permeable sediments. Geochim. Cosmochim. Acta 62, 613−631.

Johnson, A.I., 1967. Specific yield — compilation of specific yields for various materials. U.S. Geol. Surv. Water Supply Paper 1662-D, 74.

King, F.H., 1899. Principles and conditions of the movements of groundwater. U.S. Geol. Surv. Nineteenth Annual Report Part 2, 86−91.

Knochenmus, L.A., Robinson, J.L., 1996. Descriptions of anisotropy and heterogeneity and their effect on groundwater flow and areas of contribution to public supply wells in a karst carbonate aquifer system. U.S. Geol. Surv. Water-Supply Paper 2475, 47.

Krumbain, W.C., 1934. Size frequency distributions of sediments. J. Sediment. Petrol. 4, 65−77.

Kruseman, G.P., de Ridder, N.A., 1990. Analysis and Evaluation of Pumping Test Data. Bulletin 11, Institute for Land Reclamation and Improvement, Wageningen, 41−46.

Logan, J., 1964. Estimating transmissivity from routine production tests of water wells. Groundwater 2, 35−37.

Meinzer, O.E., 1923. The occurrence of groundwater in the United States with a discussion of principles. U.S. Geol. Surv. Water- Supply Paper 489, 321.

Meyboom, P., 1967. Groundwater studies in the Assiniboine river drainage basin — part II: hydrologic characteristics of phreatophytic vegetation in south-central Saskatchewan. Geol. Surv. Can. Bull. 139, 64.

Moench, A.F., 1994. Specific yield as determined by type-curve analysis of aquifer-test data. Groundwater 32, 949−957.

Moench, A.F., 1995. Combining the Neuman and Boulton models for flow to a well in an unconfined aquifer. Groundwater 33, 378−384.

Neuman, S.P., 1972. Theory of flow in unconfined aquifers considering delayed response of the water table. Water Resour. Res. 8, 1031−1045.

Neuman, S.P., 1987. On methods of determining specific yield. Groundwater 25, 679−684.

Nwankwor, G.I., Cherry, J.A., Gillham, R.W., 1984. A comparative study of specific yield determinations for a shallow sand aquifer. Groundwater 22, 764−772.

Philip, J.R., de Vries, D.A., 1957. Moisture movement in porous materials under temperature gradients. Trans. Am. Geophys. Union 38 (2), 222−232.

Rasmussen, W.C., Andreasen, G.E., 1959. Hydrologic Budget of the Beaverdam Creek Basin, Maryland. U.S. Geol. Surv. Water-Supply Paper 1472, 106.

Richard, G., Allen, 1998. Crop Evapotranspiration, Guidelines for Computing Crop Water Requirements. FAO Irrigation and Drainage Paper 56. FAO-Food and Agriculture Organisation of the United Nations, Rome.

Rushton, K.R., 2003. Groundwater Hydrology. Conceptual and Computational Models. John Wiley, 416 pp.

Schad, H., Teutsch, G., 1994. Effects of the investigation scale on pumping test results in heterogeneous porous aquifers. J. Hydrol 159, 61−77.

Şen, Z., 1988. Analytical solution incorporating non-linear radial flow in confined aquifers. Water Resour. Res. 24 (4), 601−606.

Şen, Z., 1989. Non-linear flow toward wells. J. Hydraul. Div. Am. Soc. Civ. Eng. 115 (2), 193−209.

Şen, Z., 1990. Nonlinear radial flow in confined aquifers toward large diameter wells. Water Resour. Res. 26 (5), 1103−1109.

Şen, Z., 1995. Applied Hydrogeology for Scientists and Engineers. CRC Lewis Publishers, Boca Raton, 444 pp.

Şen, Z., 2013a. Hydraulic conductivity variation in a confined aquifer. J. Hydrol. Eng. 10.1061/(ASCE)HE. 1943-5584.0000832.

Şen, Z., 2013b. Philosophical, Logical and Scientific Perspectives in Engineering. Springer, 247 pp.

Şen, Z., Wagdani, E., 2008. Aquifer heterogeneity determination through the slope method. Hydrol. Process. 22 (12), 1788−1795.

Todd, D.K., 1980. Groundwater Hydrology. John Wiley and Sons., 535 pp.

Veihmeyer, F.J., Hendrickson, A.J., 1931. The moisture equivalent as a measure of the field capacity of soils. Soil Sci. 32, 181−194.

Vogel, H.J., Roth, K., 2003. Moving through scales of flow and transport in soil. J. Hydrol. 272, 95−106.

Walton, W.C., 1970. Groundwater Resource Evaluation. McGraw-Hill, New York, 664 pp.

Wenzel, L.K., 1942. Methods of determining permeability of water bearing materials with special reference to discharging well methods. U.S. Geol. Surv. Water-Supply Pap 887.

Younger, P.L., 1993. Simple generalized methods for estimating aquifer storage parameters. Q. J. Eng. Geol. 26, 127−135.

Copyright © 2015 Elsevier Inc. All rights reserved.

3.1 GROUNDWATER WELLS

Wells are the main hydraulic structures for groundwater, oil or gas abstractions from the potential geological formations. In general, any vertical shaft or inclined excavation that reaches subsurface water or oil from the Earth's surface is a well. The purpose of water wells is to extract groundwater from the saturation zones for domestic, agricultural, industrial or any other use. The choice of well shape and dimensions depend on the piezometric level (shallow or deep), subsurface geological layers (permeable, semipermeable or impermeable), recharge possibilities (direct, indirect), purpose (water supply, oil and gas exploration, geothermal energy production), etc. In many occasions economics and politics also play important roles in well location determination. Technological aspects in well excavation are outside the scope of this book, however, it is discussed elsewhere (Driscoll, 1987; Anderson, 1967).

Prior to well excavation, it is advisable to study the hydrological, hydrogeological, subsurface geological, geophysical, economic and social aspects with preliminary office works, and subsequent reconnaissance field surveys. These activities provide some insight into the general topography, geomorphology, geology, hydrology, and composition of subsurface layers. In such a study, geographical, topographical, and geological maps are indispensable means, which can be prepared by aerial photographs, satellite images, and digital elevation models (DEM), drilling logs, and existing water supply information. Ideal water well should have circular cross-section, which is advantageous from the following points of view:

1. The groundwater pressure at any point along the well circumference on the horizontal plane is radial, and therefore, the tangential stresses are equal. As a result, the pressure distribution is radially uniform. This generates only compression strength on the well casing and screen. The geological formations are more resistant to compression than tensile stress, and consequently, any possible collapse on well face is at minimum risk in a circular cross-section, particularly in unconsolidated formations.

2. For any given cross-sectional area, the circular cross-section has the minimum circumference than any other shape, and hence, during the well drilling, the friction between the drilling bit and the well shaft lateral surface will be minimum. In addition, the casing is cheaper and all drilling techniques work on circular shapes. Other shapes are impossible, especially in deep well drillings.

3. Flow lines are radially straight-lines and the circular circumference is a potential line (Chapter 2). This property means that the flow net geometry is completely known in circular cross-section wells provided that the aquifer is homogeneous and isotropic.

3.1.1 Well Types

First classification of the wells is based on their diameter and there are large and small diameter wells. Large diameter wells are dug usually in unconfined shallow aquifers; small diameter wells penetrate big depths and they are suitable for deep confined aquifers. Figure 3.1 shows large and small diameter wells.

From ancient times, man has been using groundwater from hand-dug large diameter wells for domestic and agriculture purposes. For many centuries groundwater has been withdrawn from unconfined aquifers by means of large diameter wells. These wells are used mainly for individual exploitations and their excavation requires some minimum conditions, such that the position of the water table is close to the Earth's surface. The wells are generally circular in cross-section, but rectangular, square, or irregular shapes are not uncommon in some areas (see Chapter 2). Large diameter wells are convenient to extract significant quantities of water from low permeability geologic formations such as unconsolidated glacial and alluvial deposits at shallow depths. Performance of a large diameter well poses special problems of analysis and interpretations. The larger the diameter more will be the well storage, and hence, at the time of need

large quantities can be abstracted initially with no difficulty. These wells cannot be deep and the maximum depth observed in practice is about 100 m. They are mostly dug in unconfined aquifers and the actual depth of large diameter wells at any time is a function of the groundwater table elevation position.

A large diameter well pierces from the aquifer a volume in which free water is ready for abstraction from the well storage only. This gives rise to a difference (drawdown), s_w, between the initial piezometric level and the water level in the well. It is logical to conclude that the larger the diameter the smaller is this difference and smaller is the groundwater flow rate.

Advanced oil drilling technology gave ways to drill wells down to big depths especially for oil, natural gas, groundwater, mineral resources and/or subsurface geological exploration studies. Hence, drilled wells became to be used extensively in practice. They have small diameters and circular cross-sections only. Small diameter wells have usually diameters of 12 inches (≈ 30 cm) and the well depth may reach at many places down to 3000 m. They provide water abstraction from a multitude of aquifers along a hydro-stratigraphic sequence. There are different methods in drilling a small diameter well as explained by Driscoll (1987) and Huismann (1972). Large and small diameter wells have the following comparative advantages:

1. In large (small) diameter wells water is practically available (not available) for abstraction in the well storage for direct usage.
2. The excavation of large (small) diameter wells does not require (require) sophisticated machines for digging (drilling).
3. For large (small) diameter wells digging (drilling) and operation skilled personnel are not required (required).
4. Large diameter wells are deepened gradually after dry (drought) periods, whereas the small diameter wells are drilled once and for all.

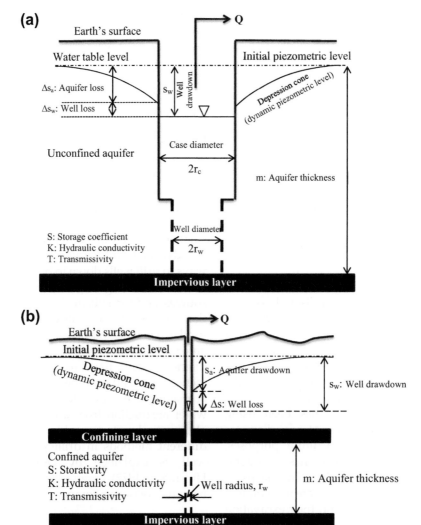

FIGURE 3.1 a) Large diameter, (b) small diameter wells.

5. In large (small) diameter well piezometric level is (not) close to Earth's surface. Accordingly, in large (small) diameter wells suction (submersible) pumps are usable for groundwater haulage to the surface.

6. Large (small) diameter well is rather cheap (expensive) to construct and maintain.

On the other hand, from disadvantage points of view they can be compared along the following points:

1. Large (small) diameter well is (not) susceptible to contamination or pollution.

2. Large (small) diameter well occupies large (small) space for the well and digging (drilling) material.

3. Large (small) diameter well is dangerous (safe) for small children and animals.
4. Large (small) diameter well is (not) subject to flood sedimentation and debris flow.
5. Large (small) diameter well may get dry in short (long) time periods during drought periods.

3.1.1.1 Early Time Well Diameter–Discharge Relationship

Well diameter–discharge relationship is valid during early time periods of pumping. In large diameter wells, the pump discharge, Q, initially comes from the well storage (Figure 3.2).

Abstraction of discharge, Q, generates drop in the static piezometric (water) level, H_o, in the well down to a dynamic water level, $H_w(t)$. In the meantime, pump water originates from the well storage only, and hence, initially the volume, $V_w(t)$, can be calculated as,

$$V_w(t) = \pi r_w^2 [H_o - H_w(t)] \qquad (3.1)$$

From the definition of discharge as volume per time, t, the well storage discharge, $Q_w(t)$, can be written from Eq. (3.1) as,

$$Q_w(t) = \frac{\pi r_w^2 [H_o - H_w(t)]}{t} \qquad (3.2)$$

Physically, the pump discharge during early times comes from the well storage only, and hence, initially $Q = Q_w(t)$; mathematically as

$t \to 0$, $Q_w(t) \to Q$. This expression indicates that discharge increases in a directly proportional manner with the well radius. For large time durations, however, Eq. (3.2) shows that the contribution from the well storage goes to zero since mathematically as $t \to \infty$, $Q_w(t) \to 0$. It means that the effect of well radius fades away with time. As the well storage fades away, the aquifer discharge, $Q_a(t)$, starts to contribute and for large times, as $t \to \infty$, $Q_a(t) \to Q$. Under the light of these two limiting conditions one can deduce that, in general,

$$Q = Q_w(t) + Q_a(t) \qquad (3.3)$$

The graphical representation of all these conditions indicates that as the well storage contribution to the pump discharge decreases, aquifer discharge starts to take on (see Figure 3.3).

Equation (3.3) is valid also for small diameter wells, but since the well diameter is small its square will be smaller, and hence, the well storage contribution to the pump discharge becomes negligible right at the beginning of the pumping. This is the main reason why in small diameter wells, under the same environmental conditions, the drop in the initial piezometric level is bigger, and hence, the well losses are also bigger. There is an inverse relationship between the well loss and diameter.

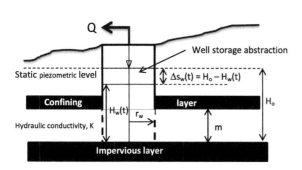

FIGURE 3.2 Large diameter well discharges.

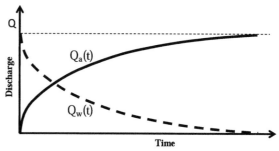

FIGURE 3.3 Large diameter-aquifer discharge variations by time.

EXAMPLE 3.1 STEADY STATE DISCHARGE CALCULATION

A well of 1.5 m radius fully penetrates a confined aquifer composed of material with hydraulic conductivity 1.2×10^{-4} m/s and the thickness, 16 m. If the hydraulic gradient in the aquifer at the well surface is equal to 0.02 during a steady state condition, what is the pumping discharge from this aquifer?

Solution 3.1

Under the light of the given information it is useful to sketch a well-aquifer configuration as in Figure 3.4.

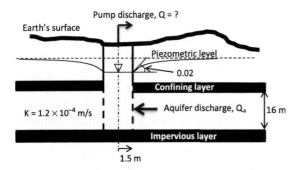

FIGURE 3.4 Large diameter well.

At the steady (or quasi) state, practically there is no more drawdown increment in the aquifer. The discharge calculation can be obtained from the hydraulic gradient at the well surface and the hydraulic conductivity of the aquifer. In the steady case, since the groundwater balance exists ($Q = Q_a$) according to the discharge definition, it is equal to flow area times the Darcy velocity.

$$Q = (2 \times 3.14 \times 1.5 \times 16) \times (1.2 \times 10^{-4} \times 0.02)$$
$$= 3.62 \times 10^{-4} \text{ m}^3/\text{s}.$$

3.1.1.2 Early Time Drawdown–Discharge Relationship

Şen (1986a) suggested a graphical method for pump discharge calculation in large diameter wells by means of early time drawdown measurements. A close logical reasoning on Eq. (3.2) leads for infinitesimally small time durations, dt, that piezometric level drop will also be small, $ds_w(t)$. Accordingly, one can rewrite Eq. (3.2) as,

$$Q = \pi r_w^2 \frac{ds_w(t)}{dt} \qquad (3.4)$$

Since, Q is a constant discharge it is possible to integrate this expression according to the separation of variables, which leads to,

$$s_w(t) = \frac{Q}{\pi r_w^2} t$$

This proves that there is a direct straight-line relationship between the water level drops (drawdown) and early time periods, where the slope of the straight-line is equal to $Q/\pi r_w^2$. As a result, the initial drawdown measurements can be plotted against the corresponding times on an ordinary paper to see the scatter of points. An eye controlled best straight-line is then fitted through the scatter points and the slope, α, of the straight-line gives the discharge estimation as,

$$Q = \alpha \pi r_w^2 \qquad (3.5)$$

EXAMPLE 3.2 LARGE DIAMETER DISCHARGE CALCULATION

Water supply to a small village has been withdrawn through a large diameter well from a shallow aquifer. After the pump start, water level

TABLE 3.1 Early Time–Drawdown Records

Time (min)	1	3	5	8	10	13	15
Drawdown (m)	0.09	0.25	0.57	0.75	0.89	1.15	1.38

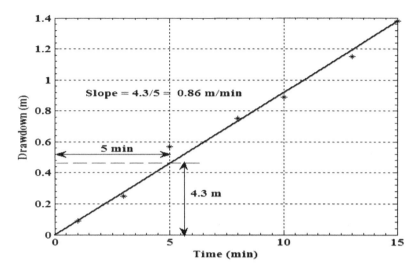

FIGURE 3.5 Early time–drawdown straight-line.

drops (drawdowns) are recorded during the first 15 min as in Table 3.1. The well radius is measured as 1.1 m. Calculate the well discharge.

Solution 3.2

The first step is to plot the scatter points as time versus drawdown on an ordinary paper and then to match the best eye fixed straight-line through the points as in Figure 3.5.

The slope of the straight-line is calculated as 0.86 m/min, and hence, its substitution into Eq. (3.5) yields,

$$Q = 0.86 \times 3.14 \times 1.1^2 = 3.27 \, m^3/min$$

3.1.1.3 Piezometers and Observation Wells

For continuous groundwater monitoring, piezometers and observation wells (OWs) must be located at suitable locations. In general, they must remain within the radius of influence distance from the main (pump) well (MW). Piezometers are fine pipes that are blind along the length and open at two ends. The bottom may end up just after its entrance into confined aquifer, but in unconfined aquifers it must enter the saturation layer such that it always remains below the depression cone dynamic level.

On the other hand, OWs help to measure groundwater level fluctuations by time. These wells allow entrance of groundwater along their whole periphery. They are either perforated pipes or drilled wells in strong geological layers. OWs serve for measuring piezometric level (hydraulic head) changes. Their diameters are rather small, just enough for measuring instrument (sounder probe) entrance. In Figure 3.6 OW position is shown in an aquifer. In practice, most often nearby existing wells are used as OW to avoid extra cost for drilling.

If both instruments (piezometer and OW) have the same piezometric levels (water levels), then there is no groundwater movement between them (zero hydraulic gradient). Such a set up of two instruments helps to decide also about the groundwater damming during the construction work. If observation wells are too close and less than three times the aquifer thickness, then the measurements can be influenced strongly from the anisotropy due to stratification. As a general rule, the distance to OW is

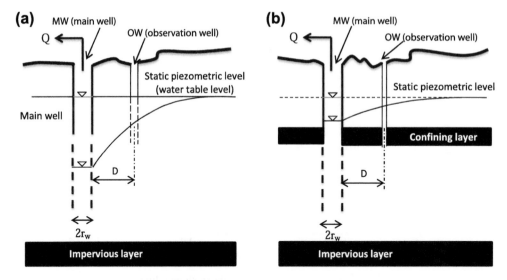

FIGURE 3.6 **Main and observation wells.** (a) Unconfined, (b) confined aquifer.

taken about 250–300 m in case of unconfined aquifers, which can be around 500–750 m in confined aquifers. If more than one OW is possible, then it is better to locate them at different distances and directions from the MW, so that the isotropy features of the aquifer can be assessed.

The number of OWs (piezometers) depends on test objectives and available resources for test program and also on the availability of the project budget.

1. In many cases measurements from a single well may be considered as sufficient, and frequently the tests are conducted in MWs for time–drawdown measurements. Such a situation leads to erroneous aquifer parameter estimations especially in the small diameter well cases, where the well losses are very high.
2. In order to increase the reliability in the aquifer parameter estimations, it is necessary to have at least one OW. If the aquifer heterogeneity and anisotropy are to be checked then more than two OWs are advised in practice. For instance, at least three OWs at

different distances are needed for regional groundwater flow direction determination.
3. Depending on the subsurface geological composition with possibilities of dykes, faults, boundaries, surface water interferences, the number of OWs can be increased accordingly.

3.1.1.4 Well-Aquifer Configuration

Depending on the relative situation of any well with respect to the piezometric level, it belongs to a different category. In Figure 3.7 different aquifers and well positions are shown in the vertical cross-section.

By consideration of the water level in each well, one can categorize them according to whether this level is below the confining layer, between the confining and impermeable layers, or above the Earth's surface:

1. Well A has its water level within the unconfined aquifer, which is the most frequently encountered case in shallow (large diameter) wells.
2. Well B is an artesian aquifer in confined aquifer, because the water level in the well is

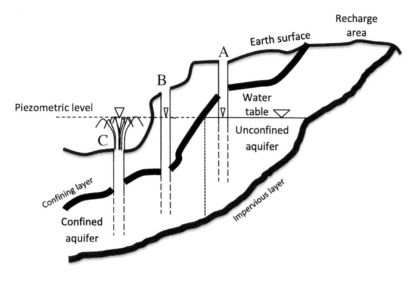

FIGURE 3.7 **Well-aquifer configurations.**

above the upper confining layer, which is the case mostly in deep (small diameter) wells.

3. Well C has the flush of water over the Earth's surface. It is referred to as flowing well, which is rarely the case in practice. It is the most preferable case for geothermal energy production through deep wells, provided the groundwater has more than 200 °C temperature.

3.1.2 Well Penetration

For water abstraction by pumps, the well should penetrate into the aquifer partially or completely. Depending on the amount of penetration, pump discharge quantity changes. Logically, more the penetration more will be the water abstraction. However, there are other restrictions, such as the subsurface geological composition, water storage and release capacity, and economy.

3.1.2.1 Full Penetration

The common depth between the well and whole saturation zone implies full penetration

(Figure 3.8). The amount of discharge is directly proportional with the penetration length. The shapes of flow and equipotential lines on a vertical cross-section depend on the well penetration.

In confined aquifers full-penetration provides planar-radial groundwater flow, which helps in easy calculations because the flow and equipotential lines are perpendicular to each other. In case of unconfined aquifer, the flow and equipotential lines have curvatures even though the well fully penetrates.

3.1.2.2 Partial Penetration

In practical studies, partial penetration occurs unintentionally most often from the lack of knowledge about the true saturated thickness of aquifer due to economic restrictions, (Figure 3.9).

These wells are common in areas where confined aquifers have large thickness. In a partially penetrating well the flow lines are forced to converge toward the partial penetration well entrance (screens with length, L_s). A water molecule has to travel longer distances than the fully penetrating wells, and therefore,

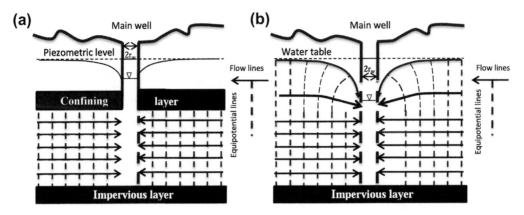

FIGURE 3.8 **Full penetration well.** (a) Confined aquifer, (b) unconfined aquifer.

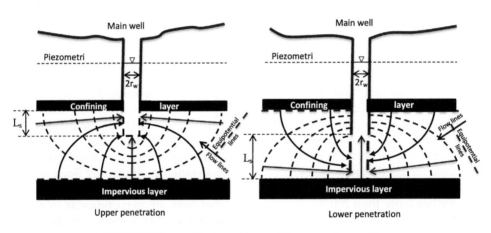

FIGURE 3.9 **Confined aquifer partial penetration aquifer.**

there are bigger drawdowns under the same conditions. In the vicinity of the well the flow lines have a three-dimensional radial but nonplanar rotational shapes provided that the aquifer is isotropic and homogeneous. However, with increase in distance from the well, the flow domain converges to two-dimensional radial flow. For a given discharge, Q, the drawdown around a partially penetrating well is more than that for a fully penetrating well. The analysis of the partially penetration case is more difficult than fully penetrating wells. Practical experiences have shown, as a rule of thumb,

that beyond a radial distance almost equal to twice the saturation thickness, the drawdown is approximately the same with a fully penetration case. In practical applications, one tries to minimize the deviations due to a fully penetrating well by considering the following points:

1. If horizontal bedding is strong in the flow domain and the observation wells are fairly close to the main well the aquifer is assumed to end at the bottom of the main well, and hence, fully penetrating well formulations are used directly on the basis of well

penetration length. This approximation leads to underestimations.

2. Observations may be taken at such radial distances that the effects of partial penetration become negligible and the streamlines are substantially the same as if the well were fully penetrating. This distance, D, (see Figure 3.6) can be calculated in terms of the aquifer thickness, m, horizontal, K_h, and vertical, K_v, hydraulic conductivities as,

$$D = 2m\sqrt{K_h K_v} \qquad (3.6)$$

3. Jacob (1944) suggested the use of corrected drawdown, s_c, by measuring the drawdown at top and bottom of the aquifer separately at a radial distance using a pair of observation wells. The corrected drawdown is calculated as the arithmetic average of top and bottom drawdowns.

In certain cases, the observed drawdown in partially penetrating wells may be adjusted for partial penetration according to the theory and empirical formulations. To this end, various researchers developed methods to correct for partial penetration in order to apply the formulations of full penetration.

Kozeny (1933) and Muskat (1937) developed a dimensionless formula, which relates the partial penetration well discharge, Q_P, ratio to full penetration well discharge, Q, as,

$$\frac{Q_P}{Q} = p\left[1 + 7\beta^{1/2}\cos\left(\frac{\pi}{2}p\right)\right] \qquad (3.7)$$

in which β is referred to as the well slimness, and is defined as a ratio,

$$\beta = \frac{r_w}{2L_s}$$

where r_w is the well radius and L_s is the screen length (see Figure 3.9). Equation (3.7) is valid for penetration ratios $p = L_s/m < 0.5$ and $L_s/m > 30$, which implies that $\beta > 0.01667$. Graphical representation of Eq. (3.7) for $\beta = 0.01667$ is shown in Figure 3.10 for quick calculation of the discharge ratios from the penetration ratio.

Huismann (1972) developed an equation relating the additional drawdown due to the partially penetrating well in terms of the pump discharge, Q, hydraulic conductivity, K, aquifer

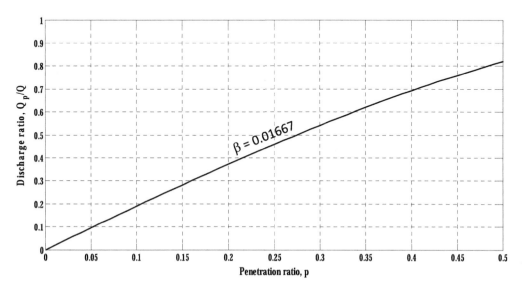

FIGURE 3.10 Penetration ratio–discharge ratio relationships.

TABLE 3.2 α Values

	e = 0	e = 0.05	e = 0.10	e = 0.15	e = 0.20	e = 0.25	e = 30	e = 0.35	e = 40	e = 0.45
p = 0.1	0.54	0.54	0.55	0.55	0.56	0.57	0.59	0.61	0.67	1.09
0.2	0.44	0.44	0.45	0.46	0.47	0.49	0.52	0.59	0.89	
0.3	0.37	0.37	0.38	0.39	0.40	0.43	0.50	0.74		
0.4	0.31	0.31	0.32	0.34	0.36	0.42	0.62			
0.5	0.25	0.26	0.27	0.29	0.34	0.51				
0.6	0.21	0.21	0.23	0.27	0.41					
0.7	0.16	0.17	0.20	0.32						
0.8	0.11	0.13	0.13	0.22						
0.9	0.08	0.12								

thickness, m, the penetration ratio, p, screen length, L_s, and the well radius, r_w, as,

$$\Delta s_w = \frac{2.3Q}{2\pi Km} \frac{1-p}{p} \log \frac{\alpha L_s}{r_w} \qquad (3.8)$$

in which α is the function of p and the amount of eccentricity, e, which is defined as a dimensionless ratio, $e = 0.5(1 - L_s/m) = 0.5\,(1 - p)$. Table 3.2 presents α values for sets of p and e values.

EXAMPLE 3.3 DRAWDOWN CALCULATION

A confined aquifer of 80 m saturation thickness with transmissivity 951 m²/day is tapped through a fully penetrating well at a constant rate, 0.05 m³/sec. The well has 1.5 m diameter and the resulting drawdown after a long abstraction period is 14.7 m. By assuming no well loss, find the drawdown value if the well is screened between 19 and 35 m below the top of the aquifer.

Solution 3.3

The penetration percentage, p, and the eccentricity, e, can be calculated as,

$$p = \frac{35 - 19}{80} = 0.20$$

and

$$e = 0.5(1 - 0.2) = 0.4$$

Hence, from Table 3.2 one can take $\alpha = 0.89$. According to Eq. (3.8) the additional drawdown due to partial penetration is

$$\Delta s_w = \frac{2.3 \times 0.05}{2 \times 3.14 \times \left(\frac{951}{24 \times 60 \times 60}\right)}$$
$$\times \frac{1 - 0.2}{0.2} \log \frac{0.89 \times (35 - 19)}{0.75}$$
$$= 19.59 \text{ m}$$

This gives a total drawdown within the aquifer at the well face, $s_w = 14.70 + 19.59 = 34.29$ m.

3.1.2.3 Skin Effect, Aquifer Loss and Other Well Losses

In any MW the drawdown is composed of aquifer and well loss components corresponding to energy losses (see Figure 3.1). Aquifer loss appears as a result of porous medium resistance against groundwater flow, and as the pump discharge increases so does the aquifer loss.

The well loss can be considered in two parts as linear and nonlinear losses. The linear part

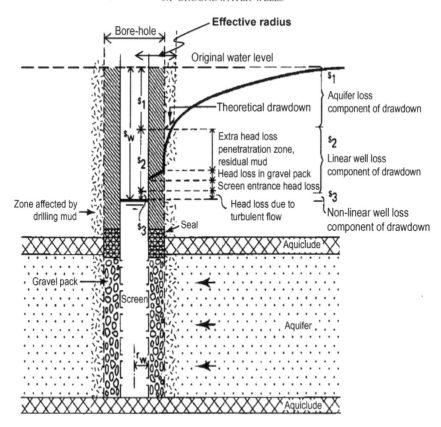

FIGURE 3.11 **Skin effect and losses.** (http://content.alterra.wur.nl).

consists of hydraulic head losses due to compaction of the aquifer material during drilling, head losses due to plugging of the aquifer with drilling mud, which reduces the permeability near the bore hole; head losses in the gravel pack; and head losses due to the screen resistance against water entrance into the well.

The nonlinear well loss includes the friction losses that occur inside the well screen and in the suction pipe where the flow is turbulent and the head losses that occur in the zone adjacent to the well. The collective effect of these well losses appears in an additive manner within the well drawdown. As a result of these nonlinear losses, the drawdown, especially in the small diameter MW is always greater than

the loss due to the groundwater flow in the aquifer. Petroleum engineers recognize the concept of "skin effect" to account for the head losses in the vicinity of a well. The theory behind this concept is that the aquifer is assumed to be homogeneous up to the wall of the bore hole, while all head losses are assumed to be concentrated in a thin, resistant "skin" against the wall of the bore hole (Figure 3.11, http://content.alterra.wur.nl).

Gravel pack is necessary for aquifer stabilization in the vicinity of wells against fine sand and silt entrance. For this purpose, a gravel belt is constructed artificially between the screen and aquifer. The gravel pack should have larger particles than the surrounding formation. This supports the development of larger void formation closer to the screen, which allows water entrance almost free of head loss. The gravel pack supports the following points:

1. Sand free water entrance into the well during pumping operation,
2. Highest permeability generation in the well vicinity,
3. Low water entrance velocity.

A characteristic standard grain size curve is given in Figure 2.1. Therein, D_s are the particle size smaller than s percentage. The uniformity coefficient has already been defined in Chapter 2, Section 2.2. Another important quantity is the pack-aquifer ratio, PA, which is the ratio of $(D_{50})_G$ size of the gravel pack to the $(D_{50})_A$ size of the aquifer material.

$$PA = \frac{(D_{50})_G}{(D_{50})_A} \qquad (3.9)$$

Aquifer loss is due to Darcy flow through the aquifer material, which causes steady energy loss as water flow approaches the main well. The coarser (finer) the medium, the smaller (bigger) is the aquifer loss. Linear well loss consists of the head losses in the transitional zone between the aquifer and the gravel pack, and screen losses. Turbulent flow loss is due to the very high gradients next to the well lateral surface, where energy losses are proportional nonlinearly with groundwater velocity, which is the greatest at the well surface. As obvious from Figure 3.11, the total drawdown, s_w, in the main well is equal to the difference between the static (initial) and the dynamic (final) water levels. This drawdown is the summation of aquifer, s_1, the linear well, s_2 and the turbulent flow, s_3, losses.

$$s_w = s_1 + s_2 + s_3 \qquad (3.10)$$

Well screen and the gravel pack together help to provide open access after well development so that there will not be clogging due to fine grains and the well is not impeded by sand. Groundwater flow toward the well carries unconsolidated sand and gravel in addition to friable materials. A proper well design should provide the following features:

1. Minimum entrance velocity,
2. Maximum open area of screen,
3. Minimization of the blockages,
4. Corrosion resistance screen material.

Prior to water abstraction for supply purposes, well development stage must be performed with high abstraction rates. After the well development the gravel pack should have the following features:

1. Sand free well-aquifer operation,
2. Low entrance velocity,
3. Resistance to chemicals.

Screen area should provide entrance velocity that is less than that required to move the finest particles of the surrounding formation into the well. Smith (1963) suggested that an entrance velocity from 1 ft/s (30 m/s) to 0.25 ft/s (7.6 cm/s) barely lifts sand from 0.25 to 0.50 mm diameter, while a velocity of 0.06 ft/s (1.83 cm/s) barely lifts a clay or silt. In practice, it is advised to use 0.1 ft/s (0.3 cm/s). The screen should also retain physically the required proportion of the formation with adequate strength to prevent collapse of the well. The head loss through a well screen is a function of the screen length, screed diameter, percentage screen opening area, opening contraction coefficient, roughness, water velocity in the screen, and discharge.

In order to protect side collapse of the wells they must be reinforced by solid materials, which can be iron, steel, concrete pipes, or stone masonry. The well losses comprise the turbulent losses about the well screen and frictional losses in the casing. One can seek to minimize the well losses so as to increase the efficiency, but the length of the casing is governed by the aquifer depth.

3.2 FIELD MEASUREMENTS AND TESTS

In general, the preliminary field work includes well inventory, topographic and surface geological assessments, identification and drilling locations in order to obtain direct subsurface data. It is also possible to use geophysical prospecting techniques for the subsurface exploration. At this point a significant advice is that one should not depend only on the instrumental measurements, but more significantly on the verbal, linguistic, and descriptive data from the local settlers and administrators who are expected to care for current and future groundwater problems. Additionally, sample collection for groundwater quality determinations and regional quality assessments are important activities during field visits. Laboratory analyses of the samples provide significant clues about the groundwater quality variations (Chapter 5).

In hydrogeology, aquifer properties may have different values depending on the number of samples taken in the field. Although their averages give representative values, for practical purposes, it would be better to appreciate temporal and spatial variations (Şen, 2009). The variations must remain within certain relative error limits ($\pm5\%$ or $\pm10\%$) for adopting arithmetic averages as representative values in the applications. If the two successive measurements (or the maximum and minimum values among all data

lot) are m_i and $m_{i+1}(m_i > m_{i+1})$ then the relative error percent, α, can be evaluated as,

$$\alpha = \frac{|m_i - m_{i+1}|}{m_i} \times 100 \qquad (3.11)$$

Provided that this value is less than ±5 or ±10, temporally the flow is in steady state or quasi-steady state case; spatially the aquifer property has regional homogeneity.

3.2.1 Field and Office Procedures

The main purpose of groundwater measurements and observations in the field is to collect data for aquifer parameter determination. Well inventory with headlines as indicated in Table 3.3 is very helpful for keeping different informative knowledge on the same sheet collectively.

The well inventory sheet is ready for filling with field information and should be completed by keeping in mind the following points:

1. The well locations must be shown even on an approximate sketch (better on an actual map) with the significant nearby features (houses, roads, bridges, lakes, hills, buildings, etc.).
2. Relevant information can be obtained from local settlers or other well owners through short interviews.
3. If outcrops are available, rock types can be specified at least approximately, in addition to possible dykes, faults, folds, fractures, etc.

TABLE 3.3 Well Inventory Sheet

Well Number	Location		Elevation	Well Type	Well Diameter	Piezometric Level	...	...	Purpose of Use
	Longitude	Latitude							
1									
2									
...	...	...	...	...	...	...	...	...	...
n									

4. Existing well diameters, distances to other wells can be recorded with special features along each direction.
5. If wells are already pumped then information about the discharge, pump duration, maximum drawdown and the time to recovery after the pump stop can be inquired.
6. Possibility of using some nearby wells can be considered as observation wells within acceptable radial distance around the selected main well.
7. Geometrical dimensions including the measurements of static water level, total well depth, and aquifer saturation thickness must be appreciated in some way.
8. Groundwater samples can be collected from a set of representative wells.

3.2.2 Aquifer (Pump) Test Planning

Any test record cannot produce satisfactory estimates of either aquifer properties or well performance parameters unless the data collection system is carefully addressed through a proper design. For successful and reliable test procedure and parameter estimations the following points are important to care for:

1. Estimation of the possible maximum drawdown during an aquifer (pump) test with constant discharge and a set of drawdowns for the well (step-drawdown) test with few discharge values (Section 3.15),
2. Estimation of the maximum pumping rate that is suitable for the environmental and hydrogeological conditions at well location. For instance, the well owners may not allow for long pumping durations and/or high discharge withdrawals. This is especially the reality in semiarid and arid regions,
3. Making decisions on the best method to measure the pump discharge,
4. Planning observation wells at convenient locations from the main well,

5. Preparation for drawdown measurements at observation wells,
6. Revision of all preparations before conduction of tests,
7. Measurement of all initial heads several times to ensure the steady-condition according to Eq. (3.11).

Economy is also a key concern; installation of pumping and observation wells may be cost prohibitive for many recharge studies. Aquifer tests require careful planning and useful guidelines are provided by Walton (1970).

3.2.3 Test Measurements

During any aquifer or well test field measurements must be recorded accurately, because the application of the analytical methodologies depends on the accuracy of time—drawdown data and the results of subsequent analyses depend on the following points:

1. The pump discharge must be maintained as constant, because almost all test evaluations require this as one of the basic assumptions.
2. Preferably time—drawdown measurements must be taken from two or more observation wells, if possible, at different radial distances.
3. Drawdown measurements can be taken at appropriate time intervals. The aquifer tests should include the information as in Table 3.4 for proper assessment of the groundwater potential under the existing hydrogeological units.

In some publications test duration is advised as 72-hour for confined and 24-hour for unconfined aquifers. In practice, most often tests are completed within a day's time (12 hours),

4. If applicable, during the test duration, some of the effective factors can also be measured such as the barometric pressure, stream, pond and lake levels, tidal oscillations, etc.

TABLE 3.4   Proper Aquifer Test Data Recording Guidance

Sheet No.	Area Name	Remarks	Date of Test

Type of test: Aquifer/pump (delete as convenient)

Location:

Performed by:

Aquifer type: Confined/unconfined/leaky/perched (delete as convenient)

Well type: Large diameter/small diameter/full penetration/partial penetration/nonpenetrating

Casing type: Iron/mason/stone (delete as convenient)

Test type: Aquifer, pump, recovery, step-drawdown

Well diameter:

Casing diameter:

Distance to observation well, if any:

Total depth of the well:

Datum elevation:

Drawdown measurement well: main/observation (delete as convenient)

Current time (hour/min)	Time Since Pump Start (min)	Piezometric Level (m)	Drawdown (m)	Discharge (m^3/s)
	(1:1:10)			
	(10:2:20)			
	(20:5:30)			
	(30:10:60)			
	(60:15:120)			
	(120:20:240)			
	(240:30:300)			
	300:60: If necessary continue with 60-minute increments			

Note-1: (a:b:c) implies start from "a" with increments "b" and end by reaching "c".
Note-2: Discharge measurements should be repeated at least three times and preferably five times during each test.

5. It is better to take drawdown measurements after the completion of aquifer or well test during the recovery period, when the discharge is zero.
6. In case of step-drawdown test at least four or five successively, increased discharge rates must be considered for time−drawdown measurements (Section 3.15).

EXAMPLE 3.4 DRAWDOWN CALCULATIONS

In the first two columns of Table 3.5 time and depth to groundwater level measurements are given in an observation well, which is 25 m away from the main well.

Hence,

TABLE 3.5   Time–Drawdown Data

Time (min)	Depth to Water Level (m)	Drawdown (m)	Time (min)	Depth to Water Level (m)	Drawdown (m)
0	32.23	0	70	34.33	2.1
1	32.29	0.06	80	34.62	2.39
2	32.36	0.13	90	34.94	2.71
3	32.44	0.21	110	35.31	3.08
4	32.52	0.29	130	35.75	3.52
6	32.62	0.39	150	36.21	3.98
8	32.73	0.5	180	36.67	4.44
10	32.84	0.61	210	37.14	4.91
15	32.97	0.74	240	37.62	5.39
20	33.12	0.89	270	38.11	5.88
25	33.27	1.04	300	38.62	6.39
30	33.45	1.22	360	39.14	6.91
40	33.64	1.41	420	39.67	7.44
50	33.85	1.62	480	40.25	8.02
60	34.08	1.85	540	40.88	8.65

1. Calculate the drawdown values.
2. Calculate the relative error for the last two drawdown measurements and decide whether the flow reached the steady state case. Consider ±5% relative error limits.

Solution 3.4

The first point is the initial (static) piezometric (water) level, which corresponds to zero time. In this table, it is at 32.23 m from the Earth's surface. Furthermore, the records in Table 3.5 indicate that as the time increases the distance to water surface also increases. In the aquifer assessment calculations drawdown values are necessary.

1. Drawdown, s, is defined as the difference between the dynamic water levels and the static (initial) water level in the observation well. Under the definition of drawdown, it is necessary to subtract from each dynamic water level the initial (static) piezometric (water) level value, 32.23 m. This leads to the third and sixth columns in Table 3.5.

2. The relative error between the two last drawdown is $100 \times (8.65 - 8.02)/8.65 = 7.28$, which is greater than the acceptable ±5%, hence, the pumping test is not complete, because it did not reach the steady (quasi) state condition.

3.2.4 Discharge and Drawdown Measurements

Pump discharge measurements can be achieved in several ways. Pumping discharge rate may change during the test due to hydraulic head or pump rounds per minute (rpm) variations. In order to adopt average constant discharge, it is necessary to take few measurements during the test duration. Depending on the discharge rate (low or high) one of the following convenient measurement methods can be used.

In cases of rather low discharges few measurements are taken to fill a certain container of known volume, V, during a certain time period, t, hence, the division of volume to time gives directly the discharge value. If this procedure is repeated few times then the arithmetic average is taken as the constant pump discharge in the calculations. This is the most frequently used method in the field.

It is also possible to use "V" notch weirs, where the heads are measured, and then, according to V-notch discharge formulation in the hydraulic textbooks, the pump discharge can be determined (Compton and Kulin, 1975). Similarly, rectangular notch weirs can also be used for head measurement leading to its conversion to water velocity through the convenient hydraulic formulation. At higher discharges Parshall flume (drop in floor) can be used for head measurement. Finally, one can also suggest the use of cut-throat flume (flat floor) and its head measurement (ASTM D 5242-92, 2001).

At high discharge rates water meters are convenient tools, if available. They are mostly impeller-driven water meters. Another alternative is to employ a circular orifice weir, where the water velocity, v, is expressed in terms of weir head, h, as (Chaudhary, 2007),

$$v = \sqrt{2gh} \qquad (3.12)$$

Drawdown measurements can be achieved through different and simple methods, but the most widely used instrument for this purpose is the sounder, which gives sound or lamp signal when the electrode at the end of a measuring tape touches the water surface. Drawdown measurements at main (pump) wells are difficult to achieve due to either turbulence, seepages from the well periphery, peripheral leakages,

collapses and alike. If time–drawdown measurements are recorded at the main well, the theoretical methodologies may yield reliable transmissivity estimation, whereas storativity cannot be calculated reliably.

3.2.5 Aquifer Test

This is also called as pumping test, but in this book "aquifer test" will be preferred, because such tests are performed in order to know the aquifer hydraulic properties such as transmissivity and storage coefficient. Groundwater specialists try to deduce from the field measurements qualitative (linguistic) and quantitative (numerical) information about the hydraulic characteristics of the aquifer geological formation. Aquifer tests are essential in effective groundwater management systems for proper planning, monitoring, observation and essential interpretations. Apart from the aquifer parameter determination, the following points are also significant for an efficient groundwater resources management (Chapter 6):

1. Determination of well efficiency and yield.
2. Determination of radius of influence for spatial effects of pumping and possible well interference calculations (Chapter 6).
3. Temporal water quality change during pumping (Chapter 5).
4. Optimum depth determination for pump location in the well.

Prior to the application of aquifer test some of the points must be taken into consideration for successfully determining the aquifer parameters and thereof interpretations.

1. Experts (in the geology and hydrogeology of the region or similar regions) with previous experience may help for successful completion of the aquifer test time–drawdown data evaluation.
2. Under the guides of preliminary reconnaissance surveys and well inventory

completion, it is possible to decide about the convenient well locations for aquifer (pump) and well (step-drawdown) tests in the study area.

3. A suitable well must be chosen for pumping close by observation wells, if possible.

4. Static water level state must be confirmed prior to the start of any pumping test with avoidance of other nearby pumping well interference. The radius of influence, R, of such wells can be calculated according to the following expression.

$$R = \sqrt{2.25T \frac{t}{S}} \qquad (3.13)$$

where T, S, and t are the transmissivity, storativity, and time duration, respectively.

3.2.6 Preliminary Time–Drawdown Assessment

As already explained in Section 3.2.3 drawdown is the difference between the static and dynamic piezometric levels. In general, piezometric surface (groundwater table) is not constant but fluctuates depending on the recharge or discharge from the aquifer. If it is stable then the piezometric surface has the same water level at different points, which is called static piezometric level (water table). Under static water level the piezometric surface is assumed as horizontal without any lateral flow at least in the vicinity of MW. In order to get useful records for aquifer parameter determination, this level is agitated by pumping from the MW. When the pump discharge is balanced by the aquifer discharge then the steady state situation is arrived. Such a case is rather rare in nature and it is never possible in unconfined or confined aquifers except when they are in connection with lakes, rivers, seas, and the like. In confined and unconfined aquifers the steady state case is expressed as quasi-steady state, which can be decided after the two successive drawdown measurements are

very close to each other within the error limits according to Eq. (3.11). Leaky aquifers may reach to true steady state case.

As time increases the drawdown also increases but in a nonlinear manner, which can be depicted by plotting field time–drawdown data on ordinary, semilogarithmic and double-logarithmic papers. In Figure 3.12 representative plotting are presented for the same time–drawdown data. Without any information about the theoretical and analytical models, one can fit by naked eye continuous lines or curves through the scatter points. These lines provide additional interpretation facilities.

Prior to any theoretical model aquifer test evaluation, the scatter points on these three graphs are worthy of investigation for gaining logical linguistic information. In practice, one can ask various questions for better understanding and evaluation of aquifer-well configuration. In general, whatever is the type of plotting paper, there is a directly proportional relationship between the time and drawdown. Another question is whether there is holistic (Section 3.6.1), partial (Section 3.6.2), or rather haphazard changes (Section 3.7). The following interpretations are not complete and many others can be added depending on the expert views by different hydrogeologists, groundwater specialists, engineers, operators, and anyone who works in this domain of interest.

1. Ordinary paper interpretations
   (Figure 3.12(a))
   There are two continuous curves fitted by eye to the scatter points, which provide the following comparative logical and verbal information:
   a. The initial part implies either the well has large or small diameter as explained in Section 3.1.1.2 or the aquifer is highly prolific with very high transmissibility.
   b. There may appear either systematic (for several consecutive measurements), or haphazard (random), or just very local

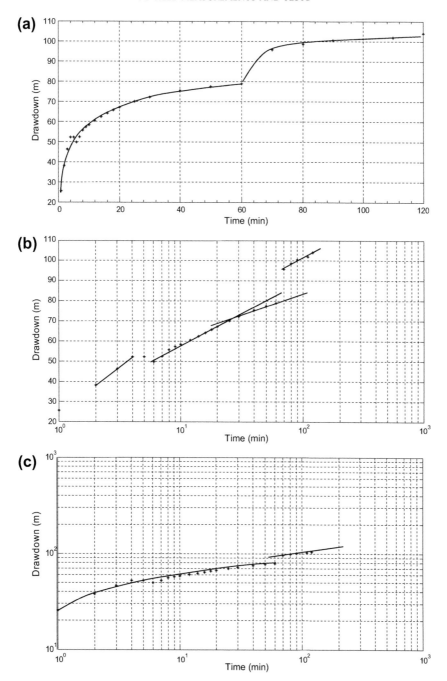

FIGURE 3.12 Time–drawdown data on (a) ordinary; (b) semilogarithmic; (c) double-logarithmic papers.

several deviations from a general trend. In case of systematic deviations depression cone expands with ease (difficulty) depending on the deviation cluster below (above) the general trend. If the deviations constitute local and partial groups above (below) the holistic trend then there are comparatively lower (higher) transmissivity zones within the depression cone. In case of haphazard deviations the aquifer material has heterogeneous and anisotropic composition of fine and course voids (pores, fissures, fractures, and solution cavities).

c. Any nonlinear trend portion implies aquifer contribution and the start of the nonlinear portion implies aquifer contribution initiation.

d. If late time scatter points are almost around horizontal line then there is internal or external recharge, which may imply leaky aquifer existence.

e. According to Eq. (3.4) during small times the slope of time−drawdown plot has a straight-line trend, the slope of which is equal to $Q/\pi r_w^2$ (Section 3.1.1.2).

A close inspection of time−drawdown plot in Figure 3.12(a) indicates that there are two successive parts prior to and after 60 min. The first part has two points off with equal drawdowns, which can be interpreted as wrong measurements. After 60 min, there is another systematic part. These systematic parts imply that the response of the aquifer is of confined type because the large times do not have a horizontal end (no leaky aquifer effect) with almost homogeneous and isotropic aquifer features (because the sequence of points falls more or less on a continuous curve without significant deviations). The second part is above the first one, which implies existence of less permeable geological formation at a certain distance within the depression cone. The flow does not reach steady state situation even after 120 min, because there are successive slight slopes at longer times. However, it may be regarded as quasi-steady state flow.

2. Semilogarithmic paper interpretations (Figure 3.12(b))
   In this figure only drawdown values on the vertical axis are real measurements, whereas time domain is in logarithmic scale. In general, on the semilogarithmic papers one tries to identify linear parts, because they may indicate transmissivity changes due to major fracture, fault, dyke, and similar subsurface geological feature effects. The following general interpretations can be deduced linguistically.

   a. Visually one searches for general linear or nonlinear trends among the scatter of points, and also looks for the continuity (or discontinuity) of such trends. Any linear trend during late times indicates applicability of Cooper-Jacob's (CJ) straight-line method (Section 3.9). Mathematically, any straight-line on a semilogarithmic paper implies an exponential function (see Eq. (3.46)).

   b. If the groundwater flow abide by the Darcy's law then the theoretical formulations suggest that for large times there is a linear trend (Section 3.9).

   c. If there are few straight-lines then there are local permeability changes, which may be due to faults, extensive fractures, and local groundwater sources, and the like.

As for Figure 3.12(b), there are few linear parts, which imply that the subsurface geological composition is heterogeneous and anisotropic. The slope of each partial line is different. As will be shown later in this chapter through the analytical derivation part that the aquifer transmissivity is a function of the slope. The smaller the slope the higher is the transmissivity (see Eq. (3.48)).

3. Double logarithmic paper interpretations (Figure 3.12(c))

Double logarithmic plots provide opportunity to magnify close scatter points and to converge far away scatter points. Such scatter diagrams bring time–drawdown data to almost equal footing for interpretations. On the other hand, double-logarithmic paper plots are requirements for theoretical type curve matching to time–drawdown data for aquifer parameter estimations. The following points are worth noticing:

a. As for the previous paper types again a general trend is sought visually through the scatter points. The reader may observe that on double-logarithmic paper the deviations are comparatively smaller.

b. On double-logarithmic paper, Eq. (3.4) yields to a straight-line during small times with slope equal to 45°, provided that the well has a large diameter. Any deviation from the 45° straight-line at early times implies nonlinear flow (Şen, 1986a).

c. Time–drawdown data scatter on double-logarithmic paper gives opportunity for decision about the aquifer type according to the type curves as will be explained later in this chapter.

d. If the large time scatters fall along a horizontal line then there is leakage contributions to aquifer and equilibrium between aquifer inflows and outflows (Section 3.5). Any horizontal line on a double-logarithmic paper indicates leakage equilibrium and such a line does not provide information about the aquifer parameters (Section 3.13).

e. If the general trend has increasing form with continuously decreasing local slope without reaching horizon at large times, then the type of aquifer is confined (Section 3.6.1).

f. If the trend has continuously increasing slope with comparatively lower rates during middle times than early times and again higher slopes at late times then it is unconfined aquifer (Chapter 4).

The closer the scatter points to continuous curve the more is the aquifer homogeneity and isotropy. Although the aforementioned explanations are valid for time–drawdown and distance–drawdown measurements, it is also possible to make necessary and similar interpretations from recovery or any other test types.

3.2.7 Aquifer Parameter Identification

Main aquifer parameters are transmissivity, T, and storage coefficient, S, (for all aquifer types), additionally leakage factor, L, (for leaky aquifers), and specific yield, S_y, and delayed yield, D_y (unconfined aquifer). Time–drawdown measurements in the field help for the identification of these parameters by processing with theoretical models. They assist to evaluate aquifer, well or pump properties, which are also useful for future predictions of discharges, drawdowns or distances between wells for the groundwater resources planning, design, operation, and management (Chapter 6). The following information can be obtained from any properly completed aquifer tests:

1. Aquifer parameter estimations (transmissivity, storage coefficient, specific yield, leakage factor, etc.),

2. Hydrogeological composition and flow quantities within the depression cone such as nature of impermeable barriers and recharge boundaries, radius of influence, drawdown at any time and distance from the main well,

3. One can follow the drawdown change by time, and hence, aquifer parameter may change by distance (Section 3.7),

4. Determination of well and aquifer characteristics such as the well losses, safe yield, specific drawdowns and the specific capacity of the well.

For the identification of aquifer parameters in the presence of multitude of prevailing geological environments, it is necessary to select suitable models, which take into account the background assumptions, initial and boundary conditions. Specific solutions are obtained from the implication of the theoretical models. The common assumptions adapted for groundwater flow to a well in any aquifer are as follows:

1. Aquifer material is isotropic and homogeneous.
2. The aquifer is areally extensive with uniform thickness and horizontal layer.
3. The flow is laminar and Darcy law is applicable.
4. The pump discharge remains constant during the aquifer test.
5. The well has full penetration.
6. The well diameter is circular and infinitesimally small.
7. There is instantaneous response from the aquifer to pumping discharge.
8. There is no well loss.
9. The aquifer parameters are independent of temporal and spatial changes, i.e., they are constants.
10. The groundwater is compressible and aquifer material has elastic behaviors.

Literature is full of analytical models developed by various researchers for the determination of aquifer parameters depending upon nature of aquifer response through time- or distance—drawdown measurements. For example, theoretical solutions about unconfined aquifers are dealt by Boulton (1954, 1955, 1963), Neuman (1972), Streltsova (1972, 1973). Confined aquifers are studied by Theis (1935), Jacob (1946, 1950), Walton (1970), Şen (1982) and Marie and Hollett (1996). The pioneering work in the development of leaky aquifer theory comes from Hantush and Jacob (1955a, b, 1960) and Hantush (1956, 1957, 1959, 1960, 1964, 1967). Some other important works regarding determination of aquifer characteristics are presented by Ferris et al. (1962), Kruseman and de Rider (1990), Rushton and Holt (1981), Şen (1986c), Corbett (1990) and, Harlow and LeCain (1993). Many of these methodologies are

explained in detail by various textbook authors (Walton, 1970; Davis and De Wiest, 1966; De Wiest, 1965; Fetter, 1983; Freeze and Cherry, 1979; Şen, 1995; Batu, 1998; Rushton, 2003). In the following sequel some of these methods are explained for use in practical applications with additional innovative methods.

3.3 AQUIFER TEST ANALYSES

During the last three decades, additional progress has been made in aquifer test analyses for hydraulic parameter computations. The reliability in the estimation of these parameters from available field data affects groundwater development and management studies. Any successful design and operation of groundwater abstraction task require reliable determination of aquifer parameters. The test that yields the aquifer material hydrogeological parameters is referred to as the "aquifer test," which is frequently referred to as the "pumping test" in the literature.

An aquifer test is the most suitable means for estimating hydraulic characteristics because it digests field data that reflects not only the point behavior of well location but a certain volume of aquifer material around the well within the cone of depression. This representative volume may extend up to 500–750 m in confined aquifers. The parameter estimations obtained through an aquifer test is more representative and reliable than laboratory test results, which depend on point samplings only.

During an aquifer test, the aquifer is probed by a main well through which controlled (constant) discharge is pumped for some time. In the meantime, the response of aquifer as the piezometric (water table) level decrease is measured at a set of predetermined time instances (Section 3.2.3), preferably drawdown measurements are taken, if possible, at least in one observation well. Theoretical models are presented in the forms of type-curves on double-logarithmic paper, and/or straight-lines on semilogarithmic paper.

This book emphasizes on the practical aspects leaving aside detailed theoretical and mathematical backgrounds. Necessary formulations are presented succinctly. Although the pump discharge is kept constant in carrying out aquifer tests in the field, it takes various contributions at different rates during the abstraction. These include:

1. The well storage contribution which is more pronounced in the case of large diameter wells at small times but as the time becomes larger this contribution ceases (Section 3.1.1.1).
2. Aquifer contribution is rather small at early times but it takes over completely at large times (see Figure 3.3).
3. Unconfined aquifers generate cone of depression born internal drainage, which is referred to as the delayed yield. Its effect occurs at moderate times (Chapter 4).
4. In the case of leaky aquifer, another source of contribution comes from the over- or underlying unpumped aquifers through semipervious layers (Section 3.5).
5. Fractures are among the early response sources that contribute to pump discharge either directly after the well storage effect or indirectly in the case of fractured porous medium aquifers such as sandstones.
6. Solution cavity contributions are rather local and they cause local reductions in the drawdown rates.

The tests yield constant aquifer parameters by mechanical matching of time–drawdown data with a convenient theoretical type curve. Additionally, field work and inquiries provide means of qualitative interpretations, which are soft computations that help to assess, evaluate, and manage groundwater resources linguistically. In any aquifer test evaluation procedure, particular attention must be paid to linguistic information sources prior to any numerical treatment of data (Section 3.2.6).

3.4 RADIAL STEADY STATE FLOW IN CONFINED AQUIFERS

Steady state flow takes place after a long water abstraction from the main well. If there is no recharge (internal or external) it is not possible to reach to absolute steady state level, but in practice almost steady state situation is assumed as quasi-steady state and the formulations are derived as if the absolute steady state exists. Separate transmissivity and storativity coefficient estimations are presented in the following sequel.

3.4.1 Radial Steady State Flow in Confined Aquifers

The groundwater movement toward a fully penetrating large diameter well in a confined aquifer after reaching the steady state case provides the depression cone insignificant expansion (quasi-steady state) under constant discharge, Q, abstraction (Figure 3.13).

Steady state flow does not provide time-wise changes but spatially the groundwater flow is in equilibrium. This means that the pump discharge, Q, is equal to aquifer discharge, Q_a, [$Q = Q_a(r)$]. Consideration of the aquifer discharge in terms of a radial cylindrical lateral surface at distance, r, with Darcian groundwater flow yields the equilibrium as,

$$Q = 2\pi r m K \frac{dh(r)}{dr}$$

Mathematical separation of variables methodology leads to,

$$dh(r) = \frac{Q}{2\pi T} \frac{dr}{r}$$

The integration of this expression by considerations of boundary conditions [r_1, $h(r_1)$] and [r_2, $h(r_2)$] yields simply,

$$h(r_2) - h(r_1) = \frac{Q}{2\pi T} Ln(r_2/r_1) \qquad (3.14)$$

FIGURE 3.13 **Confined aquifer steady state flow.**

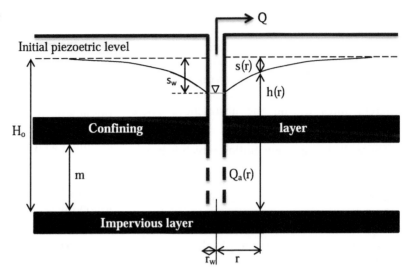

In general, the hydraulic head at any radial distance is $h(r) = H_o - s(r)$ and its substitution into Eq. (3.14) gives,

$$Q = 2\pi T \frac{s(r_1) - s(r_2)}{Ln(r_2/r_1)} \qquad (3.15)$$

This is known as the Theim (1906) equation in the groundwater literature. In practice, it helps to determine the transmissivity value of a confined aquifer through the following equation.

$$T = \frac{Q}{2\pi} \frac{Ln(r_2/r_1)}{s(r_1) - s(r_2)} \qquad (3.16)$$

This formulation can be used for large and small diameter wells. In practice, for small diameter wells the aquifer transmissivity can be determined more accurately if two observation wells are available at different distances (r_1 and r_2) with drawdown measurements, $s(r_1)$ and $s(r_2)$. The same formula provides reliable results (due to practically small well losses) in large diameter wells by using the drawdown, s_w, in the MW provided that the aquifer material has high hydraulic conductivity.

EXAMPLE 3.5 CONFINED AQUIFER TRANSMISSIVITY DETERMINATIONS

Figure 3.14 shows the general location of a fully penetrating main well (MW) and four observation wells (O1, O2, O3, O4) in a confined aquifer. The distances of each observation well to the MW of 2.2 m in diameter are given in Table 3.6. The

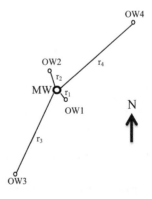

FIGURE 3.14 **Main and observation well locations.**

TABLE 3.6   Confined Aquifer Quasi-Steady-State Data

Well Number	MW	OW1	OW2	OW3	OW4
Distance (m)	1.1	25	65	105	180
Drawdown (m)	1.36	0.68	0.52	0.31	0.22

MW, main well.

quasi-steady state drawdowns after a continuous water abstraction for 900 min are also presented in the same table. The constant discharge is 0.4 m³/min. Hence,

1. Calculate all possible transmissivity estimation alternatives, make necessary interpretations as for the aquifer homogeneity and isotropy, and classify the aquifer potentiality according to the average transmissivity value.
2. Derive the radius of influence formulation and calculate its value.
3. Devise a graphical procedure for the transmissivity calculation.

Solution 3.5

1. The transmissivity estimation can be calculated from Eq. (3.16), which is applicable between any pairs of main and observation well with the substitution of given data values. The results are presented in a matrix form in Table 3.7.
   There are 10 different pairwise combinations of wells. The discharge is $0.4 \times 60 \times 24 = 576$ m³/day. Some examples for the transmissivity estimations, for instance, for MW-O3 and O2−O4 combinations are as follows.

TABLE 3.7   Transmissivity (m²/day) Matrix

	OW1	OW2	OW3	OW4
MW	947.6	1069.5	915.9	915.40
OW1		1258.1	818.8	696.9
OW2			483.0	716.2
OW3				1262.7

MW, main well.

$$T = \frac{2.3 \times 576}{2 \times 3.14} \frac{\log(105/11)}{(1.36 - 0.31)} = 915.88 \text{ m}^2/\text{day}$$

and

$$T = \frac{2.3 \times 576}{2 \times 3.14} \frac{\log(180/65)}{(0.52 - 0.22)} = 716.24 \text{ m}^2/\text{day}$$

The following conclusions and interpretations can be deduced from the transmissivity matrix:

a. Significant differences in transmissivity values indicate the heterogeneity of the aquifer material. The minimum (maximum) transmissivity value results from calculation between wells OW2 and OW3 (OW3 and OW4). The relative error between these two extreme values is about 61.7%. In practice, any aquifers can be regarded as homogeneous for relative errors less than ±5%.

b. Difference in the transmissivity values implies anisotropy of the aquifer material, because each OW is at a different direction.

c. The average transmissivity, $\overline{T}$, of the aquifer is 808 m²/day.

d. According to Table 2.10 on the basis of this average transmissivity value, the aquifer has high potentiality, because each OW is at different direction.

2. Radius of influence, R, can be calculated from Eq. (3.16) by taking into consideration one observation well at radial distance r with drawdown, s(r), and the drawdown is zero at R. Hence,

$$R = re^{\frac{2\pi\overline{T}}{Q}s(r)} \qquad (3.17)$$

Substitution of relevant values into the last expression for different observation well distances yields results as in Table 3.8. The average radius of influence is about 2556.5 m.

3. A close inspection of Eq. (3.16) implies that there is an inverse relationship between the drawdown and the radial distance. Furthermore, such a relationship is expected to be a straight-line on a semilogarithmic paper

TABLE 3.8   Radius of Influence

Well Number	OW1	OW2	OW3	OW4
R (m)	1002	6359	1614	1251

with distances on the logarithmic axis. Hence, the plot of data in Table 3.6 on a semilogarithmic paper gives scatter of points as in Figure 3.15.

Through the scatter of points a straight-line is fixed by the naked eye. Fitting straight-line is equivalent to assuming that the aquifer is homogeneous and isotropic. Transmissivity, storativity, and radius of influence values can be adopted from the graphical method. The slope of the straight-line is $\Delta s_r = 0.49$ m. In order to calculate the transmissivity value one needs to convert Eq. (3.16) into a form with slope. The term $[s(r_1) - s(r)]/Ln(r_2/r_1)$ is the theoretical slope, and hence Eq. (3.16) can be written as,

$$T = \frac{2.3Q}{2\pi\Delta s_r} \tag{3.18}$$

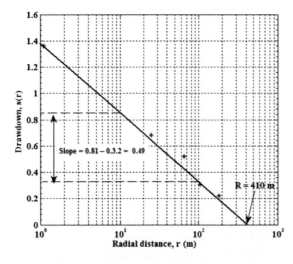

FIGURE 3.15   Distance–drawdown plots.

The transmissivity value can be calculated by substituting necessary quantities into this expression.

$$T = \frac{2.3 \times 575}{2 \times 3.14 \times 0.49} \approx 430 \ \text{m}^2/\text{day}$$

The graphical transmissivity value also shows that the aquifer is moderately potential as was the conclusion from the numerical method. Finally, the zero drawdown point on the horizontal axis gives the radius of influence as 410 m.

EXAMPLE 3.6 AQUIFER PARAMETER DETERMINATIONS

In a confined aquifer water is abstracted through a large diameter well of 2.5 m until the quasi-steady state condition is reached. The duration of withdrawal is 9 hours and 15 min with discharge of 0.5 m^3/min. The drawdowns in the MW and at 25 m OW are 5.20 m and 0.5 m, respectively. The volume of depression cone is 12,000 m^3 and the saturation layer thickness is 15 m. Hence,

1. What is the volume of abstraction water during this period?
2. What is the transmissivity of the aquifer? Interpret this parameter.
3. What is the storage coefficient of this aquifer? Interpret this parameter.

Solution 3.6

The well is in a confined aquifer, the groundwater flow is in radial directions, the quasi steady state groundwater flow situation is reached and the well fully penetrates the aquifer.

1. The water volume, V_w, abstraction from the well during the stated period, t, can be calculated easily as $V_w = Qt = 0.50 \times (9 \times 60 + 15) = 277.50 \ \text{m}^3$,

2. Transmissivity coefficient can be calculated from Eq. (3.16) as,

$$T = \frac{2.3 \times 0.5 \times \log(25/1.25)}{2 \times 3.14 \times (5.2 - 0.5)} \approx 168 \text{ m}^2/\text{day}$$

According to Table 2.10 this aquifer has moderate potentiality.

3. Storage coefficient can be calculated from Eq. (2.15) as $S = 277.5/12{,}000 = 0.0231 = 2.31 \times 10^{-2}$. The value is within the confined aquifer range of storage coefficient.

EXAMPLE 3.7 HYDRAULIC CONDUCTIVITY CALCULATION

A water pond of 10 m in radius penetrates fully the aquifer of 15 m thickness. As a result of spontaneous water withdrawal during 5 min, the water level drop is 13 cm. Hence,

1. How many cubic meters is the water withdrawal?
2. After a long time, the hydraulic gradient at the well periphery is 0.02. Accordingly calculate the transmissivity value around this well.

Solution 3.7

In case of spontaneous water withdrawal one can assume that during the abstraction duration there is no contribution from the aquifer to the well. This implies that the water initially comes from the well storage.

1. It is possible to calculate first the water volume, V_w, that is abstracted from the well storage during 5 min. If the drawdown is denoted by s_w, then, $V_w = \pi r^2 s_w$. Substitution of relevant numerical values into this expression yields $V_w = 3.14 \times 10^2 \times 0.13 = 40.84 \text{ m}^3$. Since discharge is volume per time one can calculate that $Q = V_w/t = 40.82/5 = 8.164 \text{ m}^3/\text{min}$.

2. It is first necessary to find the well entrance velocity, v_e, in the well vicinity. After a long period of time steady state groundwater flow takes place when the pump discharge is equal to entrance discharge. Basic discharge definition leads to, $v_e = Q/2\pi rm = 8.164/2 \times 3.14 \times 10 \times 1.5 = 0.0867 \text{ m/min}$. Finally, from the Darcy law by considering the hydraulic gradient at the well surface, $(dh/dr)_w$ one can write, $K = v_e/(dh/dr)_w = 0.0867/0.02 = 4.33 \text{ m/s}$.

3.4.2 Hydraulic Conductivity Variability

There are many natural cases where hydraulic conductivity variations occur within the aquifer (Wood and Nuemann, 1931; Muir-Wood and King, 1993; Rojstaczer and Wolf, 1992). Detailed accounts of radial variability are provided by Gilbert (1890) and Kitterod (2004). Şen (1989b) studied vertical permeability change and its effect on the aquifer test analysis. Radial variability of conductivity in the case of plants has been studied by plant biologists (Frensch and Steudle, 1989).

Similar to a confined aquifer, consideration of the aquifer discharge in terms of a radial cylindrical lateral surface at radial distance, r, with Darcian groundwater flow and hydraulic conductivity radial change, K(r) it is possible to write,

$$Q = 2\pi rmK(r)\frac{dh(r)}{dr}$$

Mathematically, after the separation of variables integration leads to,

$$h(r_2) - h(r_1) = \frac{Q}{2\pi m} \int_{r_1}^{r_2} \frac{dr}{rK(r)} \qquad (3.19)$$

The integration on the right hand side can be taken only after the definition of the radial hydraulic conductivity variation function.

3.4.2.1 Linear Hydraulic Conductivity Change

Let us consider linear radial variation of the hydraulic conductivity can be written as,

$$K(r) = K_w - \alpha r \quad (r > r_w) \qquad (3.20)$$

where K_w is the hydraulic conductivities at the well surface and α is defined as the hydraulic conductivity coefficient, which has the dimension of inverse time, $[1/T]$. The substitution of Eq. (3.20) into Eq. (3.19) yields (Şen, in press),

$$h(h_2) - h(h_1) = \int_{r_1}^{r_2} \frac{Q}{2\pi T} \left(\frac{1}{r} - \frac{\alpha}{K_w - \alpha r} \right) dr$$

where $T = mK_w$ is the aquifer lateral well surface transmissivity. The integration gives,

$$h(r_2) - h(r_1) = \frac{Q}{2\pi T} \left[Ln\left(\frac{r_2}{r_1}\right) - Ln\left(\frac{1 - \beta r_2}{1 - \beta r_1}\right) \right]$$

where, $\beta = \alpha/K_w$, represents the hydraulic conductivity change rate with dimensions $(1/L)$, and βr_1 and βr_2 are dimensionless radial distances. This expression can be written in terms of the drawdown, $s(r) = H_o - h(r)$, as,

$$T = \frac{Q}{2\pi} \frac{Ln[(1/r_1 - \beta)/(1/r_2 - \beta)]}{s(r_1) - s(r_2)} \qquad (3.21)$$

For $\beta = 0$ it reduces to the classical confined aquifer case with homogeneous and isotropic layer given in Eq. (3.16).

EXAMPLE 3.8 RADIAL HYDRAULIC CONDUCTIVITY CHANGE

A fully penetrating 2.4 m diameter main and three observation wells, (OW1, OW2, OW3) are in a confined aquifer with 12 m saturation thickness. The distances to each observation well are given in Table 3.9.

The measured quasi-steady state hydraulic heads are presented in the same table after continuous water extraction for 700 min with constant discharge 0.4 m^3/min. The static water level is at 20 m from the top of confining layer. Calculate whether there is a change in the aquifer hydraulic conductivity surrounding the main well.

Solution 3.8

The two numerical equations can be obtained from Eq. (3.21) by substitution pairs of observation wells such as OW1−OW2 and OW1−OW3 or OW2−OW3. The substitution of values from OW1 and OW2 gives,

$$T = \frac{0.4}{2 \times 3.14 \times (6.2 - 5.8)} Ln\left(\frac{1/25 - \beta}{1/65 - \beta}\right)$$

Additionally, consideration of OW1 and OW3 yields

$$T = \frac{0.4}{2 \times 3.14 \times (6.2 - 5.1)} Ln\left(\frac{1/25 - \beta}{1/110 - \beta}\right)$$

Since, these two expressions are equal to each other one can write after the necessary simplification that

TABLE 3.9   Confined Aquifer Quasi-Steady State Data

Well Number	MW	OW1	OW2	OW3
Distance (m)	1.1	$r_1 = 25$	$r_2 = 65$	$r_3 = 110$
Hydraulic heads (m)	9.3	$h(r_1) = 13.8$	$h(r_2) = 14.2$	$h(r_3) = 14.9$
Drawdown (m)	10.7	$s(r_1) = 6.2$	$s(r_2) = 5.8$	$s(r_3) = 5.1$

MW, main well.

$$0.4 \times \mathrm{Ln}\left(\frac{1/25 - \beta}{1/110 - \beta}\right) = 1.1 \times \mathrm{Ln}\left(\frac{1/25 - \beta}{1/65 - \beta}\right)$$

The solution of this equation yields $\beta \approx 0.0086$, which is a very small value and it indicates that the aquifer has a small linear hydraulic conductivity increase toward the main well. The substitution of this value in one of the above expression yields $T = 0.244 \, \mathrm{m^2/min} = 0.224 \times 60 \times 24 = 351 \, \mathrm{m^2/day}$, a productive aquifer, with moderate potentiality according to Table 2.10.

One can compare this transmissivity value with the one obtained by constant hydraulic conductivity assumption from Eq. (3.16) as,

$$T = \frac{0.4}{2 \times 3.14 \times (6.2 - 5.8)} \mathrm{Ln}\left(\frac{65}{25}\right)$$
$$= 0.1522 \, \mathrm{m^2/min} = 219 \, \mathrm{m^2/day}$$

The constant hydraulic conductivity approach underestimates the transmissivity value at the well vicinity by 42% with respect to the hydraulic conductivity variation case.

3.4.3 Storativity Estimation

The storage coefficient, S, estimation from quasi-steady state flow is presented by Şen (1987), who depended on the storage coefficient definition as the ratio of two volumes given in Eq. (2.15). The water withdrawal volume can be written as $V_w = Qt$ where Q is the pump discharge and t is the time required for steady state groundwater flow case. The following steps provide a straightforward graphical procedure, which gives an estimate of depression cone volume V_D and then its substitution into Eq. (2.15) yields the estimate of the storage coefficient. As mentioned earlier Eq. (3.16) represents a straight-line on a semilogarithmic paper with drawdown versus distance from the main well center (see Figure 3.16). The slope, Δs, and the

two intercept values s_L and r_U are read off from this figure.

The slope, Δs, of the straight-line is related to the aquifer transmissivity (Eq. (3.18)). The revolution volume of the same straight-line about the vertical drawdown axis gives the depression cone volume, V_D. The necessary steps for the storage coefficient calculation are as follows:

1. Plot the observed late drawdown data from observation wells located at various distances on a semilog paper with drawdowns on the vertical linear axis and the corresponding distances on the horizontal logarithmic axis (Figure 3.16). This is one of the major steps in the CJ distance–drawdown straight-line method (Section 3.9).
2. Draw the best-fit straight-line through the scatter points of data.
3. Extend this line in both directions until the intersection points are obtained on the vertical and horizontal axes.
4. Determine the slope, Δs, and then the intersection values, s_L and r_U, of this line on the drawdown and the distance axes, respectively. Clearly, r_U corresponds physically to the radius of influence and the drawdown value beyond this distance is practically equal to zero. The drawdown intersection at the well lateral surface has the distance, r_L from the main well center.
5. The transmissivity can be calculated from Eq. (3.18).
6. Calculate the area, A, of the triangle defined by the two axes and the fitted straight-line in Figure 3.16 as,

$$A = \frac{1}{2}(r_U - r_L)s_L$$

This is logically related to the depression cone volume.

7. Consider the revolution of this area below the fitted line around the vertical axis but with

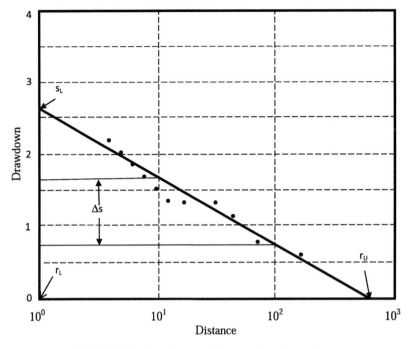

FIGURE 3.16 Schematic distance–drawdown plots.

back transformation of distances to linear scale, which yields the depression cone volume estimate as,

$$V_D = \pi \frac{(r_U - r_L)^2}{Ln\left(\frac{r_U}{r_L}\right)} s_L$$

8. The substitution of this expression into Eq. (2.15) gives the storage coefficient estimate as,

$$S = \frac{Qt}{\pi(r_U - r_L)^2 s_L} Ln\left(\frac{r_U}{r_L}\right) \qquad (3.22)$$

EXAMPLE 3.9 GRAPHICAL METHOD FOR STORAGE COEFFICIENT CALCULATION

The validity of the methodology developed for the identification of storage coefficient from quasi-steady state flow in confined aquifers and for perfectly steady state flow in leaky aquifers is presented through applications to field data already given by Kruseman and de Ridder (1990) (Tables 3.1 and 3.10). First Theim's method (Eq. (3.15)) is applied to data from pumping test "Oude Korendijk", which are reproduced and presented in Table 3.10 from Table 3.1 given by the same authors. The pumping discharge is $Q = 788 \text{ m}^3/\text{day}$ and the quasi-steady state flow is reached after $t = 830$ min. Determine graphically the storage coefficient of the aquifer.

Solution 3.9

The plots of distance–drawdown data on a semilogarithmic paper are shown in Figure 3.17 with all the necessary numerical values (Δs, r_U, r_L, s_L) for the graphical storage coefficient determination. These numerical values are shown in the same table. Although the storage coefficient estimates of the proposed method are somewhat larger than the estimates from Theis and Jacob methods, in their order of magnitude they show

TABLE 3.10   Aquifer Test Data and Results

r_1 (m)	r_2 (m)	s_1 (m)	s_2 (m)	Δs(m)	r_U (m)	r_L (m)	s_L (m)	V_D ($\times 10^3$ m³)	T (m²/day)	S ($\times 10^{-4}$)
30	90	1.088	0.719	0.64	810	0.1	3.00	687	370	6.69
0.8	30	1.236	1.088	0.73	1000	0.1	2.87	979	396	4.86
0.8	90	2.236	0.716	0.68	850	0.1	2.88	722	389	6.22
Averages									385	5.94

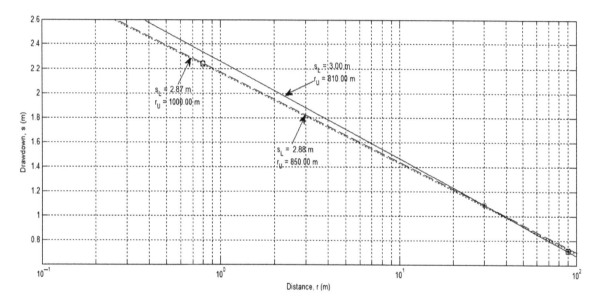

FIGURE 3.17   **Semi-logarithmic plot of distance-drawdown.**

very good agreement with Theis (1.6×10^{-4}), and Jacob (TD, DD, and CD method results as 1.7×10^{-4}, 4.1×10^{-4}, and 1.7×10^{-4}, respectively) method estimates, (Kruseman and de Ridder, 1990).

The method is also applied to the transient Qude Korendijkk data given by Kruseman and de Ridder (1990; Table 3.1). The plots of distance–drawdown values from two piezometers at distances 30 and 90 m after 140, 300, 600, and 830 min from the aquifer test start are shown in Figure 3.18, with four almost parallel straight-lines.

The implementation of the steps in the previous section gives rise to relevant data as shown in Table 3.11 for the various cases.

It is interesting to notice in this table that changes in the time duration lead to storage coefficient values with the same order of magnitude. Moreover, they also compare very well with the results obtained from other techniques. It seems from the sixth column that the estimated storage coefficient changes with time and such changes are due to the heterogeneities of the aquifer portion within the depression cone and explained

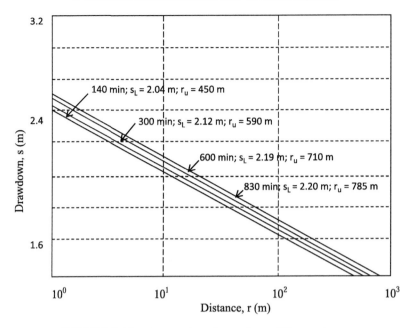

FIGURE 3.18 **Distance-drawdown plot at different times.**

TABLE 3.11   Storage Coefficient Determination

Time (min) (1)	Δs (m) (2)	r_U (m) (3)	r_L (m) (4)	s_L (m) (5)	S ($\times 10^{-4}$) (6)	r_L (m) (7)	S ($\times 10^{-4}$) (8)
140	0.76	490	1.0	2.04	3.62	10.0	3.76
300	0.76	590	1.0	2.12	4.53	10.0	2.97
600	0.76	710	1.0	2.19	6.23	10.0	4.13
830	0.76	785	1.0	2.20	7.12	10.0	4.78
Averages					5.35	10.0	3.91

in detail by Şen (1994). A critical point in the application of the method appears in choosing the lower distance value. Logically, it might be taken as equal to the well diameter. However, the choice of any small distance has no significant practical effect on the final storativity estimation. To illustrate this, r_L, is chosen as equal to 1.0, and the corresponding drawdown as well as the storativity calculations are presented in the last two columns. It is obvious that in spite of the significant difference in the r_L values, the storativity values do not deviate significantly by order of magnitude from one another.

3.5 RADIAL STEADY STATE GROUNDWATER FLOW IN LEAKY AQUIFERS

The same assumptions as for the confined aquifer are valid, with the additional one that the unconfined aquifer water table is not affected from the leakage and the leakage flow lines are perpendicular to the horizontal aquifer layer.

3.5.1 Transmissivity Estimation

The general well-leaky aquifer configuration is presented in Figure 3.19, where upper unconfined aquifer provides leakage to the lower semiconfined aquifer without its water table change by time or distance.

In leaky aquifer steady state flow equilibrium case, the main well discharge, Q, can be expressed as the summation of confined aquifer discharge, $Q_a(r)$, and semiconfined layer leakage discharge contribution, $Q_L(r)$, at radial distance, r, as,

$$Q = Q_a(r) + Q_L(r)$$

The confined aquifer and leakage discharges can be written from the geometry of flow sections and the Darcian law as,

$$Q_a(r) = 2\pi rmK\frac{dh(r)}{dr}$$

and

$$Q_L(r) = (2\pi rdr)K'\frac{H_o - h(r)}{m'}$$

where m' and K' are the thickness and hydraulic conductivity of the semiconfining layer, respectively. In the last expression the term in the parenthesis on the right hand side is the annulus area of the vertical leakage and the ratio is the hydraulic gradient that causes vertical groundwater flow from the unconfined aquifer into the confined layer. Hence, the constant pump discharge becomes,

$$Q = 2\pi rmK\frac{dh(r)}{dr} + (2\pi rdr)K'\frac{H_o - h(r)}{m'}$$

(3.23)

Division on both sides by $2\pi T$ and consideration that $s(r) = H_o - h(r)$ and $dh(r) = -ds(r)$ leads to,

$$\frac{Q}{2\pi T} = \frac{1}{L^2}rdrs(r) - r\frac{ds(r)}{dr}$$

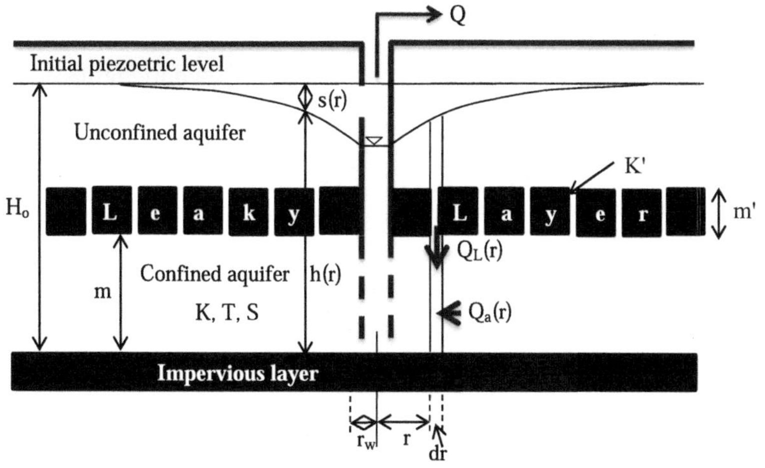

FIGURE 3.19 Leaky aquifer steady state flows.

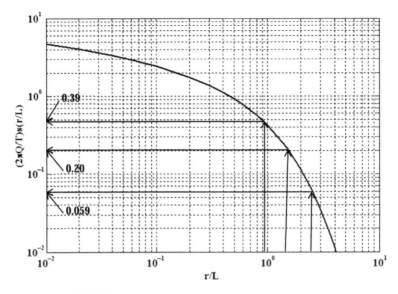

FIGURE 3.20　Dimensionless leaky aquifer function.

Herein, L is a constant value, which has length dimension and it is referred to as leakage coefficient,

$$L = \sqrt{\frac{m'mK}{K'}} \qquad (3.24)$$

The solution of this expression has been presented by de Glee (1930) by taking into consideration the condition that as $r \rightarrow 0$, $[rds(r)/dr] \rightarrow -Q/2\pi T$,

$$s(r) = \frac{Q}{2\pi T} K_o\left(\frac{r}{L}\right) \qquad (3.25)$$

Herein, r/L is a dimensionless radial and $K_o(.)$ is a modified Bessel function of the second kind. Appropriate tables are given in many textbooks (Bear, 1981; Kruseman and de Ridder, 1990). The transmissivity estimation formulation is,

$$T = \frac{Q}{2\pi s(r)} K_o\left(\frac{r}{L}\right) \qquad (3.26)$$

In practice, the transmissivity estimation can be achieved provided that there is an observation well at distance r with the steady state

drawdown, s(r), measurement and also if the leakage coefficient is known. On the other hand, one can arrange Eq. (3.26) in the following dimensionless form,

$$\frac{2\pi T}{Q} s(r) = K_o\left(\frac{r}{L}\right) \qquad (3.27)$$

This expression indicates that there is a nonlinear relationship between dimensionless drawdown $[(2\pi T/Q)s(r)]$ and the modified Bessel function. These two dimensionless variables are related to each other as shown in Figure 3.20

EXAMPLE 3.10 LEAKY AQUIFER TRANSMISSIVITY DETERMINATION

A long duration water abstraction with $0.9 \text{ m}^3/$s has been applied to a leaky aquifer and the drawdowns are measured at three observation wells (Table 3.12). The aquifer and overlying semipervious layer thicknesses are 35 and 3 m and hydraulic conductivities are 9 m/s and 5 m/s, respectively. Find the leakage coefficient and then

TABLE 3.12   Well Data

Well Number	OW_1	OW_2	OW_3
Distance, r, (m)	13	21	33
Drawdown (m)	1.06	0.87	0.25
r/L	0.945	1.527	2.400
$(2\pi Q/T)s(r/L)$	0.39	0.20	0.059
T (m^2/day)	15.36	24.60	23.96

the transmissivity values from each observation well. Give an average transmissivity value.

Solution 3.10

Since the geometric and hydraulic properties of the aquifer and semipervious layer are given, it is possible to calculate the leakage coefficient from Eq. (3.24) as,

$$L = \sqrt{\frac{3 \times 35 \times 9}{5}} \approx 13.75$$

This helps to calculate r/L values, which are given in the fourth row of Table 3.12. Entering each one of these values on the horizontal axis in Figure 3.20 provides the corresponding dimensionless drawdown values from the vertical axis, which are available in the fifth row of Table 3.12. One can calculate the transmissivity values for each observation well from Eq. (3.26) and the final results are given in the last row.

Finally the average transmissivity is 21.31 m^2/day and it indicates low aquifer potential according to Table 2.10 criterion.

3.6 UNSTEADY STATE FLOW IN AQUIFERS

The main difference from the steady state flow case is additional temporal discharge contributions from the aquifer material due to water compressibility. The discharge change is equal to the water storage change in the aquifer.

3.6.1 Unsteady Radial Flow in a Confined Aquifer Test

The simplest unsteady flow solution is for confined aquifer radial flow case, where the groundwater flow takes place horizontally and radially toward a fully penetrating well. The first solution has been obtained by Theis (1935) through the similarity work to heat transfer of Carlsaw and Jeager (1959).

It is almost impossible to encounter confined aquifer steady state flows in nature. In addition to the steady flow case assumptions, unsteady groundwater flow analytical solutions require the following assumptions:

1. The groundwater level also changes with time, h(r, t).
2. Either the flow line or equipotential line or both change their position with time.
3. The discharge at different radial sections and instances has different values, Q(r, t).
4. Previous points imply that the specific discharge magnitude, q(r,t), changes with time.
5. There is not always continuous external recharge but even if there is, it may not be enough to balance the amount of water abstracted from the aquifer.
6. There are changes in the storage within the aquifer due to water expansion. The water movement alone is not enough to represent the groundwater flow; additionally the storage capacity should also be taken into consideration.

Some of the unsteady state flows can be considered as quasi-steady state, which is an approximation to the steady state flow. Such a terminology is useful for practical purposes. Field conditions may require considerably long time periods to reach the steady state flow during field tests. Figure 3.21 shows a fully

FIGURE 3.21 **Unsteady flows to a well in a confined aquifer.**

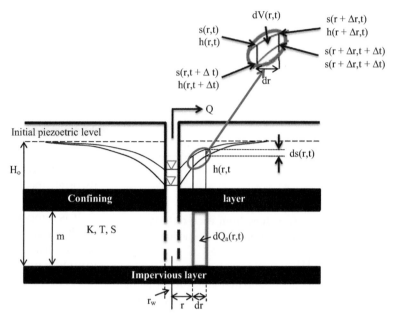

penetrating well in a confined aquifer with saturation thickness m, hydraulic conductivity K, and storage coefficient, S.

The groundwater movement mechanism has two parts due to hydraulic head temporal and spatial falls.

1. Annulus domain in the confined aquifer has discharge variation during infinitesimally small time, dt, and radial distance, dr. This discharge change is shown by $dQ_a(r, t)$ in the figure,
2. The second part is imaginary annulus between the two successive depression cones as in the figure, where four corners' drawdown and hydraulic head notations are shown explicitly. The discharge variation within the circular annulus inside the aquifer comes from the annulus volume. Its volume is $dV(r, t) = 2\pi r dr[dh(r, t)]$. The water amount that contributes into the aquifer is equal to the multiplication of this volume by the storage coefficient, $S2\pi r dr[dh(r, t)]$. This can be

converted to discharge contribution after division by dt, and hence, the discharge change in this imaginary volume is equal to $S2\pi r dr[dh(r, t)]/dt$.

Since the aquifer discharge change comes from the imaginary volume change between two successive depression cones then one can write,

$$\partial Q_a(r, t) = \frac{S2\pi r \partial r \partial h(r, t)}{\partial t}$$

If the discharge in the radial cylindrical surface can be written by taking into consideration the Darcy law then this last expression takes the following form,

$$\frac{\partial}{\partial r}\left[2\pi r m K \frac{\partial h(r, t)}{\partial r}\right] = \frac{S2\pi r \partial h(r, t)}{\partial t} \quad (3.28)$$

After the necessary arrangements, one can obtain simply the following general expression for radial unsteady flow to fully penetrating well in a confined aquifer.

$$\frac{\partial^2 h(r, r)}{\partial r^2} + \frac{1}{r} \frac{\partial h(r, r)}{\partial r} = \frac{S}{T} \frac{\partial h(r, t)}{\partial t}$$

If the initial hydraulic head at time zero is H_o, then $s(r, t) = H_o - h(r, t)$ and the last expression can be rewritten in terms of drawdown as,

$$\frac{\partial^2 s(r, t)}{\partial r^2} + \frac{1}{r} \frac{\partial s(r, t)}{\partial r} = \frac{S}{T} \frac{\partial s(r, t)}{\partial t} \qquad (3.29)$$

In general, there are two main approaches that are frequently used for practical solutions and application of Eq. (3.29). These are the Theis type curve method and its large time version known as CJ straight-line method (Section 3.9). The former depends on whole time−drawdown measurements from the field whereas the latter uses only late time−drawdown data. For quick answers, the CJ method is preferable, but in case of extensive data availability Theis method is recommended through the type curve matching procedure application.

3.6.1.1 Theis Model

Theis (1935) felt that the Eq. (3.29) has the analogy with heat flow in an isotropic and homogeneous conductive medium of uniform thickness and solved it under the relevant assumptions for a confined aquifer as,

$$W(u) = \int_{-\infty}^{u} \frac{e^{-x}}{x} dx \qquad (3.30)$$

where u is the dimensionless time factor and $W(u)$ is the well function, which are defined explicitly as,

$$u = \frac{r^2 S}{4tT} \qquad (3.31)$$

and

$$W(u) = \frac{4\pi T}{Q} s(r, t) \qquad (3.32)$$

The well function is the exponential integral (Jahnke and Emde, 1945) and can be expressed explicitly in the form of an infinite series as,

$$W(u) = -0.5572 - Ln(u) + u - \frac{u^2}{2.2} + \frac{u^3}{3.3} - \frac{u^4}{4.4} + \cdots \qquad (3.33)$$

The change of $W(u)$ with u (or $1/u$) on a double-logarithmic paper is called "type curve," which is a basic tool for aquifer parameter calculations. This curve is the focus of all the theoretical mathematical derivations, which is not necessary to know in detail. However, in practice, the use of type curve and its matching with the field time−drawdown data lead to simple aquifer parameter calculations. The type curve is presented on a double-logarithmic paper in Figure 3.22.

For practical convenience, $1/u$ is plotted versus $W(u)$, because in this case the time factor in Eq. (3.31) appears in the numerator, which provides convenience to the field plot of drawdown versus time. If [$1/u$ and $W(u)$] and [t and $s(r, t)$] pairs are known numerically then S and T can be determined from Eqs (3.31) and (3.32).

3.6.1.2 Theis Type Curve Matching Method

Type curve matching is a rather mechanical procedure through which the most convenient matching position is sought for the field data. As a general rule, any type curve leads to aquifer parameter estimations only after a proper application of the matching procedure. In any matching procedure the following steps are necessary:

1. The well function variation with the dimensionless time factor is drawn on a double logarithmic paper as in Figure 3.23, which is referred to as the type curve sheet. In practice, it is preferable to draw these curves on a transparent paper.
2. From the time−drawdown data, drawdown is plotted against time on the same scale double-logarithmic paper as for the type curve sheet. This plot is referred to as the field data sheet.

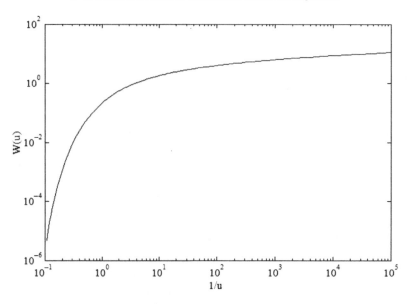

FIGURE 3.22   Theis type curve.

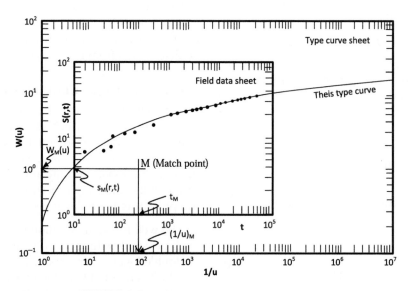

FIGURE 3.23   Theis curve and field data matching.

3. The type curve sheet is superimposed on the field data sheet and translated horizontally and/or vertically, keeping the corresponding coordinate axes (well function corresponds to drawdown) of the two sheets parallel, until a position of the best fit between the data plots and the type curve is reached. At this position the two sheets are kept without movement, preferably they are fixed by stickers.

4. An arbitrary point is chosen on the common overlapping area between the two sheets. This point, M, is called "match point."
5. The coordinates of the match point are read from the two sheets. As a result, four readings, namely, two abscissa, time, t_M, and the reverse of the dimensionless time factor, $(1/u)_M$; and two ordinates, drawdown, $s_M(r, t)$, and corresponding well function, $W_M(u)$, values are read off from the matched papers.
6. The substitutions of these four values into Eqs (3.31) and (3.32) yield aquifer parameter estimations.

All these steps are shown in Figure 3.23 with the notations mentioned in the previous steps.

TABLE 3.13   Field Data

Time (min), t	Drawdown (cm)	Time (min), t	Drawdown (cm)
1	40.2	30	152.0
2	60.4	40	161.2
3	73.8	50	169.6
4	83.0	60	175.8
5	97.0	80	185.4
8	106.8	100	193.0
10	113.4	130	200.0
15	127.0	160	208.4
20	134.0	190	225.0
25	144.0	290	220.0
		400	222.0

EXAMPLE 3.11 HOLISTIC THEIS TYPE CURVE AND AQUIFER PARAMETER ESTIMATIONS

The time–drawdown data from an observation well at 155 m radial distances from the main well center are given in Table 3.13. The pump discharge is $1.2\ m^3/min$ and the aquifer saturation thickness is 24.5 m. Hence,

1. Estimate aquifer parameters with interpretations.
2. What are the drawdown values at radial distances 45 and 62 m?

Solution 3.11

1. Field data and Theis type curve sheets are matched individually under the light of the abovementioned matching procedure steps (see Figure 3.24).

The match point values on the field data sheet are $s_M = 100\ cm$ and $t_M = 20\ min$; on the Theis type curve sheet the corresponding match point values are $W_M(u) = 3.95$ and $(1/u)_M = 91.2$, respectively. The substitution of all relevant values into Eq. (3.32) yields after the necessary arrangements,

$$T = \frac{QW_M(u)}{4\pi s_M} = \frac{0.3 \times 3.95}{4 \times 3.14 \times 1} = 135.79\ m^2/day$$

Substitutions into Eq. (3.31) yield the storage coefficient estimation as,

$$S = \frac{4t_M T}{r^2}\left(\frac{1}{u}\right)_M = \frac{4 \times 20 \times 135.79}{60 \times 24 \times 155^2} \times 91.2$$
$$= 2.8 \times 10^{-2}$$

According to the criterion given in Table 2.10 the aquifer is moderately potential since $50 < T < 500\ m^2/day$. The S value confirms that the aquifer is confined, because all $S < 10^{-1}$.

3.6.1.3 Partial Type Curve Matching Method

The actual meanings of constant transmissivity, T, and storativity, S, estimates from time–drawdown data for heterogeneous aquifers are often questioned (Şen and Wagdani, 2008,

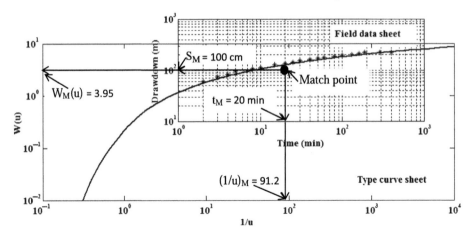

FIGURE 3.24 **Theis type curve matching to field data.**

2013). Constant discharge aquifer tests yield integrated drawdown curves for large undefined volumes of the aquifer (Tumlinson et al., 2006). Analysis of these integrated drawdown curves provide T and S estimates that have uncertain meanings with respect to the actual properties of the aquifer stressed during an aquifer test. The uncertainty typically manifests itself as some type of average T and S within the changing volume of the depression cone by constant pumping as time increases. The uncertainty increases when one considers whether drawdown data for a pumping well adequately represent some type of localized average aquifer response to spatially distributed heterogeneities within the cone of depression. Aquifer coefficients vary during early-, medium-, and late-time periods as the expansion of depression cone covers successive volumes of the aquifer. The representativeness of T and S estimates derived from good matching procedure is discussed by Tumlinson et al. (2006).

Butler and Liu (1993) concluded that constant-rate aquifer tests are not very effective for the characterization of lateral variations in flow properties. Schad and Teutsch (1994) compared T and S values from several small- and large-scale aquifer tests in a braided stream environment. They suggested that the effective length

scale of the heterogeneity structure can be estimated from their aquifer test data but not true effective T and S values. Meier et al. (1999) suggested that the straight-line method will provide a good approximation of the effective T in multi-lognormal and non-multi-lognormal T fields when constrained to late-time data.

Logically, a single aquifer value cannot be valid for the whole aquifer material around the pumping well, because the subsurface geological layers are neither homogenous nor isotropic. The only reliable conclusion when the field data fail to match the classical type curve is that the type curve theory does not fully represent actual field conditions. The aquifer parameter values derived from such an analysis may be seriously in error. Furthermore, insights about other features of aquifer response may be missed. This leads to the conclusion that in practice aquifer parameters are not distributed uniformly (Rushton, 2003).

Variable aquifer parameters are expected due to the complexity in the aquifer material. In holistic Theis type curve matching as in the previous subsection, the aquifer is assumed as homogeneous and isotropic. In order to identify partially the heterogeneities in any time—drawdown data, this section proposes application of partial matching method (PMM) based on the Theis approach. This procedure may yield a set of aquifer parameter estimations from the same time—drawdown data. If the relative errors between the PMM estimations are not more than 10% then the aquifer material can be accepted as homogeneous and isotropic. Otherwise, the aquifer parameter estimations from PMM should be preferable in practice.

The objective of this section is to summarize an effective approach, namely, PMM by Şen et al. (in press). It helps to identify possible variations in the aquifer parameters instead of constant values. This method leads to a few storativity and transmissivity estimations for different portions of the same time—drawdown data. The basis of the PMM procedure is quite simple, where well-known Theis type curve is matched partially to any given time—drawdown data. Its application can be achieved through following steps.

1. Plot drawdown versus time on a double logarithmic paper and match a group of scatter points to the most convenient part of the classical Theis type curve and calculate the aquifer parameters from Eqs (3.31) and (3.32).
2. Repeat the similar procedure for other possible partial matches and calculate new transmissivity and storativity values.
3. Calculate arithmetic or better weighted average of the partial aquifer parameters leading to overall and single T and S values.

EXAMPLE 3.12 PARTIAL THEIS TYPE CURVE AND AQUIFER PARAMETER ESTIMATIONS

The time—drawdown data is from Pakistan Chaj Doab area with the main and six observation wells (Figure 3.25) and their time drawdown data (Table 3.14), but herein, a single observation well, OB6, is considered among six wells as presented by Ahmad (1998). Aquifer test data from six observation wells (Figure 3.25) are presented and analyzed by Ahmad (1998) using the classic Theis and straight-line method. Fit convenient number of partial Theis type curve to time—drawdown data as recorded in OW-6.

Solution 3.12

The natural geological formation of the area is alluvial clay and gravelly coarse sand. Impervious clay layer also exists in some subarea around the wells. Hence, it is not possible to expect constant aquifer parameters, because the geological composition is rather complex. The following points are considered during the partial matching calculations and interpretations:

1. Aquifer material is assumed as heterogeneous and isotropic within the cone of depression during the whole pumping period.
2. The parameters are assumed to change according to the cone of depression expansion during the pumping period.

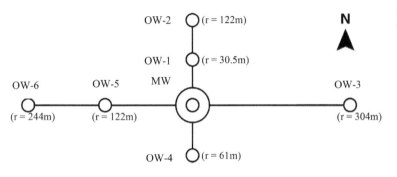

FIGURE 3.25 **Main and observation well plan positions.**

TABLE 3.14   Chaj Doab Area Aquifer Test Data

Wells	OW1	OW2	OW3	OW4	OW5	OB6
Time, t (min)	Radial distances, r, (m)					
	30.48	121.92	304.8	60.96	121.92	243.84
	Drawdown, s(r, t) (m)					
1	0.35	0.00	0.00	0.02	0.00	0.00
2	0.41	0.00	0.00	0.09	0.00	0.00
3	0.45	0.03	0.00	0.14	0.01	0.00
4	0.47	0.04	0.00	0.16	0.01	0.00
5	0.48	0.04	0.00	0.19	0.01	0.00
6	0.49	0.05	0.00	0.20	0.02	0.00
7	0.50	0.05	0.00	0.23	0.02	0.00
8	0.51	0.05	0.00	0.24	0.02	0.00
9	0.51	0.06	0.00	0.25	0.02	0.00
10	0.52	0.06	0.00	0.25	0.03	0.00
12	0.52	0.06	0.00	0.26	0.03	0.00
14	0.53	0.06	0.00	0.24	0.04	0.00
16	0.53	0.07	0.00	0.24	0.04	0.00
18	0.54	0.08	0.01	0.25	0.04	0.00
20	0.54	0.08	0.01	0.25	0.05	0.00
22	0.55	0.08	0.01	0.25	0.05	0.00
24	0.55	0.08	0.01	0.26	0.05	0.00
26	0.54	0.08	0.01	0.26	0.05	0.00
28	0.55	0.08	0.01	0.26	0.05	0.00
30	0.56	0.08	0.01	0.26	0.06	0.00
35	0.57	0.09	0.01	0.27	0.06	0.00
40	0.57	0.09	0.01	0.29	0.06	0.00
45	0.58	0.10	0.01	0.29	0.07	0.00
50	0.58	0.09	0.01	0.30	0.08	0.00
55	0.59	0.10	0.01	0.30	0.09	0.00
60	0.59	0.11	0.01	0.31	0.09	0.00
70	0.60	0.11	0.02	0.30	0.09	0.00
80	0.61	0.12	0.02	0.32	0.10	0.00

TABLE 3.14 Chaj Doab Area Aquifer Test Data (*cont'd*)

Wells	OW1	OW2	OW3	OW4	OW5	OB6
90	0.62	0.12	0.02	0.33	0.10	0.00
100	0.62	0.13	0.02	0.34	0.10	0.00
110	0.63	0.13	0.02	0.34	0.11	0.00
120	0.63	0.14	0.02	0.34	0.11	0.00
140	0.64	0.14	0.02	0.35	0.11	0.01
160	0.65	0.15	0.03	0.36	0.12	0.01
180	0.66	0.16	0.03	0.37	0.12	0.01
210	0.66	0.16	0.04	0.37	0.12	0.01
240	0.67	0.16	0.04	0.37	0.12	0.01
270	0.67	0.16	0.04	0.38	0.13	0.01
300	0.68	0.17	0.04	0.40	0.13	0.02
330	0.69	0.18	0.04	0.41	0.13	0.02
360	0.69	0.19	0.05	0.40	0.13	0.02
420	0.71	0.19	0.05	0.41	0.13	0.02
480	0.71	0.20	0.05	0.42	0.14	0.03
540	0.71	0.21	0.06	0.44	0.16	0.03
600	0.72	0.21	0.06	0.43	0.16	0.03
720	0.75	0.23	0.07	0.45	0.17	0.04
840	0.76	0.24	0.08	0.47	0.18	0.05
960	0.78	0.25	0.08	0.46	0.18	0.05
1080	0.80	0.26	0.09	0.47	0.19	0.06
1200	0.81	0.27	0.09	0.47	0.20	0.06
1440	0.80	0.27	0.11	0.49	0.21	0.06
1680	0.82	0.29	0.11	0.50	0.23	0.08
1920	0.82	0.31	0.11	0.50	0.23	0.09
2160	0.85	0.32	0.12	0.51	0.23	0.09
2400	0.86	0.32	0.12	0.52	0.23	0.09
2640	0.86	0.34	0.12	0.53	0.23	0.10
3000	0.87	0.34	0.13	0.53	0.23	0.11
3360	0.85	0.35	0.14	0.53	0.24	0.11
3720	0.87	0.35	0.15	0.54	0.24	0.12
4080	0.87	0.35	0.12	0.54	0.24	0.12

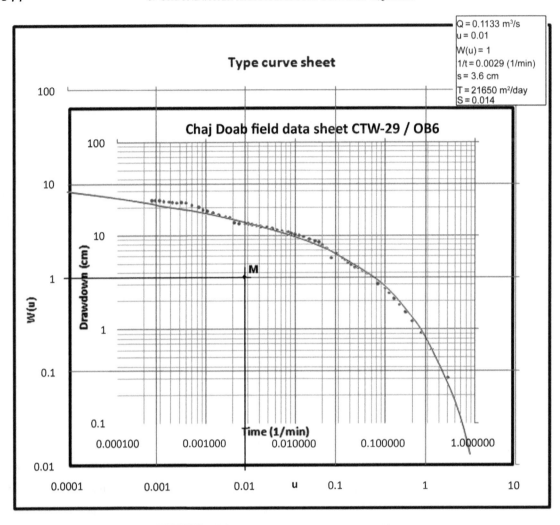

FIGURE 3.26 Holistic Theis type curve matching.

3. Transmissivity and storativity parameters are calculated separately according to convenient partial matching.
4. The partial parameter values provide information about the aquifer heterogeneity (provided that they are more than ±10% apart from each other).

Figure 3.26 represents OW6 time–drawdown scatter and the holistic Theis type curve matching

where time–drawdown data are best fitted with a single type curve. In practice, the aquifer parameters obtained from the OW6 can be regarded as representatives of homogeneous and isotropic geological composition of the aquifer material within the depression cone. A close inspection of this figure indicates that classical type curve matching cannot be representative holistically, because there are systematic derivations along some portions of the data.

This brings to the mind the application PMM procedure separately to the same time–drawdown data. Figure 3.27(a)–(d) indicate four partial matching to the OW6 time–drawdown data.

Table 3.15 indicates PMM application results for OW6 only after four PMM to the same time–drawdown data.

Since there are more than 10% deviation between the parameter values, the aquifer should be regarded as heterogeneous and anisotropic. The arithmetic averages of the transmissivity and storativity coefficients are 19,278 m^2/day and 2.3×10^{-4}, respectively. This transmissivity value implies very highly potential aquifer according to Table 2.10. There is a strict relation between partial matching time-duration and aquifer heterogeneity. By regarding i-th partial matching duration as weights, w_i, the weighted average values of the parameters, P_w, can be obtained from following equation,

$$P_w = \frac{\sum_{i=1}^{n} w_i P_i}{\sum_{i=1}^{n} w_i} \qquad (3.34)$$

Thus, four partial transmissivity and storativity coefficients are obtained and the numerical results are presented in Table 3.16. After having obtained the mean aquifer parameters the regional parameters can be calculated as the arithmetic average, leading to $T_w = 15,911$ m^2/day and $S_w = 4.04 \times 10^{-2}$. When these results are compared with the classical Theis type curve matching calculations, more realistic values are obtained as transmissivity coefficient 24.74% and storativity coefficient with 30.41% difference.

3.7 SLOPE METHOD

On the basis of core sample analyses Cardwell and Parsons (1945) suggested that the equivalent transmissive capacity of randomly distributed block heterogeneities lies between the harmonic and arithmetic means of the actual transmissive capacities of the heterogeneities. It is suggested in this section that rather than the means, the modes of the transmissivity and storativity are the most representative values (Şen and Wagdani, 2013). Bibby (1977) evaluated 122 drawdown curves for pumping wells in heterogeneous clastic sediments. He described four basic drawdown curve shapes and used the Cooper and Jacob (1946) straight-line method to ascertain the "short-term transmissive capacity" near each well. He discussed the meaning of local and regional average aquifer coefficients from early- and late-time data on the basis of the work of Cardwell and Parsons (1945). He defined weighted arithmetic, harmonic, and geometric means for the long-term transmissive capacity of a drainage volume divided into concentric rings centered on the pumping well.

Şen (1986b) devised a method, which is applicable to any type curve equation and avoids the curve matching technique in aquifer parameter determination from given field data. Elegancy of the method is the aquifer parameter estimations right after the second time–drawdown measurement, and subsequently, as the new record is measured, new estimates of the concerned parameters are obtained. The basic concept of the method is very simple and it derives from the fact that matching of two curves is equivalent to saying that they have the same slopes at corresponding points. Although the conventional aquifer tests tend to average these conditions in the aquifer response to pumping, the field test data will still have some local deviations from any analytically derived type curves; therefore, the slopes of type and field curves are not equal over the entire time domain.

Due to the aforementioned deviations one should physically expect some variation in these parameters. The main objective of slope model is to identify the likely variations in aquifer parameters by matching slopes of the type and field curves. Şen (1986b) gave the analytical slope, α,

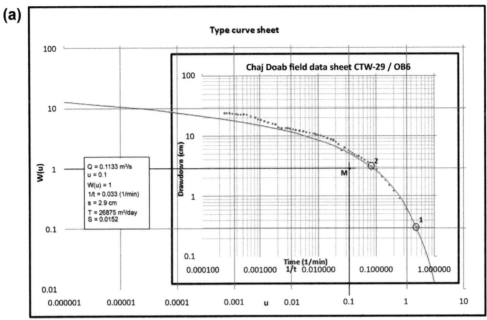

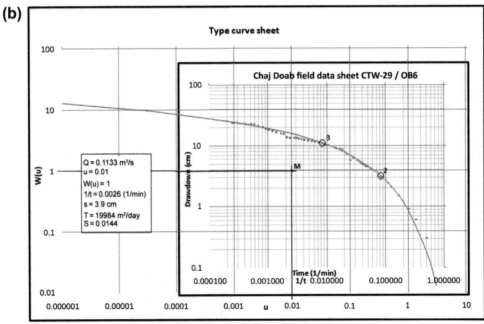

FIGURE 3.27 Partial Theis matching procedure.

(c)

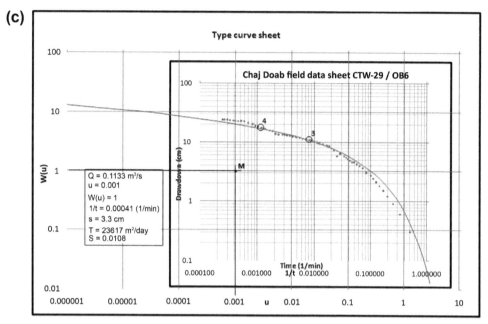

(d)

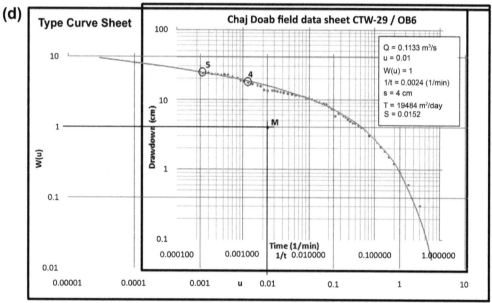

FIGURE 3.27   (*continued*).

TABLE 3.15   PMM Matching Procedure Values

Well Number	Group	Time (min)	U_M	W (U_M)	$(1/t)_M$	s_M (cm)	T	S
OW6							(m²/day)	(−)
	1st	0–12	1.00×10^{-1}	1	3.30×10^{-2}	2.9	26.875	1.52×10^{-2}
	2nd	12–110	1.00×10^{-2}	1	2.60×10^{-3}	3.9	19.984	1.44×10^{-2}
	3rd	110–960	1.00×10^{-3}	1	4.10×10^{-4}	3.3	23.617	1.08×10^{-2}
	4th	960–4080	1.00×10^{-2}	1	2.40×10^{-3}	4	19.484	1.52×10^{-2}
Arithmetic averages							22.490	1.39×10^{-2}

PMM, partial matching method.

TABLE 3.16   PMM Parameter Estimations

Distance (m)	Group	Time Duration (min)	Transmissivity (m²/day)	Storativity ($\times 10^{-2}$)	Geometric Mean T_w (m²/day)	Geometric Mean S_w ($\times 10^{-2}$)
122	1st	12	26,875	1.52	19,962	1.47
	2nd	98	19,984	1.44		
	3rd	450	23,617	1.08		
	4th	3620	19,484	1.52		

PMM, partial matching method.

expression for the Theis curve on a double logarithmic paper as,

$$\alpha = -\frac{e^{-u}}{W(u)} \qquad (3.35)$$

This expression helps to convert the Theis type curve table into a Theis slope values as presented in Figure 3.28.

The processing of the aquifer test data can be achieved through the following steps without type curve matching:

1. After the second time–drawdown measurement, calculate the slope between the two successive points in double logarithmic scale as, $\alpha_i = Ln(s_i/s_{i-1})/Ln(t_{i-1}/t_i)$ where $i = 2, 3, ..., n$ and n is the number of drawdown records.

2. Find the u_i value corresponding to this slope (α_i) from Figure 3.28.

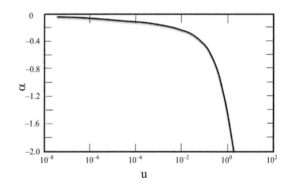

FIGURE 3.28   Type curve slope.

3. Knowing the slope and the u_i value find the well function value from Eq. (3.35) as,

$$W_i(u) = \frac{e^{-u_i}}{\alpha_i} \qquad (3.36)$$

4. Calculate local T_i and S_i values from Eqs (3.32) and (3.31), respectively.

5. Repeat the previous steps with the next time—drawdown record.

Finally, sequences of estimation are obtained for each aquifer parameters. This approach gives way to the following benefits:

1. Aquifer parameter determination even from early time—drawdown data.
2. Any late arrival at the test site means missing data for holistic type curve matching. The slope method provides an efficient way for parameter estimations even from few time—drawdown measurements.
3. Opportunity to appreciate heterogeneity in aquifer parameters.
4. Rather than constant aquifer parameters a sequence is obtained for each parameter. It is possible to obtain the frequency distribution of the parameters. Hence, the professional can appreciate the minimum, the maximum, the average, and the most frequently occurring (mode) aquifer parameter values.

EXAMPLE 3.13 TEMPORAL VARIATION OF AQUIFER PARAMETERS

The application procedure presented herein has already been given by Şen and Wagdani (2013) in full detail. To show the efficiency of the proposed methodology, a classical textbook confined aquifer test is selected from Todd (1980). A fully penetrating well in a confined aquifer is pumped at a uniform rate of 2500 m^3/day. Drawdowns, s_i, variations with time, t_i, are measured in an observation well, which is 60 m away from the pumping well (Table 3.17). Find and interpret the heterogeneity of aquifer parameters.

Solution 3.13

After a convenient matching point on Theis type curve [$W(u_M) = 1.00$, and $u_M = 1.0 \times 10^{-2}$], and on the field sheet [$s_M = 0.18$ m and $(r^2/t)_M = 150$ m], constant aquifer parameters are calculated from Eqs (3.32) and (3.31) as $T = 1110$ m^2/day and $S = 2.06 \times 10^{-4}$, respectively.

TABLE 3.17   Aquifer Test Data

Time (min)	Time (day)	Drawdown (m)	Time (min)	Time (day)	Drawdown (m)
0	0	0	18	0.01250	0.67
1	0.00069	0.2	24	0.01667	0.72
1.5	0.00104	0.27	30	0.02083	0.76
2	0.00139	0.30	40	0.02778	0.81
2.5	0.00174	0.34	50	0.03472	0.85
3	0.00208	0.37	60	0.04167	0.90
4	0.00278	0.41	80	0.05556	0.93
5	0.00347	0.45	100	0.06944	0.96
6	0.00417	0.48	120	0.08333	1.00
8	0.00556	0.53	150	0.10417	1.04
10	0.00694	0.57	180	0.12500	1.07
12	0.00833	0.60	210	0.14583	1.10
14	0.00972	0.63	240	0.1667	1.12

TABLE 3.18   Slope Matching Calculation Results

Time (15 days)	Drawdown (m)	α_i	u_i	$W(u_i)$	T_i (m²/day)	S (×10⁻⁴)
0	0					
0.0007	0.2	0.74	0.2517	1.05	889.25	1.7
0.0010	0.27	0.37	0.0471	2.60	1818.31	1.0
0.0014	0.30	0.56	0.1633	1.51	941.36	2.4
0.0017	0.34	0.46	0.0857	1.98	1109.06	1.8
0.0021	0.37	0.36	0.0305	2.71	1386.56	1.0
0.0028	0.41	0.42	0.0668	2.24	1037.33	2.1
0.0035	0.45	0.35	0.0315	2.74	1717.19	1.4
0.0042	0.48	0.34	0.0346	2.80	1104.84	1.8
0.0056	0.53	0.33	0.0205	3.00	1086.83	1.4
0.0069	0.57	0.28	0.0124	3.51	1193.87	1.1
0.0083	0.60	0.32	0.0232	3.09	998.57	2.1
0.0097	0.63	0.24	0.0900	3.73	1141.96	11.1
0.0125	0.67	0.25	0.0189	3.92	1122.69	3.0
0.0167	0.72	0.24	0.0905	3.77	1013.56	17.0
0.0208	0.76	0.22	0.0066	4.49	1136.68	1.7
0.0278	0.81	0.22	0.0053	4.60	1103.70	1.8
0.0347	0.85	0.31	0.0241	3.11	707.97	6.6
0.0417	0.90	0.11	0.0001	8.77	1907.42	0.1
0.0556	0.93	0.14	0.0004	7.03	1479.05	0.4
0.0694	0.96	0.22	0.0063	4.44	900.96	4.4
0.0833	1.00	0.18	0.0011	5.68	1108.51	1.1
0.1042	1.04	0.16	0.0091	6.35	1198.06	12.6
0.1250	1.07	0.18	0.0028	5.56	1019.30	4.0
0.1458	1.10	0.13	0.0004	7.41	1327.76	0.9

Note: $T = 1162.70\ m^2/day$ implies high potentiality; $S = 3.43 \times 10^{-4}$ implies confined aquifer.

The same aquifer test is subjected to slope matching calculations leading to the final results in Table 3.18. The slopes are calculated according to Eq. (3.35).

The plots of storativity and transmissivity against pumping time are given in Figures 3.29 and 3.30, respectively.

Their visual inspections lead to the following significant interpretations, which cannot be obtained by any other type curve matching procedure:

1. Depending on the heterogeneous composition of the aquifer material around the well, the

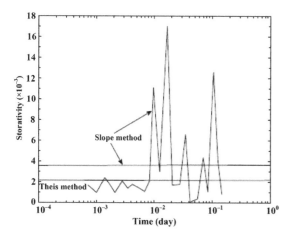

FIGURE 3.29 Temporal storativity changes.

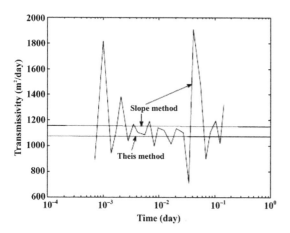

FIGURE 3.30 Temporal transmissivity changes.

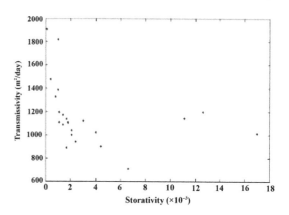

FIGURE 3.31 Storativity versus transmissivity.

4. The average slope matching aquifer parameters obtained from Table 3.18 are also shown in the same figures, and they remain over the classical estimations.
5. In general, high storativity values are attached with low transmissivity values, which imply an inverse relationship between these parameters (see Figure 3.31).

3.7.1 Statistical Considerations

The basis of any statistical assessment is first to look at the histogram or the frequency distribution of the parameter values. The frequency distributions of the storativity and the transmissivity estimations are plotted in Figures 3.32 and 3.33, respectively.

The storativity frequency diagram gives the impression of steady decrease in the frequencies with the increase in the storativity value. In other words, the aquifer is composed of dominant small storativity parts than higher storativity values. This implies, theoretically, an exponential probability distribution function (pdf) for aquifer storativity. The single parameter of this distribution is $\lambda = 3.5 \times 10^{-5}$. It is possible to calculate weighted averages of aquifer parameters with frequency weights. Figure 3.32 indicates that the aquifer is composed of

aquifer parameters assume different values with depression cone expansion.

2. The aquifer parameters have their maximum and minimum values, and hence, variation domains. The storativity (transmissivity) maximum (minimum) values are 1.7×10^{-3} (1907.42 m^2/day) and 1×10^{-4} (707.97 m^2/day), respectively.
3. Most of the time, the constant parameters from Theis type curve matching do not represent the aquifer heterogeneity, and they are biased toward low values (see Figures 3.29 and 3.30).

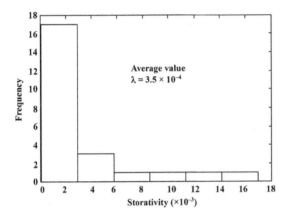

FIGURE 3.32   **Storativity frequency distribution.**

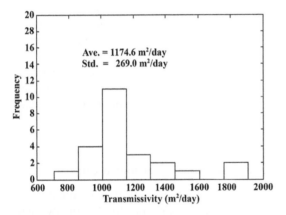

FIGURE 3.33   **Transmissivity frequency distribution.**

heterogeneities with the most frequent (mode) transmissivity values at almost $1150 \, m^2/day$. Since all the transmissivity values are greater than $500 \, m^2/day$, this aquifer is highly potential (Table 2.10).

3.7.2 Spatial Variation of Aquifer Parameters

It is also possible to calculate the change of heterogeneity with distance, r, by consideration from Eq. (3.31) for constant T and S values, which gives

$$r = \sqrt{\frac{4tTu}{S}} \qquad (3.37)$$

To find the amount of distance increment, dr, after a specific time interval, dt, it is necessary to take the derivative of this expression with respect to t,

$$dr = \frac{1}{2}\sqrt{\frac{4Tu}{tS}}dt \qquad (3.38)$$

The accumulation of distance measurements gives the radius of influence. The results from the application of Eq. (3.38) yield distances in Table 3.19. Figure 3.34 indicates the distance variation with time on semilogarithmic paper.

The mathematical expression of the straight-line in this figure can be obtained after some algebraic calculations as

$$r = 182 + 27 \log t \qquad (3.39)$$

This expression implies that the rate of expansion (change of distance by time) is higher at early time periods but steadily decreases at moderate and large time instances. Finally, it reaches practically to a constant distance, which implies quasi-steady state groundwater flow. Figure 3.35 indicates the change of drawdown with distance where an inverse relationship exists.

Initially, at small distances, the drawdown increase rate is higher than medium or big distances. It is possible to extract the maximum distance corresponding to zero drawdown at approximately 130 m, which is practically equal to the radius of influence. Considerations of Tables 3.18 and 3.19 give an opportunity to plot storativity and transmissivity coefficient changes with the distance as in Figures 3.36 and 3.37, respectively.

In general, the high (low) transmissivity (storativity) corresponds to low (high) storativity (transmissivity). On these figures, the average aquifer parameter values remain the same, similar to what have been in Figures 3.29 and 3.30.

TABLE 3.19 Distance Calculation

t (day)	dr (m)	R (m)	t (day)	dr (m)	R (m)
0	0	0	0.012500	6.32	5.89
0.001042	4.32×10^0	4.32×10^0	0.016667	1.80×10^1	7.69
0.001389	6.96	1.13×10^1	0.020833	4.35×10^0	8.12
0.001736	4.51	1.58	0.027778	5.63	8.68
0.002083	2.46	1.83	0.034772	1.07×10^1	9.75
0.002778	6.30	2.45	0.041667	5.72	9.81
0.003472	3.87	2.84	0.055556	2.19×10^0	1.00×10^2
0.004167	3.70	3.21	0.069444	7.75	1.08
0.005556	4.93	3.70	0.083333	2.90	1.11
0.006944	3.43	4.05	0.104167	1.14×10^1	1.22
0.008333	4.29	4.48	0.125000	5.79×10^0	1.28
0.009722	7.82	5.26	0.145833	1.89	1.30

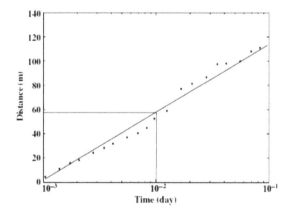

FIGURE 3.34 Distance change with time.

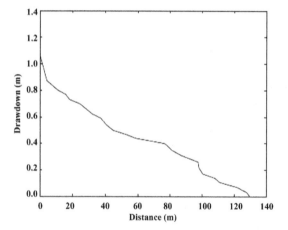

FIGURE 3.35 Drawdown change with distance.

Bibby (1977) suggested that the early-time portion of a drawdown curve probably reflects the average or effective local conditions near the well, whereas the late-time portion of a drawdown curve reflects more regional average or effective conditions. Additionally, Butler (1988) suggested that the late portions of

drawdown curves can be used to estimate sustainable aquifer yield. The estimated distances can be used as weights to derive weighted, P_W, (similar to Eq. (3.34)), harmonic, P_H, and geometric, P_G, mean transmissivity and storativity values only for the aquifer materials covered

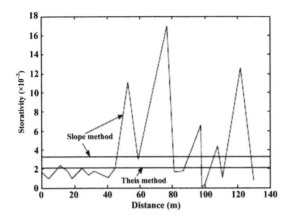

FIGURE 3.36   Storativity-distance changes.

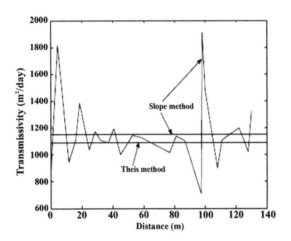

FIGURE 3.37   Transmissivity-distance changes.

by depression cone for each time step as follows (Tumlinson et al., 2006),

$$P_H = \frac{1}{\frac{1}{n}\sum_{i=1}^{n}\frac{1}{P_i}} \tag{3.40}$$

and

$$P_G = \left(\prod_{i=1}^{n} P_i\right)^{1/2} \tag{3.41}$$

The slope method provides an opportunity to calculate each one of these means in addition to other alternatives (arithmetic mean, P_A, and mode, P_M). The resultant calculations are shown in Table 3.20 with inclusion of other single parameter estimates by Theis method.

3.8   ÇIMEN METHOD

Çimen (2008) considered the Theis equation to evaluate the confined aquifer parameters by considering Eq. (3.31) and (3.32), which can be written explicitly as,

$$\frac{\partial s(r,t)}{\partial t} = \frac{Q}{4\pi T}\frac{e^{-r^2 S/4tT}}{t} \tag{3.42}$$

For large time or small radial distance, u is small and for $u \leq 0.01$ (which corresponds to the validity region of the Cooper-Jacob straight-line method) this expression reduces to,

$$t\frac{\partial s(r)}{\partial t} = \frac{Q}{4\pi T_{ap}} \tag{3.43}$$

TABLE 3.20   Single Parameter Values

Aquifer Parameters	Theis Method	Arithmetic Mean, P_A	Slope Method (Şen, 1986)		Harmonic Mean, P_H	Geometric Mean, P_G	Mode, P_M
			Weighted Mean, P_W				
			Distance	Frequency			
Transmissivity (m²/day)	1110	1162.7	1160	1180	1120	1148	1062
Storativity ($\times 10^{-4}$)	2.06	3.04	4.12	3.56	0.8	1,94	1.5

where T_{ap} is the approximate transmissivity. The error involved in adopting this expression instead of Eq. (3.42) is less than 1%. The term $t \partial s(r)/\partial t$ can be written as,

$$t\frac{\partial s(r,t)}{\partial t}\bigg|_i = t_i \frac{t_{i-1}\left[\frac{s_i(r)-s_{i-1}(r)}{t_i-t_{i-1}}\right] + t_i\left[\frac{s_{i+1}(r)-s_i(r)}{t_{i+1}-t_i}\right]}{t_{i-1}+t_{i+1}}$$

(3.44)

where i represents the time increment. One can calculate from Eq. (3.43) the transmissivity for any time as follows,

$$T_{ap} = \frac{Q}{4\pi}\frac{1}{t\partial s(r)/\partial t}$$

(3.45)

By considering this expression, the transmissivity and storage coefficient values can be achieved through the following steps:

1. Calculate $t_i(\partial s(r,t)/\partial t)_i$, values from Eq. (3.44) for each pumping time.
2. Find the approximate transmissivity values, T_{ap}, from Eq. (3.45) with the aid of the previous step.
3. Determine the approximate well function values, $W_{ap}(u)$, from Eq. (3.32) using the approximate tranmissivity values.
4. Find the approximate dimensionless time variables, u_{ap}, corresponding to these well function values from Theis type curve (Figure 3.22) that give the relationship between $W(u)$ and $1/u$.
5. Determine the approximate storage coefficient values, S_{ap}, from Eq. (3.31).
6. In order to find the exact transmissivity and storage coefficient values, multiply the approximate transmissivity values by $\exp(-u_{ap})$.
7. Repeat the previous steps from c to e for determining other $W(u)$ and u values.
8. Find the average T and S values for $u \leq 0.01$. It is necessary to continue the iterative computation until the relative error between the successive parameter estimations becomes less than $\pm 5\%$.

If the aquifer parameter values have only one value with no change, then the validity of the Theis assumptions are satisfied.

EXAMPLE 3.14 ÇIMEN SLOPE METHOD APPLICATION

The method described in the previous subsection is applied on published aquifer test data (Todd, 1980). The pumping rate is 2500 m³/day, and the drawdowns are measured in an observation well at 60 m from the pumping well. The data was also investigated with the slope-matching methods by some researchers (Şen, 1986c; Srivastava and Guzman-Guzman, 1994).

Solution 3.14

The proposed method steps and the results of the parameter estimation calculations for the Todd data are given in Table 3.21, where the aquifer parameters are calculated for the drawdown values measured at any pumping time.

This case represents that the assumptions of the Theis equation are not satisfied for the aquifer test results. In using the type curve matching to determine the aquifer parameters, the geological medium is assumed to represent an equivalent homogeneous aquifer, and consequently, the parameter estimates will be temporally and spatially invariable constants. As mentioned by Şen (1986c), there are some factors affecting the measurements during the pumping such as the systematic deviations (geological structures, hydraulic boundaries, and well design), the erratic deviations (local permeability variation, aquifer thickness, and random impermeable lenses within the aquifer), and measurement errors. However, one should always expect some variations in these parameters due to the aforementioned deviations.

The results and findings by aforementioned researchers and by Yeh (1987) with the finite-difference Newton's method are given in Table 3.22. The results are close to each other. In

TABLE 3.21 Slope-Matching Calculation for the Todd Data

Time t, (min)	Drawdown s, (m)	$t\partial s/\partial t$	T_{ap} (m²/min)	$W_{ap}(u)$	u_{ap}	$S_{ap} \times 10^{-4}$	$T = T_{ap}e^{-u_{ap}}$ (m²/min)	$W(u)$	u	$S \times 10^{-4}$	Drawdown s_{esr} (m)
0	0										
1	0.20	0.14000	0.9868	1.42857	1.564×10^{-1}	1.715	0.84394	1.42857	1.564×10^{-1}	1.467	0.20
1.5	0.27	0.13800	1.0011	1.95652	8.636×10^{-2}	1.441	0.91830	1.95652	8.636×10^{-2}	1.322	0.26
2	0.30	0.14286	0.9671	2.10000	7.393×10^{-2}	1.589	0.89817	2.10000	7.393×10^{-2}	1.476	0.30
2.5	0.34	0.17222	0.8022	1.97419	8.472×10^{-2}	1.888	0.73703	1.97419	8.472×10^{-2}	1.734	0.34
3	0.37	0.14727	0.9381	2.51235	4.772×10^{-2}	1.492	0.89438	2.51235	4.772×10^{-2}	1.423	0.37
4	0.41	0.16000	0.8635	2.56250	4.528×10^{-2}	1.738	0.82525	2.56250	4.528×10^{-2}	1.661	0.41
5	0.45	0.17222	0.8022	2.61290	4.295×10^{-2}	1.914	0.76846	2.61290	4.295×10^{-2}	1.834	0.45
6	0.48	0.16364	0.8443	2.93333	3.081×10^{-2}	1.734	0.81867	2.93333	3.081×10^{-2}	1.681	0.48
8	0.53	0.17714	0.7799	2.99194	2.900×10^{-2}	2.011	0.75761	2.99194	2.900×10^{-2}	1.953	0.53
10	0.57	0.17222	0.8022	3.30968	2.094×10^{-2}	1.866	0.78557	3.30968	2.094×10^{-2}	1.828	0.57
12	0.60	0.18000	0.7675	3.33333	2.044×10^{-2}	2.092	0.75200	3.33333	2.044×10^{-2}	2.050	0.60
14	0.63	0.17231	0.8018	3.65625	1.472×10^{-2}	1.835	0.79008	3.65625	1.472×10^{-2}	1.809	0.63
18	0.67	0.16313	0.8469	4.10728	9.324×10^{-3}	1.579	0.83907	4.10728	9.324×10^{-3}	1.565	0.67
24	0.72	0.17714	0.7799	4.06452	9.735×10^{-3}	2.025	0.77235	4.06452	9.735×10^{-3}	2.005	0.72
30	0.76	0.17222	0.8022	4.41290	6.852×10^{-3}	1.832	0.79671	4.41290	6.852×10^{-3}	1.820	0.76
40	0.81	0.17714	0.7799	4.57258	5.834×10^{-3}	2.022	0.77537	4.57258	5.834×10^{-3}	2.011	0.81
50	0.85	0.17222	0.8022	4.93548	4.052×10^{-3}	1.806	0.79895	4.93548	4.052×10^{-3}	1.798	0.84
60	0.88	0.16364	0.8443	5.37778	2.600×10^{-3}	1.463	0.84209	5.37778	2.600×10^{-3}	1.459	0.88
80	0.93	0.15429	0.8955	6.02778	1.355×10^{-3}	1.079	0.89424	6.02778	1.355×10^{-3}	1.077	0.93
100	0.96	0.17778	0.7771	5.40000	2.542×10^{-3}	2.195	0.77515	5.40000	2.542×10^{-3}	2.190	0.96
120	1.00	0.19636	0.7036	5.09259	3.460×10^{-3}	3.246	0.70114	5.09259	3.460×10^{-3}	3.235	1.00
150	1.04	0.17222	0.8022	6.03871	1.341×10^{-3}	1.792	0.80112	6.03871	1.341×10^{-3}	1.790	1.03
180	1.07	0.18000	0.7675	5.94444	1.473×10^{-3}	2.262	0.76640	5.94444	1.473×10^{-3}	2.258	1.07
210	1.10	0.17231	0.8018	6.38393	9.489×10^{-4}	1.775	0.80103	6.38393	9.489×10^{-4}	1.774	1.09
240	1.12										1.12
Average for u ≤ 0.01			0.8003			1.923	0.79697			1.915	

TABLE 3.22   Comparison of the Estimated Aquifer Parameters for the Todd Data

Method	T (m^2/day)	S × 10^{-4}	SEE × 10^{-3}
Theis curve	1110	2.06	8.85
Cooper-Jacob method	1090	1.84	35.22
Chow method	1160	2.06	22.82
Slope-matching (Şen, 1986b)	1233	1.98	21.00
Finite-difference Newton's method (Yeh, 1987)	1139	1.93	5.47
Slope-matching (Srivastava and Guzman-Guzman, 1994)	1145	2.09	18.18
Slope-matching (Singh, 2001)[1]	1125	–	–
Çimen (2008) (average for u ≤ 0.01)	1148	1.915	0.417

[1] Singh (2001) has not calculated the storage coefficient because most of the data are u ≤ 0.01.
SEE, standard error of estimation.

addition, the prediction errors are represented in the last column according to standard error of estimation (SEE).

3.9 COOPER-JACOB (CJ) STRAIGHT-LINE METHODS

Cooper and Jacob (1946) have noticed for small u values less than 0.01 that the Theis well function (Eq. (3.33)) can be approximated as,

$$W(u) = -0.5772 + 2.3 \log(u) \qquad (3.46)$$

This expression represents the time–drawdown or distance–drawdown relationships as straight-lines on semilogarithmic paper. The substitution of Eqs (3.31) and (3.32) into Eq. (3.46) leads explicitly to a semilogarithmic expression as,

$$s(r, t) = \frac{2.3Q}{4\pi T} \log \frac{2.25Tt}{r^2 S} \qquad (3.47)$$

There are three different methods of CJ straight-line aquifer parameter determination,

1. Late time–drawdown (TD) [t versus s(t, r)] method,

2. Late distance–drawdown (DD) [r versus s(t, r)] method,
3. Late time composite-drawdown (CD) [r^2/t versus s(t, r)] method.

Figure 3.38 indicates these three semilogarithmic straight-line methods.

In general, the straight-lines give two useful information for each method as slopes, [Δs_t, Δs_r or Δs_{tr}], and intercepts [t_0, r_o and $(t/r^2)_o$] corresponding to zero drawdown. The slope on a semilogarithmic paper is calculated as the difference in the drawdown values at the end and the beginning of any one complete log cycle. The steeper is the slope, the greater is the drawdown change physically implying that the specific discharge is rather small, and correspondingly, the transmissivity has also small values. There is an inverse relationship between the slope and the hydraulic conductivity as already seen through Eq. (3.18). The intercept has physical meaning as the radius of influence is only for the TD method. The analytical expression for the slope and the intercept in the late TD method application can be found from Eq. (3.47) leading to aquifer parameters estimation equations,

$$T = \frac{2.3Q}{4\pi \Delta s_t} \qquad (3.48)$$

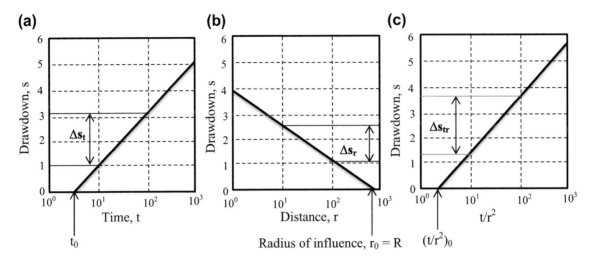

FIGURE 3.38 **Cooper-Jacob straight-line methods.** (a) Time-drawdown, (b) distance drawdown, (c) composite-drawdown.

and

$$S = \frac{2.25t_0 T}{r^2} \tag{3.49}$$

In these equations, Q and r are directly available from the field measurements but Δs_t and t_0 must be calculated in the office from the field late time—drawdown data plot on semilogarithmic paper.

The relevant equations for transmissivity and the storativity calculations can be obtained from Eq. (3.47),

$$T = \frac{2.3Q}{2\pi \Delta s_r} \tag{3.50}$$

and

$$S = \frac{2.25t_0 T}{r_0^2} \tag{3.51}$$

respectively. It is interesting to note that Eq. (3.50) is exactly the same as Eq. (3.18), which is valid for steady state groundwater flow in confined aquifer toward fully penetration wells. Finally, the relevant aquifer parameter expressions for the case of late time composite-

drawdown (CD) method application are obtainable again from Eq. (3.47) as,

$$T = \frac{2.3Q}{4\pi \Delta s_{tr}} \tag{3.52}$$

and

$$S = 2.25T \left(\frac{t}{r^2} \right)_0 \tag{3.53}$$

The CJ straight-line method is considered as a "good" approximation to Theis solution provided that the slope can be derived from data points with dimensionless time factor in Eq. (3.31) which is smaller than 0.03 and for which the approximation error is less than 1% (Sanchez-Vila et al., 1999).

EXAMPLE 3.15 UNSTEADY STATE DRAWDOWN CALCULATION

A confined aquifer of 40 m thickness has hydraulic conductivity equal to 4.5 m/day with storage coefficient 0.005. If the well is pumped at

1.6 m³/min what is the drawdown amount after 20 hours of pumping at the radial distance 42 m?

Solution 3.15

The solution can be found from the application of Eq. (3.47) as,

$$s(r,t) = \frac{2.3 \times 1.6}{4 \times 3.14 \times (40 \times 4.5)}$$
$$\times \left\{ \log \left[\frac{2.25 \times (40 \times 45) \times (20/24)}{42^2 \times 0.005} \right] \right\}$$
$$= 0.85 \text{ m}$$

EXAMPLE 3.16 TIME-, DISTANCE-, AND COMPOSITE-DRAWDOWN STRAIGHT-LINE CALCULATIONS

The aquifer test data for four OWs in Table 3.14 for Chaj Doab area in Pakistan are requested for the application of three different straight-line methods by the classical CJ method. The pump discharge is 0.113 m³/s.

Solution 3.16

For the application of TD straight-line method the observation wells' data are plotted on a semilogarithmic paper with time on the logarithmic axis (Figure 3.39).

This figure helps to find slopes and intercepts on the horizontal axis for each observation well. Four observation well time–drawdown data plots are presented out of which only three observation wells, namely, OW1, OW2, and OW3 are matched with straight-line, but the last one (OW4) time–drawdown plots does not yield to an inclined straight-line at large times, but horizontal line, which indicates that the OW4 location may have leaky aquifer condition.

The slopes of three straight-lines are calculated and given in Figure 3.39 per log cycle of time. The intercepts on the logarithmic axis are also read from the same figure. The slopes and intercept values are given in Table 3.23. Equations (3.48) and (3.49) provide the transmissivity and storativity coefficient estimations, respectively.

Numerical values of T and S indicate that the aquifer is highly potential and confined. However, the values of both parameters are comparatively

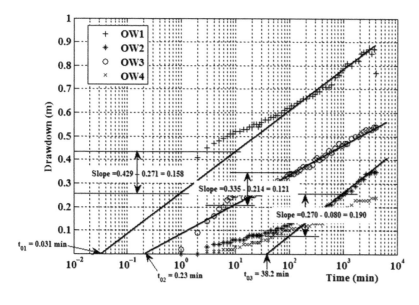

FIGURE 3.39 Time–drawdown straight-line.

TABLE 3.23   Time−Drawdown Model Calculations

Well Number	Distance, r (m)	Straight-Line Readings		Parameter Calculations	
		Slope, Δs_t (m)	Intercept, t_0 (min)	Equation (3.48) T (m²/day)	Equation (3.49) S (−)
OW1	30.48	0.158	0.031	11,310	5.89×10^{-4}
OW2	121.92	0.121	0.230	14,768	7.70×10^{-4}
OW3	304.8	0.190	38.200	9450	4.92×10^{-4}
			Average	11,842	6.17×10^{-4}

smaller than the Theis method values. The u values for the validity of CJ method for OW1, OW2, and OW3 with parameter estimates are 0.38, 0.56, and 0.06, respectively.

In order to make similar calculations as distance−drawdown plots on a semilogarithmic paper, the last row data corresponding to 4080 min are used. The scatter diagram is given in Figure 3.40.

The transmissivity and storage coefficient estimations can be obtained from Eqs (3.50) and (3.51), respectively. The numerical substitutions into these equations from Figure 3.40 yields $T = 4784.3 \, \text{m}^2/\text{day}$ and $S = 2.55 \times 10^{-1}$. These values indicate that the aquifer is next to moderately potential and in confined conditions.

Finally, the time records in the first column of Table 3.12 are converted to (t/r^2) sequence by

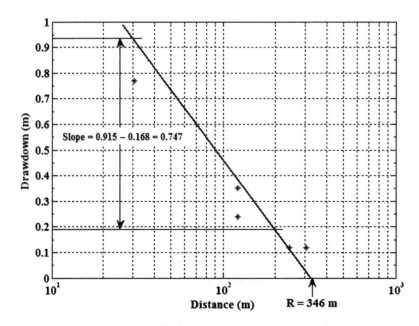

FIGURE 3.40   Distance−drawdown straight-line methods.

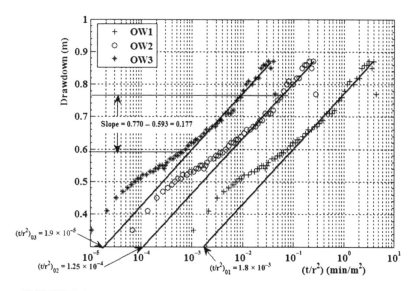

FIGURE 3.41   Time–distance–drawdown scatter diagram and straight-lines.

TABLE 3.24   Time–Distance–Drawdown Extractions

		Straight-Line Readings		Parameter Calculations	
Well Number	Distance, r (m)	Slope, Δs (m)	Intercept, (t/r^2) (min/m²)	Equation (3.52) T (m²/day)	Equation (3.53) S (−)
OW1	30.48	0.177	1.80×10^{-3}	10,096	2.84×10^{-2}
OW2	121.92	0.177	1.25×10^{-4}	10,096	2.00×10^{-3}
OW3	304.8	0.177	1.90×10^{-5}	10,096	3.00×10^{-4}
			Average	10,096	1.02×10^{-2}

taking into consideration observation wells' radial distances from the main well. The composite drawdown data from the first three observation wells (OW1, OW2 and OW3) are plotted on the semilogarithmic paper in Figure 3.41. Late (t/r^2) values are fitted by straight-lines each with slope and intercept on the logarithmic axis. Table 3.24 includes these values in the second and third columns. It is obvious that the slopes are almost equal to each other; hence, there is only one slope calculation on the same figure.

3.10 DIMENSIONLESS STRAIGHT-LINE ANALYSIS

According to CJ method, if the aquifer is almost homogeneous and isotropic, the late time–drawdown data are expected to appear around a straight-line on a semilogarithmic paper. However, in practical applications, there are numerous straight-lines, and hence, the question is "which one of these straight-lines corresponds to CJ model?" Plot of u versus W(u) in

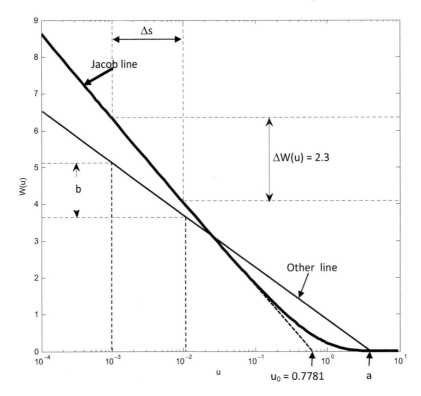

FIGURE 3.42 **Jacob straight-lines.**

Eq. (3.46) yields to an inversely proportional straight-line on a semilogarithmic paper as in Figure 3.42. The intercept on the horizontal dimensionless is $u_0 = 0.7781$ and the slope, $\Delta W(u) = 2.3$ (Şen, 2012). Any deviation from this slope invalidates CJ method applications. Otherwise, hydrogeological parameters estimations from Jacob straight-line methods may give over- or underestimations.

The verification of the CJ method is proposed by Şen (1988a) in which calculation of the dimensionless field drawdown, w_f, and time, u_f, are suggested from the already derived parameters of transmissivity T and storage coefficient, S, according to CJ straight-line method.

If the straight-line constructed by these dimensionless variables on semilogarithmic paper gives a slope of 2.3 then CJ method is valid.

Smaller slopes than the Jacob line indicate that either leakage, gravity drainage in unconfined aquifers, or recharge, or a combination of these may take place. Such slopes lead to higher transmissivity. If the slope of the dimensionless straight-line is greater than the CJ slope then either a nonlinear flow occurs due to the close location of the observation well to the main well, and/or there is a local decrease in the permeability, or limited water supply in the storage. Such slopes yield small transmissivity values. In order to arrive at a final decision the following steps must be executed:

1. Plot the large time—drawdown data on a semilogarithmic paper.
2. Calculate the aquifer parameters, T and S, by using the late time—drawdown CJ method (Eqs (3.48) and (3.49)).

3. Calculate the sequence of dimensionless field time, u_f, and drawdown, w_f, values from the fundamental dimensionless time factor and the well function definitions as in Eqs (3.31) and (3.32) as follows.

$$u_f = \frac{r^2 S}{4t_f T} \qquad (3.54)$$

and

$$w_f = \frac{4\pi T}{Q} s_f \qquad (3.55)$$

where t_f and s_f are time and corresponding drawdown measurements in the field.

4. Plot u_f versus w_f values on the same scale semilogarithmic paper as the type straight-line.
5. Draw the best straight-line through the scatter points, which is referred to as the dimensionless straight-line.
6. Compare this straight-line with the CJ line. Any difference in the slope and/or intercept on the horizontal axis suggests that CJ method is not valid and a new approach is necessary.
7. If the two slopes (dimensionless and CJ straight-lines) are practically the same (within ±5% relative error), then the validity of the CJ straight-line method is acceptable.
8. Otherwise, follow the following methodology for aquifer parameter estimations.

Any straight-line apart from the Jacob on $W(u)$—u space as in Figure 3.42 has a similar expression to Eq. (3.46) with a and b parameters.

$$W(u) = -a + b \log(u) \qquad (3.56)$$

Substitution of Eqs (3.31) and (3.32) into this expression leads to,

$$s(r, t) = \frac{bQ}{4\pi T} \log \frac{4t_0 T \exp\left(-2.3\frac{a}{b}\right)}{r^2 S} \qquad (3.57)$$

The corresponding aquifer parameter estimations can be obtained by considering two conditions as s $(r,0) = 0$ and the new slope, Δs_n, of this straight-line, which gives to new transmissivity and storativity estimations as,

$$T_n = \frac{bQ}{4\pi \Delta s_n} \qquad (3.58)$$

and

$$S_n = \frac{4t_0 T \exp\left(-2.3\frac{a}{b}\right)}{r^2} \qquad (3.59)$$

These formulations reduce to Eq. (3.48) and (3.49), respectively, for the CJ case when $a = -0.5772$ and $b = 2.3$. The application of the methodology requires the following steps:

1. Plot the time—drawdown data on a semilogarithmic paper and fit a straight-line through the large time portion of the data.
2. Determine the classical aquifer parameters according to Eqs (3.48) and (3.49). These parameters help to confirm whether the application of the CJ straight-line procedure is correct.
3. Calculate the dimensionless time factor (Eq. (3.54)) and the corresponding drawdown (Eq. (3.55)) values for the same data with the aquifer parameters found in the previous step.
4. Plot these dimensionless time and drawdown values on a semilogarithmic paper with the CJ straight-line (Figure 3.42). If the dimensionless field plots fall on the theoretical straight-line then regard the CJ parameters as valid for the aquifer.
5. In case of any difference in the intercepts and/or slope values of the dimensionless time—drawdown data straight-line from the CJ straight-line, apply the suggested methodology for aquifer parameter updates.
6. Find the intercept, a, and slope, b values from the dimensionless field data plot straight-line as in Figure 3.42.
7. Calculate the aquifer parameters from Eqs (3.58) and (3.59) for the aquifer test data at hand.

It is obvious from these steps that there are two successive semilogarithmic paper plots, one for the actual field data and the other for the dimensionless data.

Finally, it is possible to construct practical expressions between T and T_n by dividing Eqs (3.58) and (3.48) side by side, which leads to,

$$T_n = \frac{b}{2.3} T \qquad (3.60)$$

and likewise division Eq. (3.59) to Eq. (3.49) yields to,

$$S_n = 0.562S \exp\left(-2.3\frac{a}{b}\right) \qquad (3.61)$$

If $b = 2.3$ and $a = 0.7781$, then $T_n = T$ and $S_n = S$.

EXAMPLE 3.17 DIMENSIONLESS STRAIGHT-LINE FITTING METHOD

A worked example is taken from Kruseman and de Ridder (1990, p. 64) as the aquifer test data from "Oude Korendijk" (Table 3.25). The piezometer time—drawdown measurements are recorded in the field. The well site geological description, type curve application and quantitative aquifer parameter estimations are presented by them in detail. On an average, the data points fall around the Theis curve but there are significant deviations. A mechanical holistic type curve matching to overall data gives only average parameter estimates as $S = 1.6 \times 10^{-4}$ and $T = 385 \, \text{m}^2/\text{day}$.

Solution 3.17

Figure 3.43 presents classical CJ method (TD) applications for the three piezometer time—drawdown data measurements and the aquifer parameters are also given on the same figure. It is obvious that when the classical Jacob method is applied to each piezometer data separately, there is an increase both in the transmissivity and

storativity values by distance of the piezometers from the main well. The difference between the 30 and 215 m piezometer aquifer parameter estimations is more than 100%.

In order to verify the validity of the classical CJ method dimensionless time—drawdown data are produced from Eqs (3.54) and (3.55) for each piezometer and the parameter estimations are given in Figure 3.43. Dimensionless plots of Oude Korendijk data appears in Figure 3.44 almost as a single line. However, the slope and intercept on the horizontal axes are different than the classical CJ methodology and the calculation of the aquifer parameters from Eqs (3.58) and (3.59) or Eqs (3.60) and (3.61) appear as $T_n = 334 \, \text{m}^2/\text{day}$ and $S_n = 1.39 \times 10^{-4}$, respectively. These parameter values are closer to classical CJ method results, but the percentage differences between the classical CJ dimensionless approach transmissivity and storativity values are 13% and 18%, respectively.

3.10.1 Straight-Line Method Generalization

Most often CJ straight-line is used for aquifer parameter estimations from late time—drawdown data in cases of confined and unconfined aquifers. This straight-line method is attractive due to its simplicity and straightforward calculations without tedious and dubious type curve matching procedure. The warranty of CJ method application depends on the fact that the late time—drawdown measurements fall along the final portion of the Theis type curve.

Reliable test data are unavailable for the vast majority of the wells, even where reliable data are available, the nonequilibrium methods as type curve matching or fitting straight-lines are often misapplied (Misstear, 2001). It is also possible that the wide availability of software for interpreting pumping-test data hydrogeologists may depend on automatic straight-line fitting or type curve matching procedures. Even in cases of good test data availability, there

TABLE 3.25   Oude Korendijk Aquifer Test Data

Piezometer H_{30} Screen Depth 20 m		Piezometer H_{90} Screen Depth 24 m		Piezometer H_{215} Screen Depth 20 m	
t (min)	s (m)	t (min)	s (m)	t (min)	s (m)
0	0	0	0	0	0
0.1	0.04	1.5	0.015	66	0.089
0.25	0.08	2.0	0.021	127	0.138
0.50	0.13	2.16	0.023	185	0.165
0.70	0.18	2.66	0.044	251	0.186
1.00	0.23	3.00	0.054	305	0.196
1.40	0.28	3.50	0.075	366	0.207
1.90	0.33	4.0	0.090	430	0.214
2.33	0.36	4.33	0.104	606	0.227
2.80	0.39	5.5	0.133	780	0.250
3.36	0.42	6.0	0.153		
4.00	0.45	7.5	0.178		
5.35	0.50	9.0	0.206		
6.80	0.54	13.0	0.250		
8.30	0.57	15.0	0.275		
8.70	0.58	18.0	0.305		
10.0	0.60	25.0	0.348		
13.1	0.64	30.0	0.364		
18	0.68	40.0	0.404		
27	0.742	53.0	0.429		
33	0.753	60.0	0.444		
41	0.779	75.0	0.467		
48	0.793	90.0	0.494		
59	0.819	105	0.507		
80	0.855	120	0.528		
95	0.873	150	0.550		
139	0.915	180	0.569		
181	0.935	248	0.593		
245	0.966	301	0.614		
300	0.990	363	0.636		

(Continued)

TABLE 3.25   Oude Korendijk Aquifer Test Data (*cont'd*)

Piezometer H_{30} Screen Depth 20 m		Piezometer H_{90} Screen Depth 24 m		Piezometer H_{215} Screen Depth 20 m	
t (min)	s (m)	t (min)	s (m)	t (min)	s (m)
360	1.007	422	0.657		
480	1.050	542	0.679		
600	1.053	602	0.688		
728	1.072	680	0.701		
830	1.088	785	0.718		
		845	0.716		

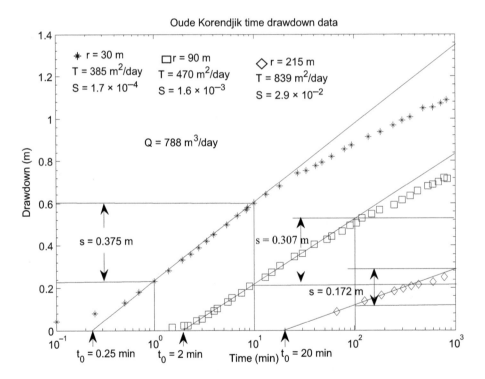

FIGURE 3.43   **Classical Cooper-Jacob method applications.**

is a tendency among some practicing hydrogeologists and engineers to adopt a prescriptive approach to the analysis, whereby the standard approaches as those Theis (1935) and CJ methods are applied to time—drawdown data with little consideration of underlying assumptions. For instance, in a CJ analysis, it is not uncommon to encounter a straight-line drawn as a best fit

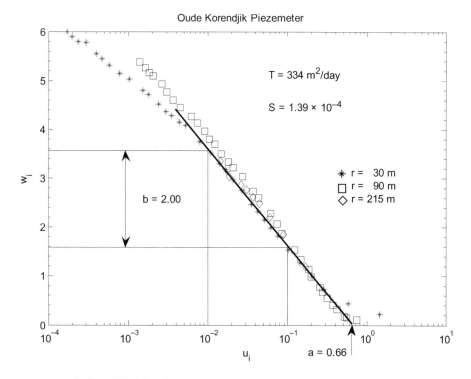

FIGURE 3.44 **Dimensionless check of the Jacob method validity.**

through a semilogarithmic data scatter that is clearly nonlinear, with no discussion of the reasons for the discrepancy between theory and the conditions at the site under investigation.

It is the main purpose of this section to present a general type straight-line model to determine aquifer properties of a confined leaky (semiconfined) aquifer using the pumping test results. The CJ method is applicable in cases where the straight-line slope on dimensionless semilogarithmic paper is equal to 2.3 only as explained in this section (Şen, in press).

Şen (1996) has presented type curve equations for leaky aquifers, which can be reduced to classical confined (Theis, 1935), and leaky (De Glee, 1930; Hantush, 1959) aquifer solutions. His study led to the general type straight-line expression as,

$$W(u, \eta, r/L) = \left(1 - \frac{1}{\eta}\right) K_0 \left(\frac{r}{L} \sqrt{\frac{\eta}{\eta - 1}}\right)$$
$$- \frac{0.577}{\eta} + \frac{2.3}{\eta} \log\left(\frac{1}{u}\right)$$

$$(3.62)$$

Herein, K_0 is zero-order Bessel function. The leakage factor, L, is defined by Hantush (1959) as and given by Eq. (3.24) and the storage coefficient ration, η, is defined as,

$$\eta = 1 + \frac{S'}{S} \qquad (3.63)$$

where S and S' are the storage coefficients of the leaky aquifer and overlying semipervious layer, respectively. Physically, for the classical leaky aquifer $S' = 0$ and $\eta = 1$. On the other hand, for

FIGURE 3.45 **Various types of straight-lines.**

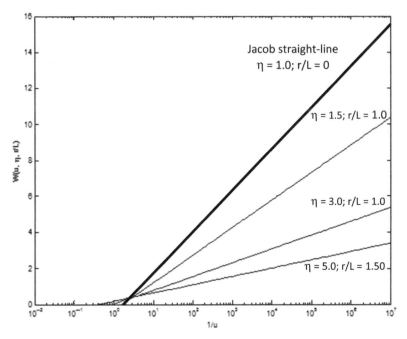

η → ∞, the aquifer is of leaky type as studied by Hantush (1956), and therefore, Eq. (3.62) reduces to Eq. (3.27), which corresponds to the expression given by De Glee (1930) for steady state flow (t → ∞) in leaky aquifers. Equations (3.62) reduces to Eq. (3.46) for η = 1, because mathematically $0K_0[(r/L)\sqrt{\infty}] = 0$ (Abramowitz and Stegun, 1972). Furthermore, η = 1 implies (S' = 0; r/L = 0) which corresponds to a completely confined aquifer case. This last expression shows many type straight-line relationships between the leaky aquifer well function, W (u, η, r/L) and u on a semilogarithmic paper for different η values. The graphical representation of Eq. (3.62) together with CJ straight-line is given in Figure 3.45. This CJ straight-line is identical with the one in Figure 3.42.

The following points provide practical information:

1. The set of straight-lines confirms the statement that not every straight-line on semilogarithmic paper implies the use of CJ method.

2. For an assigned set of parameters (S, T, Q, r) invariably the leaky aquifer straight-lines have smaller slopes than the CJ line. For very big η values the straight-line becomes horizontal implying that the aquifer is equivalent to the classical leaky aquifer. It is well known in CJ method that the aquifer transmissivity is inversely related to straight-line slope, $(\Delta s)_{CJ}$, (see Eq. (3.48)). Without distinction, whether the aquifer is leaky or not, direct use of slope value in CJ method leads to over-estimation of aquifer transmissivity.

3. The intercept of leaky aquifer straight-lines on the horizontal axis is always smaller than the CJ type straight-line intercept. The storage coefficient is directly related to the intercept value, t_0, in CJ method (Eq. (3.51)). Therefore, the substitution of intercept value in CJ formulation leads to underestimation of the storativity value.

The intercept points of general type straight-lines are defined for $[W(u, \eta, r/L) = 0, u = u_0]$ and accordingly their substitution into Eq. (3.62) yields,

$$u_0 = 0.56 e^{(\eta-1)K_0\left[\left(\frac{r}{L}\right)\left(\frac{\eta}{\eta-1}\right)\right]} \qquad (3.64)$$

which for of $\eta = 1$ yields $(1/u_0) = 1.78$ as in Figure 3.42. By considering definition of u as in Eq. (3.31), this last expression provides a practical formula for the storage coefficient estimation of the pumped aquifer at large times as,

$$S = \frac{2.25 T t_0}{r^2} e^{(\eta-1)K_0\left(\frac{r}{L}\sqrt{\frac{\eta}{\eta-1}}\right)} \qquad (3.65)$$

which reduces to Eq. (3.31) for $\eta = 1$. Since the first term on the right hand side of Eq. (3.62) is independent of u, the slope, $\Delta W(u, \eta, r/L)$, of leaky aquifer type straight-line on the semilogarithmic paper becomes,

$$\Delta W(u, \eta, r/L) = \frac{2.3}{\eta} \qquad (3.66)$$

As already mentioned by Şen (1988b, 1989a, 1990, 1995), the slope of CJ type straight-line on dimensionless plot is equal to 2.3, which results from Eq. (3.66) only when $\eta = 1$. Transmissivity estimation from the general type straight-line expression in Eq. (3.62) yields,

$$T = \frac{2.3Q}{4\pi\eta\Delta s} \qquad (3.67)$$

which reduces to Eq. (3.48) for $\eta = 1$.

EXAMPLE 3.18 GENERALIZED STRAIGHT-LINE APPLICATION

The application of methodology suggested above is performed for aquifer test data obtained from the Arabian Shield. In wadi Quaternary deposits a well is pumped with a constant discharge of 2000 m^3/day and the time drawdown measurements are collected from an observation well which is 102 m away from the pumped well. The late time–drawdown data are given in Table 3.26.

Solution 3.18

The following steps are necessary for an effective application of the suggested methodology in this section.

1. Plot the large time–drawdown data on a semilogarithmic paper with time on logarithmic axis, (see Figure 3.46).
2. The most suitable straight-line through late time–drawdown points has intercept, $t_0 = 3.9 \times 10^{-3}$ day on the time axis and the slope, $(\Delta s)_{CJ} = 0.35$ m. Substitution of these values into Eqs (3.48) and (3.49) yields classical CJ parameters as $T_{CJ} = 104.64$ m^2/day and $S_{CJ} = 9.18 \times 10^{-4}$, respectively.
3. The dimensionless time factor, u_{CJ}, and well function, $W(u_{CJ})$, are calculated with these parameters and time–drawdown values from Eqs (3.54) and (3.55). The dimensionless values are presented in Table 3.27.
4. The plot of $1/u_f$ versus w_f values on a double-logarithmic paper should fall on the classical Theis type curve if the aquifer is confined. It is obvious from Figure 3.47 that such a situation is not valid, and hence, one can conclude that the aquifer is not confined, and therefore, CJ method needs modification.
5. The same dimensionless data are plotted on the general type straight-line paper as in Figure 3.48 Fixation of the best straight-line matching yields slope value as, $[\Delta W(u, \eta, r/L)]_L = 0.24$, which is smaller than 2.3.
6. One can read from the same figure $\eta = 8$ and $r/L = 1$, which will be used in the aquifer parameter calculations.
7. Calculate the aquifer transmissivity value from Eq. (3.67), which yields for the data at hand that $T = 191$ m^2/day.
8. Calculate the storage coefficient value from Eq. (3.65). Hence, $S_L = 1.5 \times 10^{-1}$.

TABLE 3.26   Aquifer Test Data

Time (day)	Drawdown (m)	Time (day)	Drawdown (m)	Time (day)	Drawdown (m)	Time (day)	Drawdown (m)
0.0056	0.0601	0.5556	0.8097	0.0278	0.3220	2.7778	1.0717
0.0058	0.0684	0.5848	0.8180	0.0370	0.3689	3.7037	1.1185
0.0062	0.0772	0.6173	0.8268	0.0585	0.4432	5.5556	1.1929
0.0065	0.0866	0.6536	0.8361	0.0617	0.4520	6.1728	1.2016
0.0069	0.0964	0.6944	0.8460	0.0654	0.4613	6.1728	1.2109
0.0074	0.1069	0.7407	0.8565	0.0694	0.4712	6.9444	1.2208
0.0079	0.1182	0.7937	0.8677	0.0741	0.4817	6.9444	1.2313
0.0085	0.1302	0.8547	0.8798	0.0794	0.4930	7.9365	1.2425
0.0093	0.1432	0.9259	0.8929	0.0855	0.5050	9.2593	1.2546
0.0101	0.1574	1.0101	0.9070	0.0926	0.5180	9.2593	1.2676
0.0111	0.1729	1.1111	0.9225	0.1010	0.5322	11.1111	1.2818
0.0123	0.1901	1.2346	0.9397	0.1389	0.5840	13.8889	1.3337
0.0139	0.2092	1.3889	0.9588	0.1852	0.6309	18.5185	1.3805
0.0159	0.2310	1.5873	0.9806	0.2222	0.6605	18.5185	1.4102
0.0185	0.2561	1.8519	1.0057	0.2778	0.6969	27.7778	1.4465
0.0222	0.2857	2.2222	1.0354	0.3704	0.7437	55.5556	1.4933

3.11 VARIABLE DISCHARGE TYPE CURVES

Almost all of the existing analytical solutions for nonequilibrium flow toward wells assume a constant pump discharge. However, under actual field conditions, the pump discharge is found to vary either systematically or erratically around a general trend that appears usually as a function of decreasing flow with time. At the beginning of pumping, the discharge is set at a certain rate, but, due to various reasons such as the friction between the water and the pipes, increased drawdown, warming up of the motor, self-adjustment of a constant speed pump to the drawdown change in the well, etc., the initial discharge does not remain constant. These variations in the discharge are normally monitored throughout the pumping test by taking discharge measurements during the aquifer test period. However, under some conditions, the field adjustments may not be capable to maintain a constant discharge. It is rather common in some field tests that reductions in discharge during a pumping test might occur even more than 50%.

Abu-Zeid and Scott (1963) presented analytical solutions for exponential, hyperbolic, and linear discharge changes by using Laplace transformations. However, he did not present the results in the form of type curves, which are of utmost importance for any practicing hydrogeologist. The first systematic study along the topic of this section is due to Aron and Scott (1965) who presented a simplified solution for

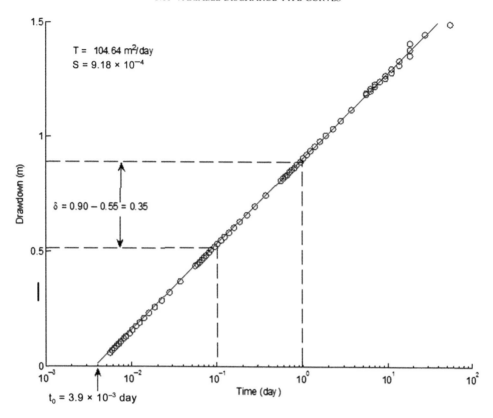

FIGURE 3.46 **Semilogarithmic plot.**

TABLE 3.27 Dimensionless Field Data

U_f	W_f	U_f	W_f
0.3948	0.0395	0.0039	0.5321
0.3750	0.0450	0.0038	0.5376
0.3553	0.0508	0.0036	0.5433
0.3356	0.0569	0.0034	0.5495
0.3158	0.0633	0.0032	0.5559
0.2961	0.0702	0.0030	0.5628
0.2763	0.0776	0.0028	0.5702
0.2566	0.0856	0.0026	0.5781
0.2369	0.0941	0.0024	0.5867

(Continued)

TABLE 3.27   Dimensionless Field Data (*cont'd*)

U_f	W_f	U_f	W_f
0.2171	0.1034	0.0022	0.5960
0.1974	0.1136	0.0020	0.6062
0.1777	0.1249	0.0018	0.6175
0.1579	0.1375	0.0016	0.6301
0.1382	0.1518	0.0014	0.6444
0.1184	0.1683	0.0012	0.6609
0.0987	0.1878	0.0010	0.6804
0.0790	0.2116	0.0008	0.7042
0.0592	0.2424	0.0006	0.7350
0.0375	0.2913	0.0004	0.7839
0.0355	0.2970	0.0004	0.7896
0.0336	0.3032	0.0004	0.7958
0.0316	0.3096	0.0003	0.8022
0.0296	0.3165	0.0003	0.8091
0.0276	0.3239	0.0003	0.8165
0.0257	0.3318	0.0002	0.8244
0.0237	0.3404	0.0002	0.8330
0.0217	0.3497	0.0002	0.8423
0.0197	0.3599	0.0002	0.8525
0.0178	0.3712	0.0002	0.8638
0.0099	0.4341	0.0001	0.9267
0.0079	0.4579	0.0001	0.9505
0.0059	0.4887	0.0000	0.9813

decreasing flow in wells. Later, Lai et al. (1973), have presented exact solutions for the drawdown in and around a large diameter well in a homogeneous, isotropic, and confined aquifer with time dependent pump discharge. Rushton and Singh (1983) have devised a curve-matching technique for large-diameter wells with decreasing abstraction rates. Zlotnik (1994) has presented a methodology for well testing with arbitrary production rate.

The radial discharge change, $\partial Q(r, t)/\partial r$, within the cylindrical annulus in the aquifer is as a result of water and aquifer compressibility variation due to hydraulic head change by time, $\partial h(r, t)/\partial t$, which gives rise to water yield equal to $2\pi r S dr \partial h(r, t)/\partial t$. This last sentence

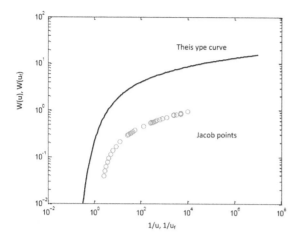

FIGURE 3.47 **Theis type curve and dimensionless field data plot.**

implies mathematically water equilibrium between these two changes as,

$$\frac{\partial Q(r,t)}{\partial r} = 2\pi r S \frac{\partial h(r,t)}{\partial t} dr \qquad (3.68)$$

The aquifer cylindrical periphery discharge can be written by the Darcy law consideration as follows,

$$\partial Q(r,t) = 2\pi r m K \frac{\partial h(r,t)}{\partial r}$$

Hence, the substitution of this expression into the previous one leads after some algebraic manipulation to,

$$\frac{\partial}{\partial r}\left[r \frac{\partial h(r,t)}{\partial r}\right] = \frac{S}{T}\frac{\partial h(r,t)}{\partial t} \qquad (3.69)$$

where T is the transmissivity of the aquifer material. It is necessary to seek the solution of this differential equation under the initial and boundary conditions, which are $\lim_{t\to 0}[h(r,t)]=H_o$, $\lim_{t\to \infty}[h(r,t)]= H_o$ and $\lim_{t\to 0}[2\pi q(r,t)]=Q(t)$. The solution of Eq. (3.69) has been presented by Şen and Altunkaynak (2003) in terms of Theis

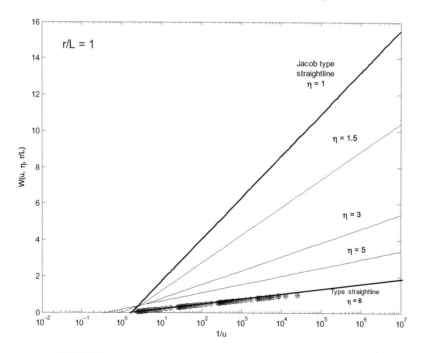

FIGURE 3.48 **Type straight-lines and dimensionless field data.**

dimensionless time factor, u, and well function, W(u), as,

$$W(u) = \int_u^\infty Q(x) \frac{\exp(-x)}{x} dx \qquad (3.70)$$

in which Q(x) is the dimensionless discharge function. This last expression is a modified Theis type curve equation which can be traced back to Carlsaw and Jeager (1959). They have used Laplace transformation in their derivation. For the constant discharge, Q(x) = 1, and Eq. (3.70) becomes identical with the classical Theis type curve for confined aquifer test (see Eq. (3.30)).

3.11.1 Exponential Discharge Variation

This discharge variation with time is given in the following form,

$$Q(t) = Q_F + (Q_I - Q_F)\exp(-c^2 t) \qquad (3.71)$$

where Q_I and Q_F are the initial and final-discharge measurements, respectively, and c is a constant. The numerical value of c can be deduced as the slope of straight-line through discharge-time data plots on a semilogarithmic paper with discharge values on the logarithmic axis. In terms of the dimensionless time factor in Eq. (3.31), the dimensionless discharge function, q(u), can be written as

$$q(u) = \frac{Q(u)}{Q_I} = \alpha + (1-\alpha)\exp\left[-\left(\frac{r}{B}\right)^2 / u\right] \qquad (3.72)$$

in which, $\alpha = Q_F/Q_I$ is named as the discharge ratio and B is an exponent constant dependent on c, T and S with the length dimension as,

$$B = \left(\frac{2}{c}\right)\sqrt{\frac{T}{S}} \qquad (3.73)$$

It is interesting to notice at this stage that Eq. (3.72) reduces to a constant discharge case for

$\alpha = 1$. On the other hand, $\alpha = 0$ implies that $Q_F = 0$, for which Eq. (3.72) becomes,

$$Q(u) = Q_I e^{-\left(\frac{r}{B}\right)^2 / u} \qquad (3.74)$$

This expression is identical to the leaky aquifer leakage discharge provided that r/B is constant, (see Şen, 1985, Eq. (18)). When $\alpha = 0$ the resulting type curves will be identical with the Hantush (1957) solution. The substitution of Eq. (3.72) into Eq. (3.70) leads simply to exponentially variable discharge type curves as,

$$W(u) = \int_u^\infty \left\{ \alpha + (1-\alpha)\exp\left[-\left(\frac{r}{B}\right)^2 / x\right] \right\}$$
$$\times \frac{\exp(-x)}{x} dx$$
$$\qquad (3.75)$$

The solution of Eq. (3.75) is given in Figures 3.49 and 3.50 as sets of type curves for different (r/B) and α value.

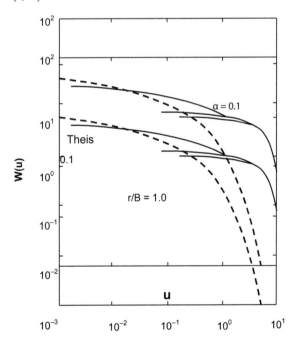

FIGURE 3.49 Exponential discharge variation type curves, (α = 0.1).

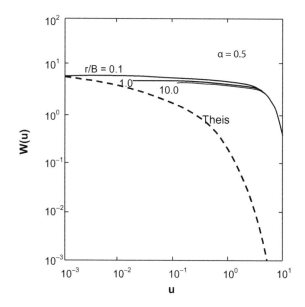

FIGURE 3.50 Exponential discharge variation type curves ($\alpha = 0.5$).

A close-inspection of these figures indicates the following significant points:

1. For small α values, the exponentially varying discharge type curves are closer to the Theis type curve for small times, i.e., at the beginning of pumping test. During moderate time periods the curvatures of the curves increase and finally, for large times they become asymptotically horizontal. Therefore, the horizontality of the late time drawdown data might apply that the pump discharge changes exponentially during the test contrary to the leaky aquifer system. In order to be confident that horizontality of the late data implies a leaky aquifer system, one must be sure that the discharge remains constant, which can be easily controlled with frequent and accurate discharge measurements during an aquifer test (Section 3.2.4).
2. For small times (big u values), all of the exponentially varying discharge type curves merge into a single curve irrespective of (r/B)

value and they are invariably above the Theis type curve. It means that for small times exponentially varying discharge case gives rise to bigger drawdowns than the constant discharge case.
3. At moderate times, the discrepancies occur and each (r/B) value begins to represent a specific curve, the extension of which appears as horizontal straight-lines as large times.

3.11.2 Straight-Line Method

It is possible to develop a simple procedure for the aquifer hydraulic parameter determination from late time–drawdown measurements. For this purpose Eq. (3.75) can be written as,

$$W(u, B) = \int_u^\infty \frac{e^{-x}}{x} dx + (1 - \alpha) \int_u^\infty \frac{e^{-x - \frac{r^2}{B^2} \frac{1}{x}}}{x} dx$$

(3.76)

As $u \to 0$ i.e., for big t values, the first integration term in this equation can be expanded in a convergent series asymptotically for $u < 0.01$,

$$\int_u^\infty \frac{e^{-x}}{x} dx = -2.3 \log\left(\frac{u}{0.561}\right)$$

(3.77)

which is the same expression as the CJ straight-line procedure. The second integration term in Eq. (3.76) becomes a constant, $C(r/B)$ for small u values depending on (r/B) value. The substitution of Eq. (3.77) and this constant into Eq. (3.76) leads to

$$W(u, B) = -2.3 \alpha \log\left(\frac{u}{0.561}\right) + (1 - \alpha)C\left(\frac{r}{B}\right)$$

(3.78)

which shows a straight-line on a semilogarithmic paper as $W(u, B)$ versus u. The change of r/B with $C(r/B)$ is shown in Figure 3.51.

Substitution of Eqs (3.31) and (3.32) into Eq. (3.78) leads to a relationship between drawdown and time as,

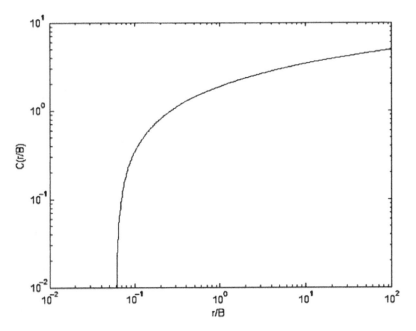

FIGURE 3.51   r/b—C(r/B) relationships.

$$s(r, t) = -\frac{2.3\alpha Q}{4\pi T}\log\frac{r^2 S}{2.25Tt} + \frac{(1-\alpha)Q}{4\pi T}C(r/B)$$

$$(3.79)$$

and

$$\frac{s(r, t)}{Q} = -\frac{2.3\alpha}{4\pi T}\log\frac{r^2 S}{2.25Tt} + \frac{(1-\alpha)}{4\pi T}C(r/B)$$

$$(3.80)$$

which also appears as a straight-line on semilogarithmic paper (time on logarithmic scale). The slope, Δs_Q, of this straight-line is

$$\Delta s_Q = \frac{2.3\alpha}{4\pi T} \qquad (3.81)$$

and the intercept, t_0, on the time axis can be found from Eq. (3.80) by substitution $s(r, t)/Q = 0$ which yields to,

$$t_0 = \frac{r^2 S}{2.25T}e^{[(\alpha-1)/\alpha]C(r/B)} \qquad (3.82)$$

These last two expressions are generalizations of Aron and Scott (1965) parameter determination equations. For $\alpha = 1$ Eqs (3.81) and (3.82) reduce to the classical CJ straight-line method on semi logarithmic paper. Provided that Δs_Q, t_0, and $C(r/B)$ are determined, then the aquifer parameters namely, S and T can be identified by using Eqs (3.81) and (3.82).

EXAMPLE 3.19 VARIABLE DISCHARGE TEST

An aquifer test has been conducted in the Saq sandstone aquifer formation, which lies in the northwestern part of Saudi Arabia. The superficial deposits covering the area consist mainly of silt and clay with variable proportions of pebbles from quartz gravel and rock of Precambrian basement. Hence this sand provides a confined aquifer situation at the aquifer test location. An aquifer test is conducted with variable discharge as shown in Table 3.28. The drawdown measurements are from the observation well which is 45 m away from the main well. The well diameter is about 20 cm.

TABLE 3.28 Discharge Variation

Time (min)	Q (m³/min)
10	5
20	4.1
30	3.4
40	2.9
50	2.5
60	2.3
90	1.9
120	1.7
Average	2.98

Solution 3.19

The drawdown versus time plots on a semi-logarithmic paper is shown in Figure 3.52. In this figure the discharge, Q, is at the time of observation as given in Table 3.28. A straight-line is usually fitted through the scatter of points during late times.

A straight-line is fitted by eye through the scatter of points during late times. The application of the methodology developed herein can be explained step by step as follows:

1. Plot the discharge variation on ordinary graph paper. The data from Table 3.28 in Figure 3.53 indicate that the discharge decrease is rather nonlinear, which will be approached by an exponential function similar to Eq. (3.71).

After taking logarithm of both sides this equation can be written as,

$$\text{Log}[Q(t) - Q_F] = \log(Q_I - Q_F) - c^2 t \qquad (3.83)$$

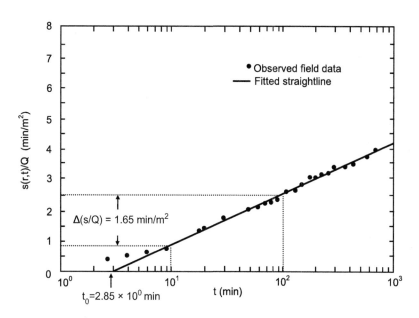

FIGURE 3.52 Semilogarithmic plot of late time–drawdown data.

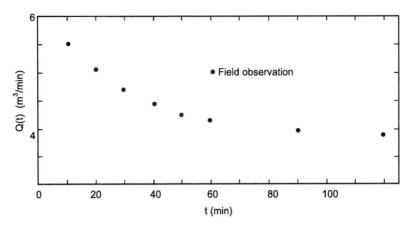

FIGURE 3.53 **Drawdown variations on mill metric graph paper.**

Hence, on a semilogarithmic paper, the relationship between $Q(t) - Q_F$ (in logarithmic scale) and t appears as a straight-line for properly chosen Q_F value. The slope of this line is equal to c^2.

2. As a first trial, choose Q_F value less than but close to the minimum measured $Q(t)$ value. Then calculate $Q(t) - Q_F$ values and plot them on the semilogarithmic paper versus time. If the points do not appear around a straight-line, choose another value smaller than previous Q_F and repeat the calculations and plotting until a good straight-line fit is obtained. For the data at hand, it is found that $Q_F = 1.5$ m^3/min gives rise to a straight-line as shown in Figure 3.54.

3. Find the intercept, Q_0, on the $Q(t) - Q_F$ axis, which appears as 4.4 m^3/min (see Figure 3.54). Consequently, $Q_I = Q_F + Q_0 = 1.5 + 4.4 = 5.9$ m^3/min, and therefore, $\alpha = Q_F/Q_I = 0.25$.

4. Find the slope of straight-line for one cycle of logarithm on the vertical axis, which yields correspondingly $\Delta t = 74$ min. Hence, the slope, c^2, is equal to $1/74 = 0.0133$, and therefore, $b = 0.12$.

5. Plot the field time–drawdown data on a semilogarithmic paper as requested and fit through the late time points the most suitable straight-line, (see Figure 3.52). Find its

intercept, t_0, on the time axis and its slope, $\Delta(s/Q)$. The results appear as $t_0 = 2.85$ min and $\Delta(s/Q) = 1.65$ m, respectively.

6. Find the transmissivity value from Eq. (3.81) by substitution of the relevant value's, and finally, $T = 0.16$ m^2/min. In practice, most of the time without checking the discharge variability, the average discharge is employed as for the constant pump-discharge. The use of average discharge, which is 2.98 m^3/min from Table 3.28, leads to $T = 0.33$ m^2/min. This value gives an overestimation of almost 50% compared with the transmissivity value of variable discharge.

7. Find the storage coefficient expression from Eq. (3.82) as,

$$S = \frac{2.25 \, t_0 T}{r^2} e^{[\frac{1-\alpha}{\alpha}]C(\frac{r}{B})} \qquad (3.84)$$

For $\alpha = 1.0$, this equation reduces to the classical CJ method. It is obvious that the storage coefficient estimation with the constant discharge case will lead to underestimation whatever the discharge ratio, $e^{[(1-\alpha)/\alpha]C(r/B)} > 1.0$. For the constant discharge, $\alpha = 1.0$, Eq. (3.84) yields $S = 1.05 \times 10^{-3}$. However, for the variable discharge case, there are two unknowns, namely, S and C(r/B). Hence, there appears a need for

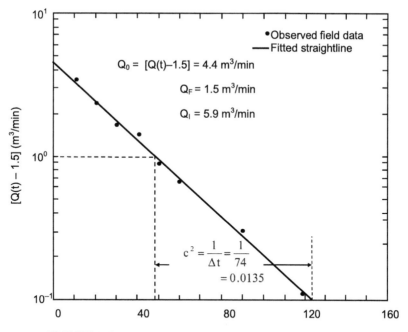

FIGURE 3.54 **Drawdown variations on semilogarithmic paper.**

another equation in order to obtain a unique solution of these two unknowns. For this purpose, Eq. (3.73) can be rearranged in such a way that,

$$S = \frac{4T}{b^2}\left(\frac{r}{B}\right)^2 \frac{1}{r^2} \tag{3.85}$$

Substitution of relevant numerical values into Eq. (3.84) and (3.85) leads to

$$S = -1.76 \times 10^{-4} e^{3c\left(\frac{r}{B}\right)} \tag{3.86}$$

$$S = 4.9 \times 10^{-2}\left(\frac{r}{B}\right)^2 \tag{3.87}$$

respectively. These two equations can be solved by trial and error method using Figure 3.51. The procedure is as follows:

1. Choose any arbitrary value of r/B; for instance, as a first try take $(r/B) = 1.0$,
2. Read the corresponding value from Figure 3.51,
3. Substitute $C(r/B)$ and (r/B) values into Eqs (3.86) and (3.87), respectively,

4. If these two equations lead to the same result by ±5% relative error, then the S value is found corresponding to (r/B) and $C(r/B)$ values for the aquifer test data at hand. Otherwise, return to step (1) and repeat the procedure until a satisfactory S value is obtained. The trial and error method for the aquifer test data at hand leads to $S = 8.5 \times 10^{-2}$, $r/B = 1.34$, and $C(r/B) = 2.05$. This storage coefficient value is bigger by one order of magnitude than the constant discharge case which was 1.05×10^{-3}. Hence, an underestimate of the storage coefficient occurs under constant discharge assumption.

3.12 QUASI-STEADY STATE FLOW STORAGE CALCULATION

Many researchers in the literature have studied determination of the aquifer parameters both in confined and unconfined aquifers (Theis,

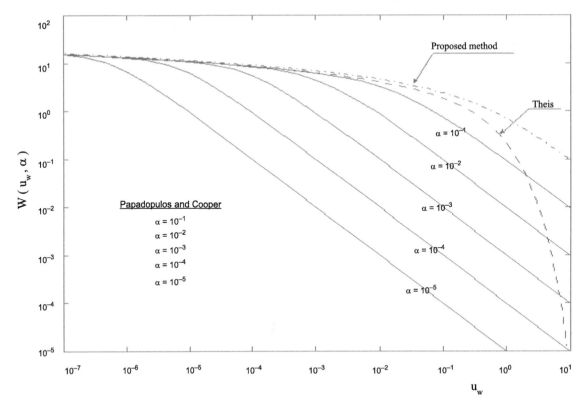

FIGURE 3.55 Comparison of models.

1935; Jacob, 1940; Hantush, 1956; Ferris et al., 1962; Boulton, 1963; Şen, 1982, 1986b) and many others. The construction of an observation well is expensive, and therefore, for determining the aquifer characteristics alternative, practical and simple applicable methods are presented for time–drawdown measurements in the abstraction well similar to that given by Papadopulos and Cooper (1967) and Şen (1982). Although, transmissivity can be estimated with a reasonable accuracy from such data, it is not possible to calculate the storage coefficient reliably.

Şen (1987) suggested a method for estimating the storage coefficient of an aquifer using pumping well data from quasi-steady state groundwater flow toward wells. The basis of the method is the storage coefficient definition as

the ratio of abstracted water volume to the depression cone volume as given by Eq. (2.15). He presented an explicit derivation of this expression for confined aquifers as,

$$S = \frac{Qt - \pi r_w^2 s_w(t)}{\frac{r_w^2 Q}{4T}\left\{\exp\left[\frac{4\pi T s_w(t)}{Q}\right] - 1\right\} - \pi r_w^2 s_w(t)} \qquad (3.88)$$

For large times the volume term Qt is much bigger than the well storage volume contribution, $\pi r_w^2 s_w(t)$, which may be neglected. Negligence of this term leads to (Birpinar and Ayhan, 2005),

$$S = \frac{4tT}{r_w^2\left\{\exp\left[\frac{4\pi T s(t)}{Q}\right] - 1\right\}} \qquad (3.89)$$

Necessary rearrangements in Eq. (3.89) by considering Eqs (3.31) and (3.32) leads to,

$$u_w = \frac{1}{e^{W(u_w)} - 1} \qquad (3.90)$$

This is a dimensionless expression, which yields the same result provided that dimensionless time factor, u_w, and well function, $W(u_w)$, remain the same. Figure 3.55 show that Eq. (3.90) approaches Theis (1935) and Papadopulos and Cooper (1967) type curves for large times.

EXAMPLE 3.20 STORAGE CALCULATION— QUASI-STEADY STATE FLOW

Extensive field pump-test data (Table 3.14) have been used for the application of the methodology for confined aquifer conditions in order to demonstrate the ability of proposed method. Calculate the storage coefficient from the Chaj Doab aquifer data in Table 3.14.

Transmissivity and storage coefficients from the aquifer data of all the observation wells are computed according to Theis method (considering all drawdowns) for pumping well (Ahmad, 1998). Three alternative CJ straight-line methods (TD, DD and CD) are applied for the treatment of field data as explained in Section 3.9. Find the storage coefficient values for the three OWs according to

quasi-steady state flow depression cone volume concept and compare them with existing methodology results.

Solution 3.20

The same observation well data (considering late drawdowns) are selected for storage coefficient estimation by the methodology proposed in this section. For this purpose, Eq. (3.88) is used for the determination of the storage coefficient of the aquifer (Eq. (3.90)). The results by using different methods are shown collectively in Table 3.29.

3.13 UNSTEADY RADIAL FLOW IN A LEAKY AQUIFER TEST

The analytical derivation of relevant mathematical expression for unsteady state flow in leaky aquifers toward fully penetrating wells is almost the same as with the confined aquifer case, except an additional contribution comes from the leakage through the upper unconfined aquifer as shown in Figure 3.56.

In addition to two already mentioned active volumes in confined aquifers, another (see Section 3.6) one will come into view as leakage volume from the unconfined aquifer. The annulus volume of this part is $2\pi r dr[dh(r, t)/dt]$

TABLE 3.29 Unconfined Aquifer Storage Coefficients

| Well Number | Theis Method | Jacob Method | | | Şen (1987) Analytical Method |
		TD	DD	CD	S
OW1	2.7×10^{-5}	1.3×10^{-3}		4.7×10^{-5}	3.0×10^{-5}
OW2	3.5×10^{-3}	9.4×10^{-3}		9.8×10^{-3}	1.9×10^{-3}
OW3	2.9×10^{-2}	7.0×10^{-3}	2.7×10^{-2}	2.0×10^{-2}	5.1×10^{-3}
Average	1.1×10^{-2}	5.9×10^{-2}	2.7×10^{-2}	9.9×10^{-3}	2.3×10^{-3}

TD, time–drawdown; DD, distance–drawdown; CD, composite-drawdown. Comparison of the storage coefficients indicates that the method proposed in this section yields results that are practically in good agreement with other methods.

FIGURE 3.56 **Unsteady flows to a
well in a leaky aquifer.**

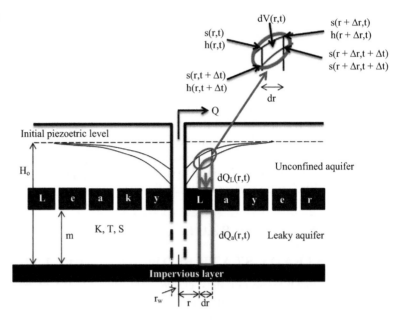

and its contribution to overall discharge is
$S2\pi rdr[dh(r, t)/dt)]$ within the unconfined
aquifer as the storage contribution from the
leaky aquifer. Hence, the third discharge contributions to Eq. (3.23) and differentiation with
respect to r gives,

$$\frac{\partial}{\partial r}\left(2\pi rmK\frac{\partial h(r,t)}{\partial r}\right) + 2\pi rK'\frac{H_o - h(r,t)}{m'}$$

$$= \frac{S2\pi r\partial h(r,t)}{\partial t} \quad (3.91)$$

After the necessary arrangements, it can be
written by taking into consideration the drawdown as $s(r, t) = H_o - h(r, t)$,

$$\frac{\partial^2 s(r,r)}{\partial r^2} + \frac{1}{r}\frac{\partial s(r,r)}{\partial r} + \frac{s(r,t)}{L^2} = \frac{S}{T}\frac{\partial s(r,t)}{\partial t} \quad (3.92)$$

Similar dimensionless time factor and well
function as in Eqs (3.31) and (3.32) are valid
with an additional leakage factor which has
been already been defined by Eq. (3.24).

The first analytical solutions for leaky aquifers
were developed by Hantush and Jacob (1955a)
and Hantush (1956, 1960). On the other hand,
Neuman (1972) and, Neuman and Witherspoon
(1969a,b) enlarged the development of leaky
aquifer theory.

Hantush type curves express the relationship
between well function and dimensionless time
factor for different r/L values as in Figure 3.57.

Each type curve in this figure has three
distinctive portions, which are labeled as A, B
and C and their explanations are as follows.

1. The first portion A coincides with the Theis
   curve, indicating that the aquifer behaves as
   confined at the early stages of water
   abstraction. There is no contribution from the
   upper confined aquifer through leakage.
   Increase in the observation well distance, r,
   and/or decrease in leakage reduces the
   duration of this portion,
2. The second part B starts at the time when a
   particular type curve starts to deviate from

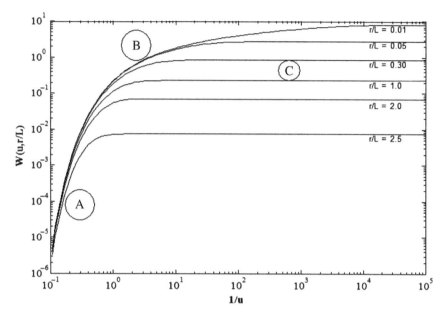

FIGURE 3.57 **Hantush leaky aquifer type curves.** *After Walton (1970).*

Theis curve downward and continues until it becomes horizontal. During this portion both the aquifer storage and the leakage contribute,
3. The third part C is completely horizontal implying a steady state flow during which all pump discharge comes from leakage water only. This means that the water source is only from the above unconfined aquifer.

In order to apply leaky aquifer (Hantush) type curves for the field data the following steps are necessary:

1. Plot the time drawdown data on log–log scale,
2. Select the most suitable Hantush type theoretical curve by considering that both the field and the theoretical curves are drawn on the same scale double-logarithmic papers,
3. Both sheets are brought together keeping their axes parallel such that most of the data points fall around a type curve. Preferably, one of the sheets should be transparent,

4. Both sheets are fixed marking a match point, M, and reading the corresponding values of field (time-drawdown), and dimensionless type curve (time factor versus well function) values,
5. Substitution these values into Eqs (3.31) and (3.32) lead to transmissivity and storage coefficient calculations,
6. Knowing the distance, r, of the observation well from the main well, the leakage factor, L, can be determined from the label, r/L of the most suitable type curve.

EXAMPLE 3.21 LEAKY AQUIFER TYPE CURVE MATCHING

The time–drawdown data from an observation well 125 m away from the main well are given in Table 3.30. The pump discharge is 0.5 m^3/min. Determine the aquifer type and then calculate all possible parameter values.

TABLE 3.30 Leaky Aquifer Data

Time (min)	Drawdown (m)	Time (min)	Drawdown (m)
2	15.5	90	31
3	18	120	31.5
4	19	180	32
5	20	240	33
10	24	300	33.5
15	26	360	34
20	27	420	34
30	27.5	480	34
40	28	570	34
50	28.5	650	34
60	29		

Solution 3.21

The aquifer type can be decided from the time–drawdown scatter plot as in Figure 3.58. It is obvious that the late time drawdown data fall on a horizontal line and this indicates that the aquifer is of leaky type.

After the decision on the leaky aquifer type, type curve matching procedure is necessary to determine the aquifer parameters, namely, leakage coefficient, transmissivity and storage coefficient and Figure 3.59 represents the final form of the matching procedure.

The set of extractable data from this figure includes $r/L = 0.1$ and since $r = 125$ m then $L = 1250$ m; $t_M = 10$ min, $s_M = 10$ m, $(1/u)_M = 3.85$ and $W_M(r, r/L) = 0.8$. The substitution of relevant values into Eq. (32) and then Eq. (31) yields $T = 4.58$ m^2/day and $S = 2.1 \times 10^{-3}$, respectively. According to Table 2.10 this aquifer has negligible potential, because $T < 5$ m^2/day.

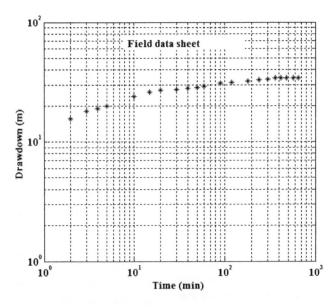

FIGURE 3.58 Time–drawdown data plot.

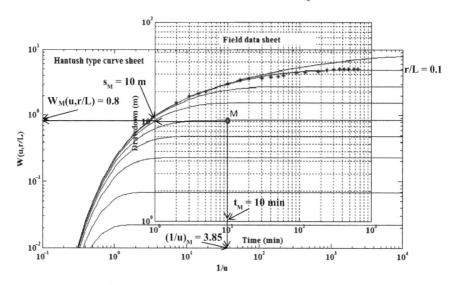

FIGURE 3.59 Leaky aquifer type curve matching.

3.13.1 Hantush Inflection Point Model

Hantush (1964) described the properties of the inflection point at which general behavior of the curve starts to deviate from that of pure confined aquifer. Hantush (1960) observed the initial time–drawdown data fall on the Theis type curve for a period $t < t_i/4$ on the semilogarithmic paper. Herein, t_i is the time at which an inflection point occurs on the leaky aquifer type curve. He found a number of relationships, which are valid whatever the value of r/L factor. He expressed the dimensionless time factor at the inflection point as,

$$u_i = \frac{r^2 S}{4Tt_i} = \frac{r}{2L} \qquad (3.93)$$

The slope at the inflection point is given as,

$$\Delta s_i = \frac{2.3Q}{4\pi T} \exp\left(-\frac{r}{L}\right) \qquad (3.94)$$

The inflection point drawdown, s_i, is taken as equal to the half of the maximum drawdown, s_m,

which implies steady state flow. The maximum drawdown expression is given in Eq. (3.25), and hence,

$$s_i = 0.5\,s_m = \frac{Q}{4\pi T} K_0\left(\frac{r}{L}\right) \qquad (3.95)$$

Another useful expression for practical calculations is given as follows:

$$K_0\left(\frac{r}{L}\right)\exp\left(\frac{r}{L}\right) = 2.3\frac{s_i}{\Delta s_i} \qquad (3.96)$$

For $r/L < 0.01$ which corresponds to $s_i/\Delta s_i > 2$ the equivalent approximation to this last expression is,

$$\log\left(\frac{2L}{r}\right) = 0.251 + \frac{s_i}{\Delta s_i} \qquad (3.97)$$

This inflection point method needs time–drawdown data from a single observation well. Apparently, the inflection point is located at half of the maximum drawdown (s_m) on the semilogarithmic plot of time–drawdown data. Aquifer pumping duration should be long enough to attain representative maximum drawdown, s_m, which ultimately is used for the

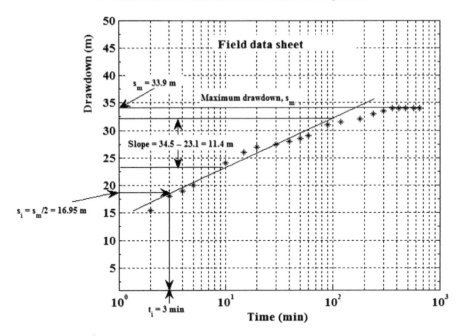

FIGURE 3.60 **Inflection point graph.**

location of the inflection point. The time drawdown data on the semilogarithmic scale yields
estimates of maximum drawdown, s_m, and the
slope, Δs_i, at the inflection point. At half of the
maximum drawdown, s_m, the values of draw-
down, s_i, and time, t_i, are read from the corresponding axes of the time–drawdown plot on
semilogarithmic scale. The substitution of s_i and
Δs_i into Eq. (3.97) gives the estimates of r/L value
by making use of Figure 3.20. The estimated
value of r/L and the slope, Δs_i, are substituted
into Eq. (3.94) in order to calculate transmissivity,
T, of the aquifer parameter. Finally, storage
coefficient, S, is determined from Eq. (3.97) by
substituting values of T, r/L and t_i.

EXAMPLE 3.22 LEAKY AQUIFER INFLECTION POINT METHOD

Subject the same time–drawdown data in
Table 3.30 to inflection point methodology and
determine the aquifer parameters.

Solution 3.22

For the application of the methodology first it
is necessary to plot the given field data on a
semilogarithmic paper with time on logarithmic
axis as in Figure 3.60.

One can read off from this figure the necessary
set of parameters as $\Delta s_i = 11.4$ m, $s_m = 33.9$ m,
$s_i = 16.95$ m and $t_i = 3$ min. Substitution of s_i and
Δs_i into Eq. (3.97) gives $\log(2L/r) = 1.738$ or $2L/r = 56.68$ or $L/r = 28.34$. Since the distance of the
observation well to the main well is 125 m, then
$L = 3642$ m. Equation (3.94) provides possibility
for calculation of the transmissivity value, which
after the substitution of relevant values turns out
as $T = 11.17$ m^2/day. Finally the storage coefficient from Eq. (3.93) becomes as $S = 2.18 \times 10^{-1}$.

3.13.2 Hantush Distance-Slope Model

This method is used when time–drawdown
data is measured at more than one observation

wells located at different distances from the main pumping well. At least three observation well data are needed. Equation (3.94) can be expressed by taking logarithms on both the sides as,

$$r = 2.3L\left[\log\frac{2.3Q}{4\pi T} - \log\Delta s_i\right] \qquad (3.98)$$

The slopes of time−drawdown data from each observation well on semilogarithmic scale are estimated at inflection points. It is obvious from this last expression that plot of these slopes versus the radial distances, r, on semilogarithmic scale should yield to a straight-line with slope,

$$(\Delta s)_i = 2.3L \qquad (3.99)$$

The intercept, $(\Delta s_i)_0$, of this straight-line at $r = 0$ is given with the following expression,

$$(\Delta s_i)_0 = \frac{2.3Q}{4\pi T} \qquad (3.100)$$

The leakage factor, L, and transmissivity, T, are estimated from Eqs (3.99) and (3.100), respectively and storage coefficient, S, is calculated by substituting the values of relevant parameters and variables into Eq. (3.93).

EXAMPLE 3.23 DISTANCE-SLOPE METHODS

Ahmad (1998) gave the semilogarithmic plots of six observation wells' time−drawdown data plots in Figure 3.61. Calculate the aquifer parameters.

Solution 3.23

It is obvious from this figure that in all six observation wells there are horizontal ends at large times, which indicate the leaky aquifer formation. The maximum drawdowns and the radial distances to the main well are given in Table 3.31.

For the determination of hydrogeological parameters, Hantush distance-slope method is

applied and the results are given in Figure 3.61. The necessary calculations are shown in Table 3.32.

3.14 LARGE DIAMETER WELL HYDRAULICS

The first study about the unsteady state flow toward large diameter well in a confined aquifer is conducted by Papadopulos and Cooper (1967) by considering the well storage. They obtained a set of type curves for large-diameter wells on the basis of Laplace transformations. The evaluation of aquifer response by Papadopulos and Cooper method requires numerical integration of an improper integral involving Bessel's function. Patel and Mishra (1983) analyzed by a discrete kernel approach unsteady flow to a large-diameter well. The discrete kernel technique is applicable only for a linear system.

Şen (1982) has used the concept of depression cone volume to derive type curve equation for large-diameter wells in confined aquifers of finite extent. The relative error between these and Papadopulos and Cooper type curves is less than practically acceptable limit of ±5.

In the following sequel type curves for large diameter wells will be derived with the effect of impervious barriers as shown in Figure 3.62.

A vertical boundary is represented by an irregular and steeply sloping barrier. A fully penetrating large-diameter well of radius r_w is located adjacent to a barrier of distance L from the well center. The total drawdown, $s_T(t)$ in the well due to existence of a barrier is given as follows,

$$s_T(t) = s_w(t) + s_i(t) \qquad (3.101)$$

where $s_w(t)$ is the drawdown in the main well without the barrier effect, and $s_i(t)$ is the drawdown from the image well. This equation is valid completely after arriving of the depression curve at the image point; otherwise $s_i(t) = 0$. Since the

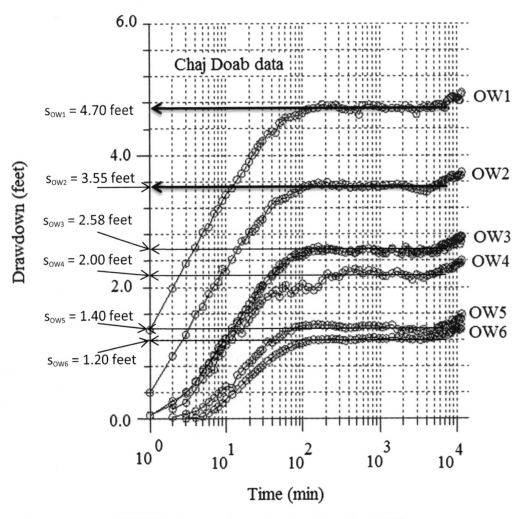

FIGURE 3.61 A set of time–drawdown data. *Ahmad (1998).*

constant well discharge, Q, comes from the aquifer discharge, $Q_a(t)$, and well storage discharge as given in Eq. (3.4) one can write that,

$$Q_a(t) = Q - \pi r_w^2 \frac{ds_w(t)}{dt} \qquad (3.102)$$

In an aquifer domain between two concentric cylinders of respective radius r and r + dr, the water volume balance equation can be written as,

$$dtdQ_a(r, t) = S2\pi rdrds(r, t)$$

where S is the storage coefficient. This equation can be rewritten succinctly as,

$$dtdQ_a(r, t) = SdV(r, t)$$

where $dV(r, t)$ is the incremental depression cone volume during an infinitesimally small time interval, dt. The integration of this last expression with respect to r and boundary conditions $Q_a(r_w, t) = Q_a(t)$; $dV(r_w, t) = 0$ and $dV(\infty, t) = dV(t)$ gives,

$$Q_a(t)dt = SdV(t) \qquad (3.103)$$

TABLE 3.31   Hantush Distance-Slope Method Characteristic Values

Well Number	Radial Distance (feet)	Maximum Drawdown (feet)	Inflection Point		
			Drawdown (feet)	Time (min)	Slope
OW1	100	4.70	2.35	3.0	1.95
OW2	200	3.55	1.76	12.0	1.65
OW3	400	2.58	1.29	4.0	1.85
OW4	400	2.00	1.00	15.0	1.10
OW5	600	1.40	0.70	20.0	1.05
OW6	800	1.20	0.60	9.0	1.50

TABLE 3.32   Aquifer Parameter Estimations

Well Number	(r/L) $(-)$	Transmissivity (ft^2/s)	Storativity $(\times 10^{-4})$
OW1	0.09	0.1904	6.17
OW2	0.30	0.1824	4.92
OW3	0.20	0.1798	4.31
OW4	0.60	0.2027	6.08
OW5	0.70	0.1922	5.00
OW6	0.50	0.1643	5.50
Averages	0.40	0.1853	5.33

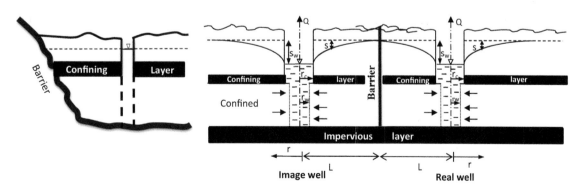

FIGURE 3.62   **Natural barriers and its mathematical abstraction.**

where $dV(t)$ is the difference between depression cones at time instances t and $t + dt$. Substitution of Eq. (3.102) into Eq. (3.103) results in,

$$Qdt = \pi r_w^2 ds_w(t) + SdV(t)$$

The integration of which with relevant boundary conditions, $s_w(0) = 0$ and $V(0) = 0$ yields,

$$Qt = \pi r w_w^2 s(t) + SV(t) \qquad (3.104)$$

where $V(t)$ shows the depression cone volume since the pumping start. The depression cone is a revolutionary surface around the main well. The groundwater flow motion equation can be written according to Darcy's law for any concentric cylinder of radius r as,

$$Q_a(r, t) = -2\pi r T \frac{ds(r, t)}{dr} \qquad (3.105)$$

where T is the transmissivity constant. This equation is approximately valid for any instantaneous time instant during the water level decline in nonequilibrium state. The integration of Eq. (3.105) for boundary conditions $Q_a(r_w, t) = Q_a(t)$; $Q_a(\infty, t) = 0$; $s(r_w, t) = s_w(t)$; and $s(\infty, t) = 0$ leads to the depression cone equation as,

$$r = r_w \exp\left\{ -\frac{2\pi T[s(r, t) - s_w(t)]}{Q_a(t)} \right\} \qquad (3.106)$$

The depression cone volume can be found as a result of the following integration,

$$V(t) = \int_0^{s_w(t)} \pi r^2 ds(r, t) - \pi r_w^2 s_w(t)$$

After the substitution of Eq. (3.106) into this last expression and the completion of necessary algebra yields,

$$V(t) = r_w^2 Q_a(t) \frac{\left\{ \exp\left(\frac{4\pi T s_w(t)}{\frac{e}{Q_a(t)}} \right) - 1 \right\}}{4T}$$
$$- \pi r_w^2 s_w(t)$$

$$(3.107)$$

Consideration of Eq. (3.31) and Eq. (3.32) after some simplifications leads succinctly to,

$$u_w = \frac{1}{e^{W(u_w)} + W(u_w)\left(\frac{1}{S} - 1\right) - 1} \qquad (3.108)$$

When u_w is plotted against $W(u_w)$, a set of similar curves to Papadopulos and Cooper (1967) type curves are obtained as shown in Figure 3.63.

So far in the derivation of Eq. (3.108) the barrier situation (Eq. (3.101)) has not been considered actively, therefore, it would be valid for large-diameter wells in extensive aquifers. Equation (3.108) is valid if the depression cone does not reach the image well. At the image point, according to Eq. (3.105) the drawdown, $s(2L - r_w, t)$, should be zero. Therefore, from the same equation, the main well drawdown at this instance can be found as,

$$s_w(t) = \frac{Q_a(t)\text{Ln}\left(\frac{2L}{r_w} - 1\right)}{2\pi T}$$

or in terms of dimensionless well function as,

$$W(u_{2L}) = 2\text{Ln}\left(\frac{2L}{r_w} - 1\right) \qquad (3.109)$$

If $W(u_{2L}) < 2\text{Ln}(2L/r_w - 1)$, then the effect of the barrier does not appear in the main well and the type curves up to this dimensionless drawdown value are obtainable from Eq. (3.103). Otherwise, the drawdown, $s_i(t)$ value at the imaginary well point becomes from Eq. (3.106) as,

$$s_i(t) = s_w(t) - \frac{Q_a(t)\text{Ln}\left(\frac{2L}{r_w} - 1\right)}{2\pi T}$$

The substitution of which into Eq. (3.101) gives the barrier-affected drawdown, $s_w'(t)$, in the main well as,

$$s_w'(t) = 2s_w(t) - \frac{Q_a(t)\text{Ln}\left(\frac{2L}{r_w} - 1\right)}{2\pi T}$$

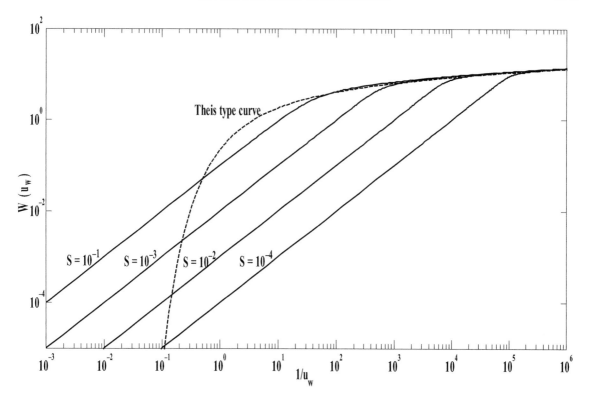

FIGURE 3.63 **Large diameter well type curves.**

Consideration of Eq. (3.31) together with Eq. (3.32) gives the barrier-affected well function in the main well as,

$$W(u_w) = 2W(u_w) - 2Ln\left(\frac{2L}{r_w} - 1\right) \qquad (3.110)$$

The substitution of $W(u_w)$ from this expression into Eq. (3.108) leads to the following expression, which is valid for the barrier effect in the large-diameter well as,

This equation provides the extensions of type curves in the domain where $W(u_{2L}) > 2Ln(2L/r_w - 1)$. In Figure 3.64 these extensions are presented for $L/r_w = 2$ only.

For any desired value of L/r_w, the type curves can be obtained easily from Eq. (3.111). A comparison of Figures 3.63 and 3.64 shows that in the case of barrier existence, the type curves are deflected upward. When the depression cone of the large-diameter well appreciably affects the imaginary well, the time rate of drawdown,

$$u_w = \frac{1}{e^{\left[\frac{W(u_w')}{2} + Ln\left(\frac{2L}{r_w} - 1\right)\right]} + \left[\frac{W(u_w')}{2} + Ln\left(\frac{2L}{r_w} - 1\right)\right]\left(\frac{1}{S} - 1\right) - 1} \qquad (3.111)$$

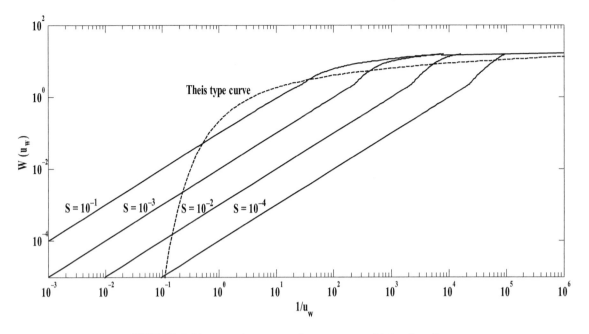

FIGURE 3.64 Large diameter well type curves with barrier effect.

according to Eq. (3.101) will increase; thus the drawdown curve of the main well is deflected upward. The effect of a barrier is evident by an increase in the rate of drawdown with departure from the Papadopulos and Cooper (1967) type curves. It is clear from Figure 3.64 that the upward deflections in the type curves for various S values but common L/r_w values arrive at asymptotic value practically at large times, but theoretically as u_w goes to zero. Another very distinctive characteristic of type curves is that as L/r_w takes large values, they converge to Theis type curve. On the other hand, the deflection observed in the field data curve can be used for finding the effect of boundary and its approximate distance from the main well.

3.15 AQUIFER DOUBLE TEST

In order to check practical homogeneity and isotropy, it is recommended here to perform at least two independent (double) aquifer tests in the same well with different discharge values. After calculating the aquifer parameters from these tests, it is necessary to look at the relative error between the two aquifer parameters according to Eq. (3.11). If the relative error is less than the acceptable limits of ±5 or ±10 then the aquifer may be regarded as homogeneous and isotropic with respect to the parameter concerned.

Groundwater flow toward wells is assumed to take place according to Darcy law, where the only causal effect is the hydraulic gradient under the assumption of constant hydraulic conductivity and transmissivity (Section 3.4.1). This assumption simplifies mathematical model derivations, but if there are spatial hydraulic conductivity variations, the use of Darcy law is possible with a suitable modification (Section 3.4.2). There is now a large body of experimental and theoretical evidence to show that water can be induced to flow through porous media under the influence of gradients other than that of hydraulic gradient. The basic question is how hydraulic conductivity gradient affects the groundwater flow? Hydraulic conductivity changes influence the groundwater flow, and any local decrease (increase) in the hydraulic conductivity causes local increase (decrease) in the hydraulic gradient as well as in the groundwater velocity. The steady-state flow formulation for confined aquifer reduces to the classical Theis formulation when there is no hydraulic conductivity gradient (Eqs (3.15) and (3.16)). The practical use of these general formulations requires double pumping tests in the same well with different pump discharges.

The geological composition affects the piezometric surface, and hence, the hydraulic gradient. If the geological structure is discontinuous, including faults, fractures, joints, dykes, solutions cavities, etc. (but not anticlines and synclines), or continuous with facial changes, then their impacts are appreciated directly on the piezometric surface. In the discontinuity locations such as fractures, the value of hydraulic gradient is comparatively very small. Hence, it is worthy to consider that the hydraulic gradient is also dependent on the aquifer material composition apart from recharge and discharge events. This is equivalent to saying that the groundwater velocity is a function of not only the hydraulic gradient, but similarly, the material properties among which the hydraulic conductivity is the most significant factor for groundwater movement. Hydraulic conductivity gradient is reflected in an induced manner in the piezometric level variations. (Şen and Baradi, 1991). Şen (1989b) studied vertical permeability change and its effect on the aquifer test analysis.

An isotropic media where permeability is equal in all directions in the aquifer is a theoretical assumption which is used for the simplification of governing groundwater equation analytical solution in porous media. This assumption is considered valid both in large-scale hydrology (Dullien, 1992) and comparatively small-scale flow through permeable sediments (Huettel et al., 1998). On the contrary, natural sediments are anisotropic, where the vertical hydraulic conductivity is different from the horizontal conductivity. Additionally, heterogeneity is the property of the aquifer medium, where the hydraulic conductivity is different from one place to another. It is observed through many field trips that natural sediments have spatial anisotropy and heterogeneity which affect both the pattern and rates of porous medium flow (Dullien, 1992; Dagan, 1984; Freeze and Cherry, 1979). Although detailed information might be available on small sampled volumes of soil, the spatial distribution of the permeability at basin scale can only be extracted as equivalent or lumped value. Classical pumping tests provide average permeability estimation for the depression cone volume only. Vogel and Roth (2003) stated that a change of scale through several orders of magnitude in length is necessary to achieve mapping of permeability at basin scale.

There are many natural cases where hydraulic conductivity variations occur within the aquifer. For instance, the release of groundwater is considered to reflect an increase in hydraulic conductivity (permeability) of the near-surface rocks as a result of intense shaking (Rojstaczer and Wolf, 1992). They mentioned about the increased permeability (hydraulic conductivity) from the fall of the level of some wells located on the ridge top immediately above the northwest end of the Loma Prieta fault rapture in California, where nearby rivers increased in flow. In such a model, a correlation might be anticipated between the level of ground shaking, and the proximity of fault rapture, to the size of the hydrological response in groundwater movement due to the hydraulic conductivity change. As stated by Wood and Nuemann (1931) changes in spring flow and well levels have sometimes become incorporated as an intensity indicator. Crust permeability may tend to be higher in a region subject to the extension than one involved in compression deformation. In both normal and strike-slip tectonic environments, enhanced or induced permeability will tend to be along vertical fractures, while in the compression tectonic environments, they are expected to lie in the horizontal plane (Muir-Wood and King, 1993). Butler and Liu (1993) concluded that constant-rate aquifer tests are not very effective for the characterization of lateral variations in flow properties.

In practical applications, two or more estimates of transmissivity are derived from alternative methods of analysis based on the same set of data. If there is a set of measurements in a number of observation wells, alternative transmissivity values are obtained for different radial distances from the main well. If classical methods such as Theim (1906) steady-state groundwater theory are applied, multiple values of transmissivity cannot be correct since this theory requires that the transmissivity of the aquifer is constant in the extensive aquifer. Hence, in the case of different transmissivity value calculations one of the basic assumptions of the Theim (1906) theory is being violated as the spatial change in the transmissivity, and hence, in the hydraulic conductivity.

EXAMPLE 3.24 AQUIFER DOUBLE WELL TEST

A fully penetrating main (MW) and two observation wells (OW1, OW2) are considered in a confined aquifer with 12 m saturation thickness (Şen, 1995). The distances of each observation well to the MW of 2.2 m in diameter are given in Table 3.33. A double pumping test is carried out on such

a well configuration. The quasi-steady state hydraulic heads after a continuous water abstraction for 700 min and then another independent one for 900 min are also presented in the same table with the constant $0.25 \, m^3/min$ and $0.4 \, m^3/min$ discharges, respectively. Interpret the given data for practically important conclusions.

Solution 3.24

If the aquifer is assumed as homogeneous and isotropic then Theim equation is applicable. It yields two transmissivity values from a double pumping test by use of Eq. (3.16) as $137 \, m^2/day$ and $125 \, m^2/day$, respectively. This means that two different pumping tests yield different transmissivity even though the aquifer is assumed to have extensive areal homogeneity and isotropy. This point confirms that the aquifer is not homogeneous because the relative error percentage difference between the two transmissivity values is about 10%. The difference between these two transmissivity values is due to the extra depression cone coverage of the bigger discharge such that it covers more extensive volume within the aquifer, which may have geological composition difference from the earlier test coverage. The best that one can use in practice is the arithmetic average, which is $131 \, m^2/day$.

3.16 RECOVERY METHOD

During an aquifer test after reaching to a more or less steady state case, the pump shut-down (Q = 0) causes rise in the piezometric level, which is referred to as the recovery test. Records

TABLE 3.33 Confined Aquifer Quasi-Steady State Data

Well Number		OW1	OW2	Transmissivity (m²/day) Eq. (3.16)
Distance (m)		$r_1 = 25$	$r_2 = 65$	
$Q_1 = 0.25 \, m^3/min$	Hydraulic heads (m)	$h_1^{'} = 13.8$	$h_2^{'} = 14.2$	137
$Q_2 = 0.40 \, m^3/min$		$h_1^{''} = 13.2$	$h_1^{''} = 13.9$	125

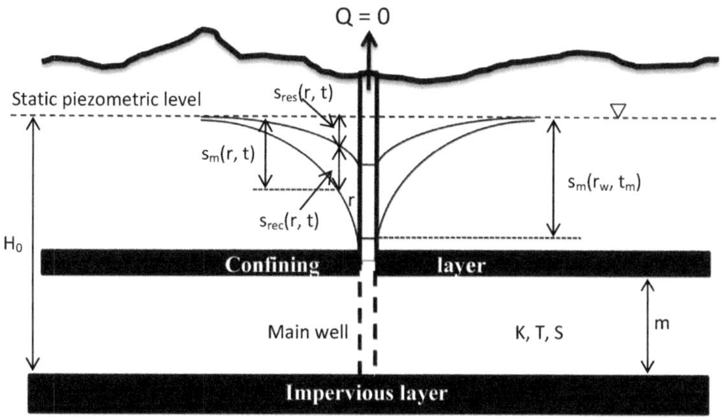

FIGURE 3.65   Recovery test setup.

of the time—drawdown both in the MW and OWs during the recovery period provide useful quantitative and qualitative information about the aquifer characteristics. The quantitative interpretation of the recovery test data is used very often for determining the hydraulic behavior of the oil and gas deposits; however, the recovery data are rarely used in the hydrogeology literature.

At the time, t_m, of pump shut down; there is the maximum drawdown, $s_m(r, t_m)$ at any radial distance. With the rise in water level this maximum drawdown has two complementary parts.

1. The recovery drawdown, $s_{rec}(r, t)$, is the difference between the water level measurement at time, t', and $s_m(r, t_m)$,
2. The residual drawdown, $s_{res}(r, t)$, is the difference between the static water level, H_o. and the water level at current time during recovery. Figure 3.65 shows the three drawdowns and they are related to each other as,

$$s_m(r, t_m) = s_{res}(r, t) + s_{rec}(r, t) \qquad (3.112)$$

A representative time—drawdown variation graph during the pumping and recovery phases is given in Figure 3.66.

The discharge is equal to zero, since there is no water withdrawal. Recovery tests provide independent transmissivity estimation, but do not yield storage coefficient estimations.

3.16.1 Theis Recovery Method

Since the flow is linear (Darcian) during the pumping and recovery periods, the principle of superposition holds. According to Cooper and Jacob (1946), during recovery period an imaginary well continues to recharge the aquifer with the same discharge. If t marks the time since the start of the pumping and t' is the time since pump shutoff, then the residual drawdown $s_{res}(r, t)$ can be written from Eq. (3.46) by difference as,

$$s_{res}(r, t) = \frac{Q}{4\pi T}[W(u_1) - W(u_2)] \qquad (3.113)$$

where u_1 and u_2 are the pumping and recovery period dimensionless time factors similar to Eq. (3.31), the substitution of which with relevant notations into this last expression gives,

$$s_{res}(r, t) = \frac{2.3Q}{4\pi T}\log\frac{t}{t'} \qquad (3.114)$$

The plot of $s_{res}(r, t)$ data from the main or an observation well versus t/t' ratio on a semilogarithmic paper appears as a straight-line. The

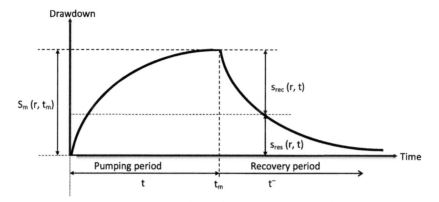

FIGURE 3.66 Graphical representations of pumping and recovery phases.

slope, Δs_r, of this line helps to determine the transmissivity from Eq. (3.114) as,

$$T = \frac{2.3Q}{4\pi\Delta s_r} \qquad (3.115)$$

This expression is similar to Eq. (3.48) for late time–drawdown CJ straight-line method.

EXAMPLE 3.25 THEIS RECOVERY METHOD

A 50 cm well is pumped at a constant rate $Q = 100 \, \text{l/min}$ for 18 hours. At this time the drawdown in the well is 1.90 m and the pump is stopped. Table 3.34 presents the residual drawdown during recovery for 60 min. Determine the aquifer transmissivity.

Solution 3.25

The values of $s_{res}(r, t)$ versus t/t' are plotted on a semilogarithmic paper as in Figure 3.67.

The residual drawdown $\Delta s_{res}(r, t)$ per logarithmic cycle of time t/t' is measured and T is determined from Eq. (3.115) as,

$$T = \frac{2.3 \times 100 \times 10^{-3}}{4 \times 3.14 \times 0.25} = 0.0732 \, \text{m}^2/\text{min}$$
$$= 105.4 \, \text{m}^2/\text{day}$$

TABLE 3.34   Residual Drawdown Data

Time Since Pump Stop (min)	Residual Drawdown $s_{res}(r, t)$ (m)	Time Ratio $t/t' = (18 \times 60 + t')/t'$
1	0.875	1080
2	0.735	541
3	0.690	361
4	0.662	271
5	0.640	217
6	0.625	181
8	0.590	136
10	0.570	109
12	0.556	91
16	0.536	68
20	0.498	55
30	0.453	37
40	0.423	28
45	0.410	25
60	0.382	19

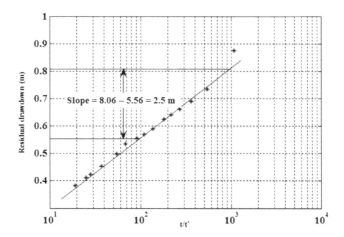

FIGURE 3.67   **Recovery dimensionless time drawdown plot.**

3.17 WELL TEST (STEP-DRAWDOWN TEST)

After drilling of a well and its completion with screens, casing and gravel pack in a certain zone around the skin area, there are impacts on the groundwater flow regime in the well vicinity. Therefore, overall parameter estimations from the known methods can no longer represent the well vicinity. These man-made products around the well affect the groundwater movement in a different pattern than the undisturbed aquifers. Generally, skin domain causes additional energy losses during a water particle movement before it reaches the well. Several expected discontinuities in the piezometric level within the well vicinity are already shown in Figure 3.9. The skin effect causes additional nonlinear losses.

Jacob (1946) suggested theoretically that the drawdown, $s_w(t)$ in a well can be expressed as the summation of linear and nonlinear energy loss terms as,

$$s_w(t) = A(t)Q + CQ^2 \qquad (3.116)$$

in which $A(t)$ is time dependent aquifer loss coefficient. The second term on the right hand side indicates nonlinear losses within the skin

effect and C is a constant. Determination of this constant requires step-drawdown test, which furnishes information about the following important questions:

1. What is the effective well radius that will give rise to minimum well loss? By definition it is the distance measured radially from the well axis to the theoretical drawdown, which equals the actual drawdown within the aquifer. This diameter is directly proportional with the transmissivity but inversely related to the storage coefficient,

2. What is the efficiency of a complete well, in other words, what portion of the total well drawdown is due to the linear aquifer flow and what percentage is a result of other factors? The smaller the percentage the more efficient is the well. The well efficiency, η, is defined usually as the ratio of the aquifer loss to the total losses as,

$$\eta = \frac{A(t)}{A(t) + CQ} \times 100 \qquad (3.117)$$

The ideal well with 100% efficiency rarely exists in practice with the exception of uncased

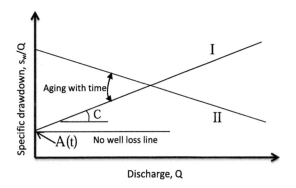

FIGURE 3.68 **Specific drawdown–discharge relations.**

wells in fractured rock aquifers or cavernous limestone. Badly designed wells will have rather small efficiency percentages,

3. How can one control the future performance of a well? Operation of the well for water supply purpose gives rise to some aging problems such as corrosion, encrustation, clogging etc., which affect the total drawdown in the well (see Figure 3.68). The lines in this figure are obtained by dividing both sides of Eq. (3.116) by Q,

$$\frac{s_w}{Q} = A(t) + CQ \qquad (3.118)$$

in which s_w/Q is the specific drawdown, which is the reverse of the specific capacity, Q/s_w.

In an ideal situation with no well loss, the relationship appears as a horizontal line parallel to Q axis at a distance $s_w/Q = A(t)$. Otherwise, it is a straight-line with intercept, A(t) and slope, C. With aging, the drawdown in the well increases compared to the original time–discharge curve, I. If the increase is more than 10% of the initial drawdown, then the well needs cleaning or rehabilitation.

Many field data fail to fit the formulation in Eq. (3.116). In order to account for some of these deviations Rorabaugh (1953) proposed an alternative formula as,

$$s_w = A(t)Q + CQ^n \qquad (3.119)$$

in which n is an exponent often greater than 2 with an average value of 2.5. His method usually provides a value for C smaller than the value calculated from the same data by CJ method.

In order to determine the characteristic shape of specific drawdown–discharge relationship, a set of sequential pumping tests, each with different discharge, must be conducted in the field. Although it is possible to conduct independent tests in the same well with different discharges yielding different drawdown values, it is rather cumbersome and requires longer time which is not a practical solution, although such a procedure is similar to aquifer double test (Section 3.15) there is no benefit for nonlinear well loss calculations.

It is preferable to conduct a continuous pumping test consisting of discharge increments for prefixed time durations. To save time, step drawdown tests are carried out with no stop between successive steps. As a result, the measured drawdown in the field during each step contains the residual drawdowns of the preceding step. At least three epochs should be included in such a test with constant discharge increment during each epoch as schematized in Figure 3.69.

The identification of parameters, namely, A(t) and C in Eq. (3.116) from given pairs of observations sequences on Q and corresponding s_w values is possible by using different methods.

For instance, Jacob (1946) method is a graphical approach, which helps to find the well loss coefficient from a general expression and data concerning two successive epochs, say, $(i-1)$-th and i-th as,

$$C_i = \frac{\dfrac{\Delta s_{wi}}{\Delta Q_i} - \dfrac{\Delta s_{w(i-1)}}{\Delta Q_{(i-1)}}}{\Delta Q_i + \Delta Q_{(i-1)}} \quad (i = 1, 2, \dots, j) \quad (3.120)$$

where j is the epoch number and other terms are self-explanatory from Figure 3.69. Accordingly, Table 3.35 is prepared for the systematic calculation of necessary steps.

The solution of Eq. (3.116) needs a graphical representation of time–drawdown data on a

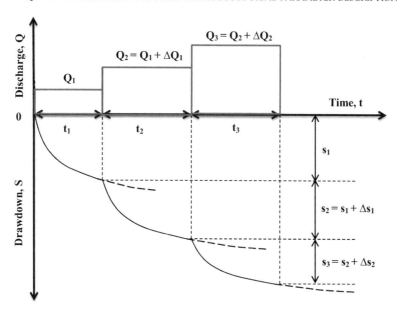

FIGURE 3.69 **Step drawdown procedures.**

TABLE 3.35 Jacob Calculations of C Values

Step No., i (1)	Q_i (m³/s) (2)	ΔQ_i (3)	Δs_{wi} (m) (4)	$\Delta s_{wi}/\Delta Q_i$ (5)	C_i (From Eq. (3.120)) (s²/m⁵) (6)
0	0	0	0	0	0
1	Q_1	Q_1	s_{w1}	$s_{w1}/\Delta Q_1$	C_1
2	$Q_2 + \Delta Q_1$	$\Delta Q_1 = Q_2 - Q_1$	s_{w2}	$s_{w2}/\Delta Q_2$	C_2
3	$Q_3 + \Delta Q_2$	$\Delta Q_2 = Q_3 - Q_2$	s_{w3}	$s_{w3}/\Delta Q_3$	C_3
4	$Q_4 + \Delta Q_3$	$\Delta Q_3 = Q_4 - Q_3$	s_{w4}	$s_{w4}/\Delta Q_4$	C_4
5	$Q_5 + \Delta Q_4$	$\Delta Q_4 = Q_5 - Q_4$	s_{w5}	$s_{w5}/\Delta Q_5$	C_5

semilogarithmic paper for genuine drawdown increments in each step. The drawdown curves of each step are extended smoothly until the end of the succeeding step ends. The incremental drawdown, Δs_i, are read off and written into column 4. The remaining columns in the table are filled accordingly, which leads to individual C estimations for each step, and then an average value can be adopted as a representative of the well loss factor.

3.18 AQUIFER CLASSIFICATION BY FUZZY HYDROGEOLOGICAL PARAMETER DESCRIPTIONS

The classical aquifer parameter estimation techniques yield single, and constant T and S values. Hence, it is not possible for practicing hydrogeologists or groundwater specialists to classify the concerned aquifer into different classifications by simultaneous consideration of both parameters. Based on the transmissivity value only, it is possible to say crisply that the aquifer has "negligible," "weak," "low," "moderate," or "high" potential (Table 2.10), in addition to its "confined," "unconfined," or "leaky" aquifer specification as mentioned in Chapter 2. Each one of these words is rather vague and they constitute fuzzy sets (Zadeh, 1965). Heterogeneous and anisotropic composition of geologic layers does not allow aquifer belongings crisply to one class only, but it may have shares in two or more sets. In order to make noncrisp classification (fuzzy), it is necessary to have a sequence of aquifer parameter estimations from a given time—drawdown data, and then to treat these sequences in a fuzzy manner into two or more subclasses. As shown in Section 3.7, slope matching procedure provides a sequence of parameter estimations, which can be treated by fuzzy logic principles (Şen, 2010).

3.18.1 Fuzzy Aquifer Classification Chart

In classical hydrogeological assessments of groundwater potentiality based on transmissivity values, Table 2.10 indicates the classifications which are crisp and at the boundaries do not have clear meanings. Tables 2.10 and 2.11 in Chapter 2 are for individual transmissivity and storage coefficient classifications, but their simultaneous consideration does not exist. In Figures 3.70 and 3.71, the fuzzy membership functions (MF) of the transmissivity and storativity are presented on logarithmic scale with five and six membership functions, respectively.

The storage coefficient classifications in Table 2.11 are not adopted as they are in Figure 3.70, but the whole domain of parameter variation is divided into three aquifer types, namely, "confined," "leaky," and "unconfined" in the

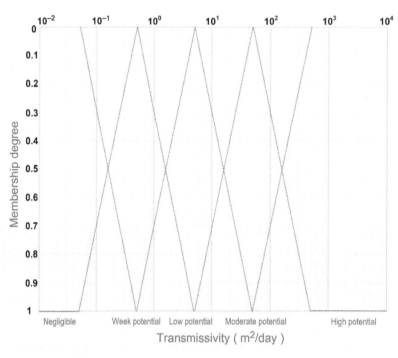

FIGURE 3.70 Transmissivity membership functions.

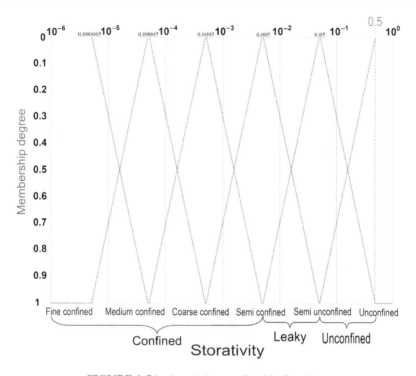

FIGURE 3.71   **Storativity membership functions.**

order of increasing storage coefficient value. On the basis of these aquifer types, it is possible to make refined classifications as "fine confined," "medium confined," "coarse confined," "semiconfined," "semiunconfined," and "unconfined" in increasing order of storativity. The diagram in Figure 3.72 is prepared for joint fuzzy classification.

The scatter domain is now divided into $5 \times 6 = 30$ simultaneous classification subdomains each with different degrees of transmissivity, T, and storativity, S. The fuzzy simultaneous classifications (fuzzy inference) of T and S are given in Table 3.36, where there are 30 joint specifications with the same number of logically deductible rules only of which few are presented below. The completion of all the rules is left to the reader.

R1: IF T is "High" and S is "Fine confined" THEN the aquifer has "high" potential
R1: IF T is "Moderate" and S is "Course confined" THEN the aquifer has "very high" potential
R1: IF T is "Low" and S is "Semi confined" THEN the aquifer has "Moderate" potential
R1: IF T is "Weak" and S is "Fine unconfined" THEN the aquifer has "very very low" potential

EXAMPLE 3.26 FUZZY HYDROGEOLOGICAL PARAMETER DESCRIPTIONS

In order to show the efficiency of the slope methodology a classical textbook confined aquifer test is selected from Todd (1980). A well

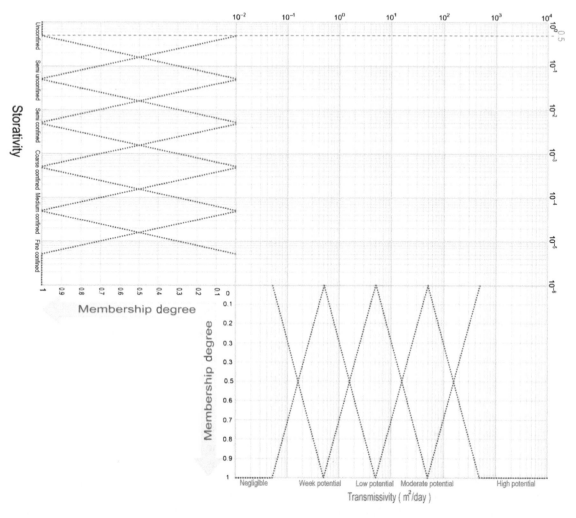

FIGURE 3.72 Joint classification templates.

penetrating confined aquifer is pumped at a uniform rate 2500 m³/day. Drawdown during the pumping period is measured in an observation well 60 m away, and time–drawdown measurements are listed in Table 3.32.

Solution 3.26

After a convenient Theis type curve matching with a matching point selection, $W(u) = 1.00$, and $u = 1 \times 10^{-2}$, and corresponding $s = 0.18$ m and $r_2/t = 150$ m values, the aquifer parameters are

calculated from Eqs (3.31) and (3.32) as $T = 1110$ m²/day, and $S = 2.06 \times 10^{-4}$, respectively. The same aquifer test is subjected to slope matching calculation steps as explained in the previous section and the final results are presented in Table 3.37.

It is obvious that for each time–drawdown measurement after the first reading there are u_i $W_i(u_i)$, T_i and S_i values, which change during the whole pumping test. On the other hand, extensive aquifer tests are performed on the

TABLE 3.36   Hydrogeological Parameter Fuzzy Inference

			Transmissivity (m²/day)				
			High	Moderate	Low	Weak	Negligible
Storativity	Confined	Fine	H	VH	H	L	N
		Medium	VH	H	L	VL	N
		Course	VMH	VH	H	M	L
	Leaky	Confined	VH	H	M	L	VL
		Unconfined	H	M	L	VL	VML
	Unconfined		VH	L	VL	VML	VMN

VMH, Very much high; VH, Very high; H, High (H); M, Medium; VML, Very much low; VL, Very low; L, Low; VMN, Very much negligible; VN, Very negligible; N, Negligible.

TABLE 3.37   Slope Matching Calculation Results

Time (day)	Drawdown (m)	α_i	u_i	$W_i(u_i)$	T_i (m²/day)	S_i
0	0					
0.0007	0.20	0.74	0.2517	1.05	889.25	0.00017
0.0010	0.27	0.37	0.0471	2.60	1818.31	0.00010
0.0014	0.30	0.56	0.1633	1.51	941.36	0.00024
0.0017	0.34	0.46	0.0857	1.98	1109.06	0.00018
0.0021	0.37	0.36	0.0305	2.72	1386.56	0.00010
0.0028	0.41	0.42	0.0668	2.24	1037.33	0.00021
0.0035	0.45	0.35	0.0315	2.74	1171.19	0.00014
0.0042	0.48	0.34	0.0346	2.80	1104.84	0.00018
0.0056	0.53	0.33	0.0205	3.00	1086.83	0.00014
0.0069	0.57	0.28	0.0124	3.51	1193.87	0.00011
0.0083	0.60	0.32	0.0232	3.09	998.57	0.00021
0.0097	0.63	0.24	0.0900	3.73	1141.96	0.00111
0.0125	0.67	0.25	0.0189	3.92	1122.69	0.00030
0.0167	0.72	0.24	0.0905	3.77	1013.56	0.00170
0.0208	0.76	0.22	0.0066	4.49	1136.68	0.00017
0.0278	0.81	0.22	0.0053	4.60	1103.70	0.00018
0.0347	0.85	0.31	0.0241	3.11	707.97	0.00066

(Continued)

TABLE 3.37 Slope Matching Calculation Results (*cont'd*)

Time (day)	Drawdown (m)	α_i	u_i	$W_i(u_i)$	T_i (m²/day)	S_i
0.0417	0.90	0.11	0.0001	8.77	1907.42	0.00001
0.0556	0.93	0.14	0.0004	7.03	1479.05	0.00004
0.0694	0.96	0.22	0.0063	4.44	900.96	0.00044
0.0833	1.00	0.18	0.0011	5.68	1108.51	0.00011
0.1042	1.04	0.16	0.0091	6.35	1198.06	0.00126
0.1250	1.07	0.18	0.0028	5.56	1019.30	0.00040
0.1458	1.10	0.13	0.0004	7.41	1327.76	0.00008
$T = 1162.70 \text{ m}^2/\text{day}$				High potential		
$S = 3.43 \times 10^{-4}$				Confined aquifer		

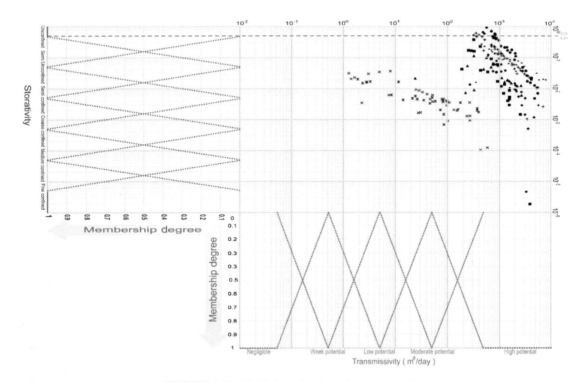

FIGURE 3.73 **Field data plot from slope test results.**

Arabian Peninsula in the same aquifer through eight wells. The application of slope method to these tests yielded a set of S and T values. In Figure 3.73 Todd data are shown by full squares and the Arabian Peninsula data by crosses.

It is obvious that even though all the test wells are in the same aquifer the aquifer parameters differ significantly. S and T estimations fall within ranges 1×10^0–10^{-7} and 1×10^0–2×10^4 m^2/day, respectively. According to Table 2.36 the initial fuzzy results with distinction between the MFs are presented in Table 3.38 based on the number of samples in each joint class.

It is obvious from this table that the aquifer is not homogeneous and the most frequently appearing portion falls under the specification of "highly potential leaky unconfined" aquifer corresponding to "VMH" joint specification according to inference in Table 3.36. The next significant portion of the aquifer has "moderate potential unconfined" condition, the combination of which gives "L, low" combined potential according to Table 2.10.

Table 3.39 presents the complete assessment of the aquifer tests in the same aquifer, including aquifer types, subtypes and fuzzy membership degrees (MDs) for S and T. It is possible to obtain

TABLE 3.38   Aquifer Test Results

	Aquifer Types		Transmissivity (m²/day)				
			High	Moderate	Low	Weak	Negligible
Storativity	Confined	Fine	2	0	0	0	0
		Medium	0	0	0	0	0
		Course	1	3	2	0	0
	Leaky	Confined	23	12	26	3	0
		Unconfined	53	20	2	11	0
	Unconfined	Fine	15	46	0	0	0

TABLE 3.39   Aquifer Classification Details

		Fuzzy Degrees			
		S		T	
Aquifer Type	Aquifer Subtype	(−)	MD	(m²/day)	MD
Confined	Fine	1.1×10^{-6}	1.0	7.5×10^3	1.0
	Medium	2.5×10^{-5}	0.35	1.0×10^4	1.0
	Course	3.5×10^{-4}	0.15	6.0×10^2	1.0
Leaky	Confined	3.5×10^{-3}	0.14	1.0×10^2	0.3
	Unconfined	3.4×10^{-2}	0.14	7.5×10^2	1.0
Unconfined	Fine	1.8×10^{-1}	0.45	8.0×10^2	1.0

from Figure 3.73 fuzzy degrees for each data point on the horizontal and vertical axes MFs. Table 3.39 indicates the membership degrees of the arithmetic averages of each fuzzy inference location from Table 3.38.

It is possible to calculate weighted average of T and S with MDs as weights, and the results appear as $T = 3726.4 \, m^2/day$ and $S = 6 \times 10^{-3}$. Hence, according to Tables 2.10 and 2.11, the overall average behavior of the aquifer is "highly potential" and "fine confined" aquifer.

References

Abo-Zeid, M.A., Scott, V.H., 1963. Non-steady flow for wells with decreasing discharges. Proc. Am. Soc. Civ. Eng. J. Hydraul. Div. 89 (HY3), 119–132.

Abramowitz, M., Stegun, I.A. (Eds.), 1972. Handbook of Mathematical Functions with Formulas, Graphs, and Mathematical Tables. Dover Publications, New York, ISBN 978-0-486-61272-0.

Anderson, K.E., 1967. Water Well Handbook. Scholin Brothers, St. Louis.

Ahmad, N., 1998. Evaluation of groundwater resources in the upper middle part of Chaj Doab area, Pakistan (unpublished Ph.D. thesis). Technical University of Istanbul, Istanbul, Turkey, p. 308.

Aron, G., Scott, V.H., 1965. Simplified solution for decreasing flow in wells. Proc. Am. Soc. Civ. Eng. J. Hydraul. Div. 91 (HY5), 1–12.

ASTM D 5242-92, 2001. Standard Test Method for Open Channel Flow Measurement of Water with Thin-Plate Weirs.

Batu, V., 1998. Aquifer Hydraulics – A Comprehensive Guide to Hydrogeological Data Analysis. Wiley Interscience, New York.

Bear, J., 1981. Hydraulics of Groundwater. McGraw-Hill Book Co., New York, 567 pp.

Bibby, R., 1977. Characteristics of Pumping Tests in Heterogeneous Clastic Sediments. Alberta Res. Council Bull. 35, Edmonton, Alberta, Canada, pp. 31–39.

Birpinar, M.E., Ayhan, G., 2005. On the application of storage coefficient determination by quasi-steady-state flow. Nordic Hydrol. 36 (2), 13.

Boulton, N.S., 1954. Unsteady radial flow to a pumped well allowing for delayed yield from storage. Int. Assn. Sci. Hydrol. 37, 472–477.

Boulton, N.S., 1955. Unsteady radial flow to a pumped well allowing for delayed yield from storage. In: Publication de L'Association Internationale d'Hydrologie, Assemblée Générale de Rome, vol. 37, 472–477.

Boulton, N.S., 1963. Analysis of data from non-equilibrium pumping test allowing for delayed yield from storage. Proc. Inst. Civ. Eng. London 26, 469–482.

Butler Jr., J.J., 1988. Pumping tests in non-uniform aquifers — the radially symmetric case. J. Hydrol 101 (4), 15–30.

Butler Jr., J.J., Liu, W., 1993. Pumping tests in non-uniform aquifers: the radially asymmetric case. Water Resour. Res. 29 (2).

Cardwell, W.T., Parsons, R.L., 1945. Average permabilities of heterogeneous oil sands. Trans. Am. Inst. Min. Pet. Eng. 160 (1), 34–42.

Carlsaw, H.S., Jeager, J.C., 1959. Conduction of Heat in Solids, second ed. Oxford Univ. Press, London and New York.

Chaudhary, M.H., 2007. Open Channel Flow, Second ed. Springer. 540 pp.

Compton, P.R., Kulin, G., 1975. A Guide to Methods and Standards for the Measurement of Water Flow. National Bureau of Standards Special Publication 421, Institute for Basic Standards, Washington, 97 pp.

Cooper, H.H., Jacob, C.E., 1946. A generalized graphical method for evaluating formation constants and summarizing well field history. Trans. Am. Geophys. Union 27, 526.

Corbett Jr., R., 1990. Day centers and the advent of a mixed model in corrections. IARCA J. 4, 26–27.

Çimen, M., 2008. Confined aquifer parameters evaluation by slope-matching method. J. Hydrol. Eng. ASCE 13 (3), 141–145.

Dagan, G., 1984. Solute transport in heterogeneous porous formations. J. Fluid Mech. 145, 151–177.

Davis, S.N., De Wiest, R.J.M., 1966. Hydrogeology. Krieger Publishing Company, 463 pp.

De Glee, G. J., 1930. Over groundwaterstromingen bij teronttrekking door middel van putten (Thesis). Waltman, J., Delft, The Netherlands, 175 pp.

De Wiest, R.J.M., 1965. Geohydrology. Krieger Publishing Company, 366 pp.

Driscoll, F.G., 1987. Groundwater and Wells. Johnson Division, St. Paul, MN, 1089 pp.

Dullien, F.A.L., 1992. Porous Media: Fluid Transport and Pore Structure, second ed. Academic Press, San Diego.

Ferris, J.G., Knowles, D.B., Brown, R.H., Stallman, R.W., 1962. Theory of aquifer tests. U.S. Geol. Surv. Water Supply Paper 1536-E, 174 p.

Fetter, C.W., 1983. Applied Hydrogeology. Prentice Hall, Upper Saddle River, New Jersey, 598 pp.

Freeze, R.A., Cherry, J.A., 1979. Groundwater. Prentice-Hall, Inc., Englewood Cliffs, New Jersey, 604 pp.

Frensch, J., Steudle, E., 1989. Axial and radial hydraulic resistance to Roots of Maize. Plant Physiol. 91, 719–726.

Gilbert, G.K., 1890. Lake Bonneville. Monogr. U.S. Geol. Surv. 1, 438 pp.

Hantush, M.S., 1956. Analysis of data from pumping test in leaky aquifers. Trans. Am. Geophys. Union 37, 702.

Hantush, M.S., 1957. Nonsteady flow to a well partially penetrating an infinite leaky aquifer. Proc. Iraqi Sci. Soc. 1, 10.

Hantush, M.S., 1959. Nonsteady flow to flowing wells in leaky aquifers. J. Geophys. Res. 64, 1043.

Hantush, M.S., 1960. Modification of the theory of leaky aquifers. J. Geophys. Res. 65, 3713.

Hantush, M.S., 1964. Hydraulics of wells. In: Chow, V.T. (Ed.), Advances in Hydrosciences.

Hantush, M.S., 1967. Flow of groundwater in relatively thick leaky aquifers. Water Resour. Res. 3 (2), 583.

Hantush, M.S., Jacob, C.E., 1955a. Nonsteady Green's function for an infinite strip of leaky aquifers. Trans. Am. Geophys. Union 36, 101.

Hantush, M.S., Jacob, C.E., 1955b. Nonsteady radial flow in an infinite leaky aquifer. Trans. Am. Geophys. Union 36, 95.

Hantush, M.S., Jacob, C.E., 1960. Flow to an eccentric well in a leaky circular aquifer. J. Geophys. Res. 66, 3425.

Harlow Jr., G.E., LeCain, G.D., 1993. Hydraulic characteristics of, and ground-water flow in, coal-bearing rocks of southwestern Virginia. U.S. Geol. Surv. Water Supply Paper 2388, 36 pp.

Huettel, M., Ziebis, W., Forster, S., Luther, G.W., 1998. Advective transport affecting metal and nutrient distributions and interfacial fluxes in permeable sediments. Geochim. Cosmochim. Acta 62, 613–631.

Huismann, L., 1972. Artificial Groundwater Recharge. Delft University of Technology, Dept. of Civil Engineering, Delft, Netherlands.

Jahnke, E., Emde, F., 1945. Tables of Functions with Formulae and Curves. Dover Publications, 382 pages.

Jacob, C.E., 1940. On the flow of water in an elastic artesian aquifer. Am. Geophys. Union Trans. 72 (Part II), 574–586.

Jacob, C.E., 1944. Notes on determining permeability by pumping test under water table condition. U.S. Geol. Surv. (Open File Report).

Jacob, C.E., 1946. Radial flow in a leaky artesian aquifer. Am. Geophys. Union Trans. 27, 191.

Jacob, C.E., 1950. Flow of groundwater. In: Rouse, H. (Ed.), Engineering Hydraulics. John Wiley, New York, pp. 321–380 (Chapter 5).

Kitterod, N.O., 2004. Dupuit-Forchheimer solutions for radial flow with linearly varying hydraulic conductivity or thickness of aquifer. Water Resour. Res. 40, 1–5.

Kruseman, G.P., de Ridder, N.A., 1990. Analysis and Evaluation of Pumping Test Data, Bulletin 11. Institute for Land Reclamation and Improvement, Wageningen, pp. 41–69.

Kozeny, J., 1933. Theorie und Berehnung der Brunnen. Wasserkraft Wasserwirtschaft 39, 101.

Lai, R.Y., Karadi, G.M., Williams, R.A., 1973. Drawdown at time-dependent flowrate. Water Resour. Bull. 9 (5), 892–900.

Marie, J.R., Hollett, K.J., 1996. Determination of hydraulic characteristics and yield of aquifers underlying Vekol Valley, Arizone, using several classical and current methods. U.S. Geol. Surv. Water-Supply Paper 2453.

Meier, P.M., Carrera, J., Sanchez-Vila, X., 1999. A numerical study of the relationship between transmissivity and specific capacity in heterogeneous aquifers. Ground Water 37 (4), 611–617.

Misstear, D.R., 2001. The value of simple equilibrium approximations for analyzing pumping test data. Hydrogeol. J. 9, 125–126.

Muir-Wood, R., King, G.C.P., 1993. Hydrological signatures of earthquake strain. J. Geophys. Res. 98 (B.12), 22,035–22,068.

Muskat, M., 1937. Flow of Homogeneous Fluids through Porous Media. McGraw-Hill.

Neuman, S.P., 1972. Theory of flow in unconfined aquifers considering delayed response of the water table. Water Resour. Res. 8 (4), 1031–1045.

Neuman, S.P., Witherspoon, P.A., 1969a. Transient Flow of Groundwater to Wells in Multiple-Aquifer Systems, Geotechnical Engineering Report. University of California, Berkeley.

Neuman, S.P., Witherspoon, P.A., 1969b. Theory of flow in a two aquifer system. Water Resour. Res. 5 (4).

Papadopulos, I.S., Cooper, H.H., 1967. Drawdown in a well of large diameter. Water Resour. Res. 3, 241–244.

Patel, S.C., Mishra, G.C., 1983. Analysis of flow to a large diameter well by a discrete kernel approach. Ground Water 21, 573–576.

Rojstaczar, S., Wolf, S., 1992. Permeability changes associated with large earthquakes: an example from Loma Prieta, California. Geology 20, 211–214.

Rorabaugh, M.I., 1953. Graphical and theoretical analysis of step-drawdown of artesian well. Proc. Am. Soc. Civ. Eng. 79.

Rushton, K.R., Holt, S.M., 1981. Estimating aquifer parameters for large-diameter wells. Ground Water 9 (5), 505–509.

Rushton, K.R., Singh, V.S., 1983. Drawdown in large diameter Wells due to decreasing abstraction rates. Ground Water 21 (6), 670–677.

Rushton, K.R., 2003. Groundwater Hydrology. Conceptual and Computational Models. John Wiley and Sons., New York, 416 pp.

Sanchez-Vila, X., Meier, P.M., Carrera, J., 1999. Pumping test in heterogeneous aquifers: an analytical study of what can be obtained from their interpretation using Jacob's method. Water Resour. Res. 35 (4), 943–952.

Schad, H., Teutsch, G., 1994. Effects of the investigation scale on pumping test results in heterogeneous porous aquifers. J. Hydrol. 159 (1–4), 61–77.

Singh, S.K., 2001. Confined aquifer parameters from temporal derivative of drawdowns. J. Hydraul. Eng. ASCE 127 (6), 466–470.

Smith, R.C., 1963. Relation of screen design to the design mechanically efficient wells. J. Am. Water Works Assoc. 55, 609–614.

Srivastava, R., Guzman-Guzman, A., 1994. Analysis of slope-matching methods for aquifer parameter determination. Ground Water 32 (4), 570–575.

Streltsova, T.D., 1972. Unsteady radial flow in an unconfined aquifer. Water Resour. Res. 8, 1059–1066.

Streltsova, T.D., 1973. Flow near a pumped well in an unconfined aquifer under nonsteady conditions. Water Resour. Res. 9 (1), 227–235.

Şen, Z., 1982. Type curves or large-diameter Wells near barriers. Ground Water 20 (3), 274–277.

Şen, Z., 1985. Volumetric approach to type curves in leaky aquifers. J. Hydraul. Eng. ASCE 111 (3), 467–484.

Şen, Z., 1986a. Discharge calculation from early drawdown data large diameter wells. J. Hydrol. 83, 45–48.

Şen, Z., 1986b. Determination of aquifer parameters by the slope matching method. Ground Water 24 (2), 217–223.

Şen, Z., 1986c. Volumetric approach to non-Darcy flow in confined aquifers. J. Hydrol. 87, 337–350.

Şen, Z., 1987. Storage coefficient determination from quasi-steady state flow. Nordic Hydrol. 18, 101–110.

Şen, Z., 1988a. Dimensionless time drawdown plots of late aquifer test data. Ground Water 26 (5), 615–624.

Şen, Z., 1988b. Analytical solution incorporating non-linear radial flow in confined aquifers. Water Resour. Res. 24 (4), 601–606.

Şen, Z., 1989a. Nonlinear flow toward wells. J. Hydraul. Div. ASCE 115 (HY2), 193–209.

Şen, Z., 1989b. Radial flow in vertically graded hydraulic conductivity aquifers. J. Hydrol. Eng. ASCE 115 (12), 1667–1682.

Şen, Z., 1990. Non-linear radial flow in confined aquifers toward large diameter wells. Water Resour. Res. 26 (5), 1103–1109.

Şen, Z., 1994. Hydrogeophysical concepts in aquifer test analysis. Nordic Hydrol. 25, 183–192.

Şen, Z., 1995. Applied Hydrogeology for Scientists and Engineers. CRC Lewis Publishers, Boca Raton, New York, 465 pp.

Şen, Z., 1996. Volumetric leaky aquifer theory and type straight lines. ASCE J. Hydraul. Eng. 122 (5), 272–280.

Şen, Z., Altunkaynak, A., 2003. Variable discharge type curve solutions for confined aquifers. J. Am. Water Resour. Assoc. 40 (5), 1189–1196.

Şen, Z., Wagdani, E., 2008. Aquifer heterogeneity determination through the slope method. Hydrol. Processes 22 (12), 1788–1795.

Şen, Z., Wagdani, E., 2013. Statistical evaluation of parameters in heterogeneous aquifers. Hydrol. Processes 15 (5), 576–581.

Şen, Z., 2009. Spatial Modeling Principles in Earth Sciences. Springer, New York, 351 pp.

Şen, Z., 2010. Fuzzy Logic and Hydrological Modeling. Taylor and Francis Group, CRC Press, Boca Raton, 340 pp.

Şen, Z., 2012. Dimensionless straight line fitting method for hydrogeological parameter determination. Arabian J. Geosci. http://dx.doi.org/10.1007/s12517-012-0783-3.

Şen, Z., 2014. Hydraulic conductivity variation in a confined aquifer. ASCE J. Hydrol., in press. EngJournal of Hydrologic Engineering, vol 19 (3), 654–658.

Şen, Z., 2014. Dimensionless straight line fitting method for hydrogeological parameter estimation. Arabian J. Geosci. 7, 819–825.

Şen, Z., Straight-Line Method Generalization for Aquifer Parameter Estimations. J. Irrig. Drain. Eng. 138 (12), 1082–1087.

Şen, Z., Dabanli, İ., Şişman, E., and Güçlü, Y.S. Hydrogeological parameter estimations by partial type curve matching methodology. Arabian J. Geosci. http://dx.doi.org/10.1007/s12517-013-1205-x

Şen, Z., Al-Baradi, A., 1991. Sample functions on indicators of aquifer heterogeneities. Nordic Hydrol. 22, 37–45.

Theim, G., 1906. Hydrologische Methoden. J.M. Gebhart, Leipzig, 56.

Theis, C.V., 1935. The relation between lowering of the piezometric surface and the rate and duration of discharge of a well using ground water storage. Trans. Am. Geophys. Union Part 2, 519–524, 16th Annual Meeting.

Todd, D.K., 1980. Groundwater Hydrology. John Wiley and Sons, 535 pp.

Tumlinson, L.G., Osiensky, J.L., Fairley, J.P., 2006. Numerical evaluation of pumping well transmissivity estimates in lateral heterogeneous formations. Hydrogeol. J. 14 (1–2), 21–30.

Vogel, H.–J., Roth, K., 2003. Moving through scales of flow and transport in soil. J. Hydrol. 272, 95–106.

Walton, W.C., 1970. Groundwater Resource Evaluation. McGraw-Hill Book Co., New York.

Wood, S.H., Neumann, F., 1931. Modified Mercalli intensity scale of 193. Bull. Seismol. Soc. Am. 21, 277–283.

Yeh, 1987. Aquifer Parameter Estimation with Newton Method.

Zadeh, L.A., 1965. Fuzzy sets. Inf. Control 8, 338–353.

Zlotnik, A., 1994. Well testing with arbitrary production rate. Hydrol. Sci. Technol. 10 (1–4), 178–194.

Unconfined Aquifers

Practical and Applied Hydrogeology
http://dx.doi.org/10.1016/B978-0-12-800075-5.00004-2

Copyright © 2015 Elsevier Inc. All rights reserved.

4.1 UNCONFINED AQUIFER PROPERTIES

An unconfined aquifer may receive recharge directly from the surface after each rainfall event or from the surface waters such as rivers, lakes, ponds, etc. Its water table surface is free to fluctuate up and down, depending on the recharge/discharge rates. Groundwater level starts to rise after a time lag from rainfall occurrence and enough infiltration process. Geologically, there is no overlying impermeable "confining layer" to isolate the groundwater system from direct recharge. The water table in such an aquifer corresponds to the top of the saturation zone, below which pores, fractures, and solution cavities (karst) are filled entirely with water. During the wet (dry) periods, the water table typically rises (drops) since there is abundant water recharge (long periods of droughts). Well penetrations into the unconfined aquifers have to extend below the water table surface in order to withdraw water from the aquifer.

A big problem associated with unconfined aquifers is their extreme susceptibility to contamination (Chapter 5). Should something spill on the land surface, it can vertically infiltrate and make its way downward towards the groundwater storage. The distinctions of unconfined aquifers from other types are due to the following properties.

1. They are shallow aquifers with depths not more than 100 m.
2. They receive direct recharge from surface waters.
3. They have physical water table (piezometric level).
4. The water table is under atmospheric pressure.
5. In general, they have two geological layers permeable above impermeable layer; however, if due to some reason (such as overpumping) the piezometric level falls below the upper confining layer of the confined aquifer, then the same confined aquifer starts to behave like an unconfined aquifer.
6. Wells in shallow unconfined aquifers have large diameters and they may be deepened as the water table approaches the well bottom during dry and drought periods.
7. They are exposed to pollution and contamination more than confined aquifers and small diameter wells.
8. They are suitable for large diameter wells, especially, if the hydraulic conductivity of the aquifer layer is low. In such cases well storage plays significant role, because after the water withdrawal from the well storage and with the drop of water level in the well, the aquifer layer starts to yield groundwater from the whole well periphery. If the well is partially penetrating, which is the most frequent case in unconfined aquifers, then groundwater entrance takes place also from the bottom of the well.
9. After each water abstraction from the main well, there is physical depression cone around the well.

10. Depression cone causes delayed yield due to gravitational force.
11. The piezometric level (water table) remains within the unsaturated layer; if it is at the earth surface level then the unconfined aquifer domain is regarded as completely saturated. However, this is rarely the case in nature. Unsaturated zone provides additional water impoundment facility (natural or artificial recharge) for water storage.
12. Unconfined aquifers are suitable geological formations for horizontal man-made drainage galleries, which are ancient groundwater structures named "qanats," "aflaj" (singular is falaj), or "collector wells," which are the main source of irrigation water apart from wells.
13. Unconfined aquifers are suitable for large diameter injection wells or small diameter injection pipes behind small groundwater recharge dams for groundwater storage augmentation.
14. Subsurface dam constructions are possible in unconfined aquifer geological environments.

4.2 QUATERNARY DEPOSIT AQUIFERS

As already mentioned unconfined aquifers have two-layer geological set ups; pervious layer above lower impervious one. For mathematical derivation of relevant expressions and groundwater calculations each layer is considered as horizontal with uniform saturation thickness in addition to homogeneity and isotropy assumptions. Among the most common media for unconfined aquifers are any types of Quaternary deposits as shortly explained in Chapter 1.

4.2.1 Alluvium Channel Aquifers

Alluviums (sedimentary deposits) at low elevations of each drainage basin are formed as a result of weathering surface water flows along channels; sediment accumulation appears at depressions and along the coastal areas (deltas). Boulders and gravels, which are the coarsest products of erosion, are moved along shorter distances from their sources and deposited in restricted areas in the upstream part of drainage basin. Fluvial gravels are widespread, especially in arid regions, where they fill the beds of intermittent streamflow surface depressions or fault zones. The alluvial fills in arid regions are known as "wadis," which are natural water courses but they are dry most of the time (Şen, 2008). At times, however, they become conveyors of flash floods that carry away large amounts of sediments, leaving marked imprints on the desert landscape. Usually, the medium has a coarse-grained type, as the fine grains are either washed away by flash floods or blown by wind during dry spells. In general, the groundwater is found in Quaternary deposits in addition to the limited amounts in the weathered and underlying fractured zones (Figure 4.1).

They make up potential groundwater reservoirs for local users. The convenient features of a channel are the flood plain, terraces, meander scrolls, swamps, and natural drainage toward the main channel with the result that silt-loaded flood water remains on the flood plain

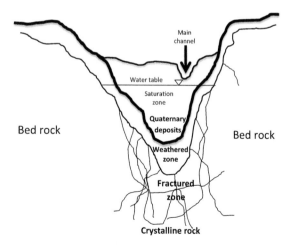

FIGURE 4.1 Wadi alluvial fill cross-section.

swamps. Evaporation of the swamps causes salt deposition, which generates salinity problems for groundwater storage.

In drainage basins the vertical cross-section at any main-channel point yields four different layers that are meaningful for groundwater studies. At the bottom, the crystalline rock that can give water only depending on the degree of fracturing, which is overlain with weathered layer that has been subject to atmospheric and various hydrological effects prior to the deposition of alluviums during geological past. In general, the alluvium is dry in the upstream. On the average, the weathered zone has about 10 m thickness. Other hydrogeological layers are within the alluvium as the saturation and unsaturation zones. Valley alluviums consist mainly of sand and gravely sand in addition to silt at some places. In alluvium aquifers wells usually yield about 5–15 l/s discharge.

Due to intense pumping, the discharge may exceed the recharge causing big drops in the groundwater level. Although the groundwater flows naturally from high elevations, but at places due to excessive and heavy groundwater withdrawal or recharge there are locally high hydraulic gradients, which may cause inverse subsurface flows. Such groundwater level depressions (molds) play the role of subsurface barrier against the natural valley downward groundwater flow.

4.2.2 Alluvial Fan Aquifers

These formations are of fluvial origin and occur where a stream leaves a steep valley and slows down as it enters a plain. They are very noticeable and abound especially in arid and semiarid regions. The growth of alluvial fans was initiated during the Pleistocene epoch when the climate was more humid with intensive rainfalls. Most often the water appears in the form of springs; otherwise it may continue its journey farther downstream where it emerges as surface flow (runoff). At mouths of stream tributaries sand and gravel deposits are formed

with some silt along both sides of large valleys (wadis) in the forms of deltas. They have typical depths of 3–15 m and receive recharge primarily from the fan-building. Alluvial fans generally form a locally thick unconfined aquifer, which overlies fine-grained valley-fill deposits (impervious layers). If they overlie impervious and/or semipervious layers, then leaky or multiple aquifer system may occur.

4.2.3 Sabkhahs (Playa) Aquifers

Any depression or low-lying closed basin without outlet is subject to occasional storm rainfall events and floods, and hence, a temporary pond or wetland comes into existence. In arid regions, evaporation of surface water from these lands leaves salt accumulation remnants at the base of the low areas. In the meantime, water infiltrates into the subsurface layers and may generate unconfined aquifers with salty water quality. Such salt concentration locations especially along the coastal areas are referred to as a "playa" in English and "sabkhahs" in Arabic. During dry periods salt accumulations appear as a crust on the surface. Much of the accumulated salts stem from evaporitic marine sediments outside the basin; many Mesozoic (Triassic, Jurassic) and Tertiary sediments are very rich in evaporates (Şen, 1995).

Water is always associated with sabkhahs as a result of flooding, runoff accumulation, capillary rise, and tidal fluctuation. The sediments that fill sabkhahs consist of sand, clay, silt, and salts in various combinations. Their flat surfaces mark the elevation to which soil moisture rises above the static water surface. Below this surface the materials are damp, wet, or saturated, but above they dry out and blow away.

4.2.4 Coastal Aquifers

These locations are either at the foothills of mountainous areas close to the coastal zones as in many regions or at the downstream ends of

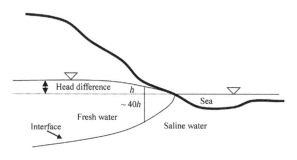

FIGURE 4.2 Coastal unconfined aquifers.

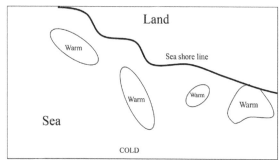

FIGURE 4.3 Temperature difference.

drainage basins that confluence with seas. In the former case the groundwater recharge is rather new with good quality but in the latter there are unconfined groundwater storages with low quality. They are in the form of unconfined aquifers, where there is an interface between the bottom sea born salt and overlying land born fresh waters as in Figure 4.2.

Groundwater resources end up in deep free surface bodies such as lakes, seas, or oceans. Fresh groundwater discharge into sea is possible mostly in fractured and karstic aquifers. Especially karstic geological formations, such as limestone and dolomite include solution cavities that confluence with the sea bottom and these outlets are the major groundwater (fresh) locations along the coastal areas. The fresh water has lighter density than the saline seawater, and consequently, it floats on the saline water. The continuous feeding from the main land gives rise to the formation of fresh water upcoming packets along the seashore at about 50–200 m away from the coastal line.

The seawater fresh sources are results of either solution cavities (especially limestone aquifers, which are referred to as the karstic terrain) or fractures, which are available in sedimentary or volcanic rocks. In addition to these features any type of structural geological elements such as faults, contacts, and unconformities may cause groundwater recharge into the sea bottom. The interface surface is located underneath the land surface and it extends toward the inland with increasing fresh water saturation.

Dispersion is another cause of hydrodynamic forces that leads to groundwater mixture. With the entrance of groundwater at the sea bottom outlet into the main saline water body due to its lighter density, it rises to the surface. Continuous support of the groundwater causes formation of fresh water upcoming with a large base area at the surface of the sea as in Figure 4.3. The locations of these fresh water patches depend on the sea surface conditions that are affected by the sea currents and surface winds.

Heat source is an effective hydrodynamic factor for groundwater mixture. In general, the groundwater resources are comparatively warmer than the seawater in winter periods, whereas in summer the opposite temperature gradients prevail. Preliminary survey by thematic satellite images may help in the identification of these patches.

4.3 FRACTURED ROCKS

As already explained groundwater is available locally under unconfined conditions in fractured rocks (see Figure 4.1). However, fractured unconfined aquifers are also available in extensive volcanic rocks such as basalts (Figure 4.4).

Fractures occur chiefly in dense crystalline or cemented rocks. Major fractures are of

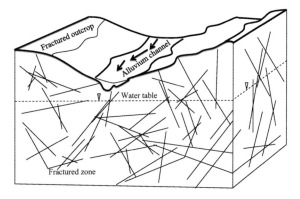

FIGURE 4.4 Fractured rock unconfined aquifers.

supercapillary size and/or fed by tributary fractures that are commonly very small in size. From a geological point of view, the generation of fractures can be attributed to the following different events.

1. Tectonic movements that cause the Earth's crust to deform,
2. Significant erosion of the overburden so that fracturing takes place along different planes, joints, or weak points,
3. Changes in rock volume due to the loss or gain of water, especially in shale or mixture with sands,
4. A change in rock volume due to temperature differences especially in the igneous rocks.

The most significant fracture features are orientation, aperture, density, alignment, and roughness. These characteristics depend on the rock resistance against various internal and external forces. For instance, in hard rocks fractures are extensive, large, and dense as compared with those in comparatively softer rocks, where fractures are of limited extent, relatively small, and less dense. Rock fractures (unless filled with cementing material) facilitate the flow of water through the rock mass. Compared with a porous medium, a fractured medium offers relatively less resistance to water movement. It is logical to say that the longer the

fractures and the smoother their walls, easier is the water transmission. Since fractures dominate the movement of groundwater, knowledge of their geological settings is indispensable in hydrology and hydrogeology.

4.4 KARSTIC MEDIA

The formation of a karstic medium requires very special sediments such as limestone, dolomite, gypsum, halite, anhydride, and other soluble rocks. Through time, because of the rock solubility and the effects of various geological processes that come into play, the terrain develops especially unique topography that is defined as "karst" (Milanovich, 1981). Limestone, no matter how hard, is dissolved by water so that caves or even river channels develop underground while drainage sinks, rifts, and shallow holes appear on the surface all of which leave the land dry and relatively barren. Especially, the fractured limestone, due to dissolution of carbonates by carbonic acid present in the atmosphere and the soil, gives rise to enlarged fractures, conduits, caverns, or caves referred to as a karstic terrain. Very often the surface water is in direct contact with the groundwater through numerous sinkholes and outlets. Figure 4.5 indicates various karstic formations with haphazard solution cavities and their connections.

Contacts of the limestone with water and carbon dioxide, CO_2, lead to chemical solution with various complicated subsurface cavern developments. In scientific terminology, the collection of phenomena characterizing regions of soluble limestone rocks is known as a "karst." It is for this reason that nonclastic formations are called karst reservoirs. The characteristics differ from others in many respects.

1. The process is chemical rather than mechanical or tectonic.
2. Wide channels develop within the karstic reservoirs and enlarge by time.

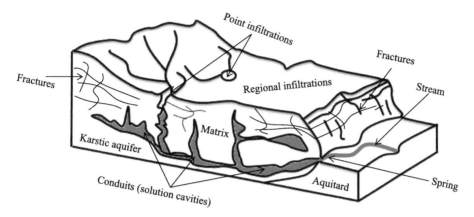

FIGURE 4.5 Karstic formation patterns.

3. Overall permeability decreases rapidly with increasing depth, especially below the water table. However, the zone above the water table has a high transmissivity.

From the hydrogeological point of view, solution cavities provide more storage capability than fractures. The general mechanism in the formation of karstic reservoirs has four successive stages as sedimentation, consolidation, fracturing, and chemical solution. Depending on the spread of these events, the region is either fully or partially karstified. Areas of extensive karstification contain widely spaced solution channels. Fully karstified areas generally have the following reservoir characteristics.

1. an interconnected network of highly permeable channels near the water table,
2. rapidly decreasing karstification with increasing depth below the water table,
3. exceptionally high permeability zones in and around the valleys,
4. a very cavernous unsaturated zone,
5. rather saline water in the lower and less permeable parts of the reservoir,
6. low storage of fresh water after long periods of dry weather.

Larger rock fragments may create blockages in the karstic network that divert water through the granular body or fine fractures, which bring a delayed flux to the main system where the subsurface flow is rapid. In a karstic terrain, the surface runoff may be carried totally or partially as subterranean flow. Springs in limestone terrain can be interconnected to topographic depressions caused by collapsed sinkholes at high elevations. The water-level fluctuation in the sinkhole indicates recharge and discharge to the karstic formations.

Karstic formations are fully developed in humid and semiarid regions, where the lakes are usually interconnected with underlying solution-cavity networks. During wet seasons, this network transports water from the surrounding terrain into the lake, that is, recharge to the lake takes place. In dry seasons, however, the water level is higher than the surrounding subsurface flow level, and therefore, the same network feeds the karstic aquifer. Sometimes the sinkholes in humid regions take all the surface water, which disappears in the underground.

4.5 HYDRAULIC STRUCTURES

There are few practically and frequently used special hydraulic structures for groundwater abstraction from unconfined aquifers.

FIGURE 4.6 Well cross-sections (a) circular, (b) rectangular, (c) irregular.

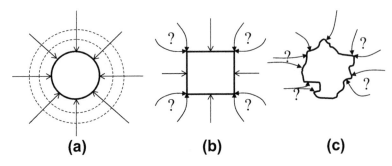

(a)   **(b)**   **(c)**

4.5.1 Vertical Large Diameter Wells

Cross section of water well is important for different functions so as to have the optimum performance during planning, design, operation, and management stages of water supply. In calculating discharge or identification of aquifer parameters, the geometrical shapes of flow and equipotential lines in the vicinity of the well provide restrictive conditions (Chapter 2). For instance, if the well boundary is smooth without any discontinuity (corners, fractures) then the flow lines will end up without any dilemma. Existence of corners on the well periphery gives rise to complications in the stream and equipotential lines, which may cause extra energy loss around the well vicinity as shown in Figure 4.6.

In regular cross sections the equipotential and flow lines take simple regular shapes. For instance, in Figure 4.6(a) the equipotential lines are concentric circles, whereas the flow lines are radial straight lines, but in Figures 4.6(b) and (c) they have irregular shapes. As has been explained in Chapter 3 the calculation of discharge depends on the streamline positions, which is rather difficult to measure or visualize around the irregular well cross sections. The longer the stream lines the more will be the energy loss. Well shapes as in Figure 4.6(c) is used commonly for temporary purposes as shallow wells that are dug by simple excavation machines. They are filled by debris and sedimentation after each flood and water is hauled from these wells by suction pumps for nearby consumption centers for domestic, husbandry, or agriculture purposes.

4.5.2 Collector Wells

As a general setup, these wells have two main parts, namely, a vertical and circular cross-sectional water storage volume acting as a collector and horizontally driven radial drains close to the collector base as horizontal groundwater conveyors from the aquifer (see Figure 4.7). The collector has groundwater seepage neither from its wall nor bottom but only from the radial conveyors through gate valves which are operated from the top.

The water collector is essentially a large diameter, usually 4–5 m, water-tight tank. After the completion of this tank from reinforced concrete with a sealed bottom by pouring a thick concrete plug heavy enough to resist the buoyancy, lateral pipes, made of steel, having 15–20% slot area, are driven horizontally into the groundwater reservoir by special hydraulic jacks. There may be more than one layer of conveyor pipes, especially if the groundwater reservoir is dissected by clay layers preventing free vertical movement.

Collector wells are constructed in alluvial deposit unconfined groundwater reservoirs. They

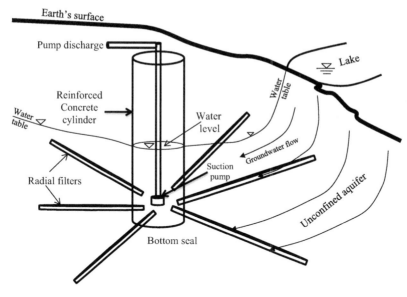

FIGURE 4.7 Collector well.

prove especially useful for large irrigation supplies with a permanent source as a result of continuous recharge from lakes, perennial streams, or water-logged areas irrigated by big canal systems. These wells yield large supplies of water from relatively thin shallow unconfined aquifers. The advantages of these wells are:

1. Availability of large filtered sterilized water supply,
2. Reduced operation and maintenance costs,
3. Suitability in thin aquifers,
4. Reduced wear of pumping due to the entrance velocities, resulting in sand free water.

Collector wells have the following differences from common well types.

1. The initial cost exceeds that of a vertical well.
2. The maintenance cost is less than other wells.
3. They have large yields under low pumping heads.
4. The maximum head difference between the water level in the well and in the adjacent

aquifer occurs at the time of well construction completion.
5. Horizontal drains in various directions have very big chance to intersect with more fractures than vertical borehole wells. Hence, the well productivity increases significantly.
6. Hydraulic head losses, that is, drawdowns around the collector wells are smaller than the drilled wells, and therefore, the sanding problem does not occur in the collector wells.
7. The saline groundwater generally lies at big depths of the saturation thickness. Use of the collector wells will not give rise to groundwater quality deterioration due to the mixing of deep lying saline water with overlying relatively fresh water.

4.5.3 Injection and Scavenger Wells

Artificial recharge is possible through injection wells after each rainfall event behind the surface water impoundment dams for groundwater recharge purposes. In general, these wells help to place surface water into the groundwater

FIGURE   4.8 a)   Static   case,
(b) upconing, (c) scavenger well role.

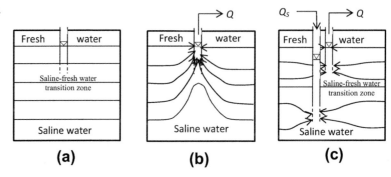

(a)   (b)   (c)

storages in the possible short time durations so as to avoid evaporation losses. These wells have additional wide range of uses such as waste disposal, oil production enhancement, mining, and salt water intrusion prevention through scavenger wells. They can be used as means to control saline water upconing in the coastal aquifers. They stop the advancement of salt water toward inlands and also stabilize the interface between the fresh and salt waters (see Figure 4.8). Through the scavenger wells nonpotable water or alike must be injected. Scavenger wells are useful to prevent the contaminated water from reaching the production well.

An injection well is a vertical pipe in the ground into which water is pumped or allowed to flow. They are frequently used in arid regions to lead the surface water into groundwater reservoirs after each storm rainfall and especially flash floods. There are different types of injection wells.

1. In arid regions, irregularly excavated shallow ditches are prepared for surface water trap and their direct use for a while by tanker transportations and in the meantime the groundwater is also recharged (Figure 4.6(c)). They may be filled with sedimentation partially or fully after runoff, but simple instruments are used for cleaning or digging another series of ditches for the next storm rainfall.

2. In some semiarid regions large diameter wells with radius reaching even to 10 m are dug for groundwater recharge along the Quaternary deposit valleys (wadis).

3. In extremely dry regions occasional flash floods bring abundant surface water (runoff), that is lost either in the desert areas or confluence areas next to seas. These huge amounts of occasional water volumes can be gained for later exploitation, if they can be injected into the unconfined aquifers in the shortest possible time. For this purpose, vertical injection pipes are used behind small surface dams at convenient locations along the valley. The main purpose of such surface dams in arid regions is for speedy groundwater recharge, but they can also be combined with vertical injection pipes for better groundwater recharge efficiency. Figure 4.9 indicates the configuration of surface dams with vertical recharge pipes within the impoundment area.

The injection pipes are located at convenient places within the dam reservoir area. If necessary, prior to dam construction, vertical geophysical sounding prospecting method can be used to explore the subsurface geological compositions and geometrical dimensions. In Figure 4.9, as the surface water level in the

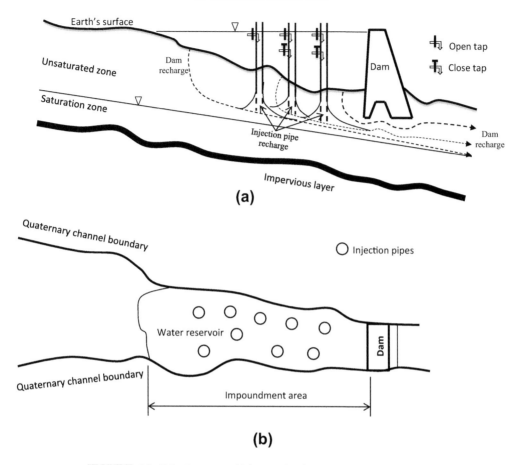

FIGURE 4.9 Injection pipes, (a) longitudinal cross-section, (b) plan view.

reservoir subsides then the lower taps are opened for clean water to infiltrate into the Quaternary deposits for groundwater recharge enhancement. The lower taps are located above the levels of dam bottom sedimentation accumulation.

Another point of worth in practice is to open a ditch along the main channel before dam construction and fill it with coarse materials that will help speedy groundwater flow downward. In this manner, unconfined aquifer will be replenished and downstream settlements will have more water at their disposal and the groundwater quality will improve.

4.5.4 Horizontal Man-Made Conduits—Qanats

A system of wells connected together by a gallery that brings water from the foothills to the plains is called "qanats" as shown in Figure 4.10. They have various local names in different countries. For instance, in Saudi Arabia, they are known as ayn (or ayun in plural) shat-at-ir in Marocco, foggariur in Algeria, falaj (or aflaj in plural) in Oman, shariz in Yemen, and karez in Pakistan, Iraq, and Afghanistan. They were constructed first around 800 BC in the Middle East. In addition to their hydrologic importance,

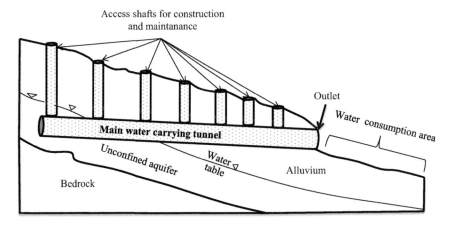

FIGURE 4.10 Qanats.

maintenance, and specific properties, a detailed account of qanats construction in the arid zones is given by Amin et al. (1983).

The first stage in the construction of qanats is to dig large diameter exploratory wells (vertical shafts) along the potential aquifer. These wells are about 300—400 m apart and may reach down to 30—50 m depth and they have large diameter of at least 0.75—1.0 m. The gradient of qanats is usually around 1:1500. Qanats are constructed as gently sloping coarse grain filled tunnels in the alluvial deposits, fills, or fans. The tunnel collects the unconfined groundwater seepage and carries it down the slope until it appears as surface water near the consumption areas. The wide spread usage of qanats in the arid or semiarid regions is due to the following reasons.

1. The groundwater flows under the gravitational force and thus there is no need for extra power source.
2. The evaporation losses are at the minimum level because the flow takes place completely in the subsurface.
3. They provide dependable and sustainable water supply for local domestic and agricultural lands. For instance, Ayn

Zubaydah qanats system was constructed mainly for supplying water to the pilgrims in the Holy places of Makkah City in about 800 and it is functional even today.

4. The groundwater flow is safeguarded against pollution.

EXAMPLE 4.1 QANATS CALCULATIONS

One of the qanats has been constructed in the form of a rectangle canal with 870 m length, 4 m width, and 6 m height. It is assumed to work as an unconfined aquifer. Hence:

1. The base slope of this qanats is 0.01; hydraulic head at the upper entrance point is 13.50 m, at the middle part 13.01 m, and at the outlet point 12.96 m. All the measurements are taken from the qanats base. Hence, at which point (upper, middle, or lower) is the groundwater velocity fastest? Why? What are the necessary assumptions?
2. The porosity, specific retention, and storage coefficient of qanats material are 0.32, 0.13, and 0.09, respectively. What is the total water volume in the qanats itself?

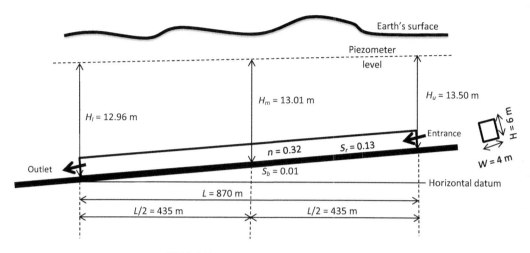

FIGURE 4.11 Representative qanats data.

3. A trace element poured from the upper point reaches the middle and lower points after 14 and 23 min, respectively. What can one say about the two parts (up−middle and middle−down) of the qanats? What type of calculation is necessary?

4. Calculate the upper and lower parts' hydraulic conductivity values.

Solution 4.1

It is possible to visualize the problem after drawing a rough sketch as in Figure 4.11, with the relevant data at hand.

1. It is first necessary to find the hydraulic heads with respect to a common datum, which is taken as the level of the outlet in this case. The slope of qanats provides additional geometric heights to middle and upper point hydraulic heads. With the data given in Figure 4.11, the additional height for the middle (upper) point is $0.01 \times 435 = 4.35$ m (8.70 m). Hence, the common horizontal level hydraulic heads at the middle and upper points of the qanats are $13.01 + 4.35 = 17.36$ m and $13.50 + 8.70 = 22.20$ m, respectively. The hydraulic gradient between upper and middle

points is $i_{U-M} = (22.20 - 17.36)/435 = 0.0111$; between the middle and lower points, $i_{M-L} = (17.36 - 12.96)/435 = 0.0101$.

In the calculations the qanats fill material is assumed homogeneous and isotropic with uniform cross-section. Since the groundwater velocity is directly proportional to the hydraulic gradient, the groundwater velocity is relatively higher at the upper part of the qanats. Such a situation may cause problems so far as the water balance (continuity) equation is concerned, because the input through the middle section into the lower part of the qanats cannot find easy passage from the lower part. In order to alleviate this situation the qanats material in the lower part must be filled with a coarser material so as to help the upcoming groundwater flow to pass easily. Such an assessment is necessary for the discharge continuity along the qanats.

2. In this situation, the qanats canal is assumed to be unconfined. It is necessary to find specific yield value for the water withdrawal calculation. Since porosity and specific retention values are given as 0.32 and 0.13, from Eq. (2.5) one can find simply the specific yield value as $S_y = 0.32 - 0.13 = 0.18$. Water

withdrawal volume, V_w, is equal to the multiplication of qanats volume, V_q, by specific yield ($V_w = V_q S_y$), and hence, $V_w = 870 \times 4 \times 6 \times 0.18 = 3758.4$ m³. Herein, the qanats is assumed as horizontal.

3. Since the time durations to cover the two parts are known from the physical definition of the velocity as, $v = L/t$ (length L, and time, t) then the upper part velocity, $v_{U-M} = 435/(14 \times 60) = 0.518$ m/s; and the velocity at the lower part is $v_{M-L} = 435/(9 \times 60) = 0.8056$ m/s.

4. Hydraulic conductivity can be calculated from the Darcy law as v/i. Finally, the upper and lower parts' hydraulic conductivities can be calculated as $K_{U-M} = 0.518/0.011149 = 56.4615$ m/s and $K_{M-L} = 0.8056/0.0101149 = 79{,}644$ m/s, respectively.

4.6 GROUNDWATER HYDRAULICS

Apart from the geological and hydrogeological differences from the confined aquifers, unconfined aquifers include nonlinearity in the flow lines, and hence, the analytical solutions become more complex. Such solutions are assumed to start with Dupuit (1863) studies.

The first aquifer parameter estimation work from field tests is due to Forchheimer (1898). The upper boundary limitation is the groundwater table (phreatic level) does not allow horizontal flow around the well within the depression cone influence area. Unsaturated zone is a subsurface domain above the water table where the pore spaces contain a combination of air and water. This zone provides ready space for groundwater recharge accumulation.

4.6.1 Dupuit−Forchheimer Assumptions

Quantitative treatment of confined aquifers with fully penetration wells is easier than unconfined aquifers because in confined aquifers the saturation zone is completely defined by impervious layers and the streamflow line geometry remains parallel and does not change with time. The difficulty in unconfined aquifers comes from spatial and temporal changes of saturation thickness near the main well. Temporal changes render the flow not only into a three-dimensional type near the well but also equipotential lines become curves leading to decreases in saturation thickness toward the well as in Figure 4.12. Thus the groundwater velocity direction is not horizontal at every point but

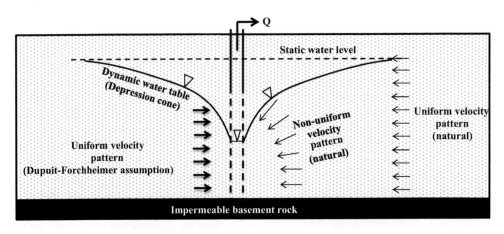

FIGURE 4.12 Flow net around a fully penetration well.

inclined especially in the well vicinity near the water table.

The hydraulic gradient also changes from one point to another within the flow domain. The inclinations and variations cause difficulties in the groundwater movement quantification. In order to avoid such difficulties Dupuit (1863) proposed the following assumptions that were later advanced by Forchheimer (1901).

1. The water is homogeneous and it has the same physical properties in every direction.
2. The flow lines are horizontal, and accordingly, the equipotential lines are vertical. This assumption implies horizontal groundwater velocity and its uniform distribution along the whole saturation thickness. It becomes close to reality if the water table hydraulic gradient is small.
3. The hydraulic gradient at every point along a vertical line is equal to the slope of the free surface. The velocity of flow is proportional to the sine and not the tangent of the water table slope; the flow is horizontal and uniform near the water table.
4. The capillary zone is negligibly small.
5. Theim (1906) assumed that in unconfined aquifers transmissivity is constant provided that the variations in saturation thickness are small compared to the initial saturation thickness.
6. The aquifer material and water are incompressible.

It is obvious that these assumptions neglect completely the vertical flow component. However, the practical value of Dupuit–Forchheimer assumptions lies in the fact that they reduce a three-dimensional flow into a two-dimensional type similar to the confined aquifer case, which is easier to deal with analytically.

Unconfined aquifers are the most frequently used types because of their hand-dug well convenience, easy reach to shallow groundwater levels, and direct replenishment (recharge) possibilities.

4.6.2 Delayed Yield

It is an important parameter of unconfined aquifers and occurs within the depression cone. When pump starts, the water is discharged initially only from the well storage and then aquifer starts to respond. After a short time period, the drawdown increments do not increase as before due to the addition of water to dynamic water table by vertical "gravity drainage," which is a delayed response, and it is also referred to as "delayed yield." This phenomenon creates difficulties at the time of aquifer parameters quantification. A typical TD curve is shown on double logarithmic paper in Figure 4.13(a) which displays delayed yield effect clearly. This curve has three different portions.

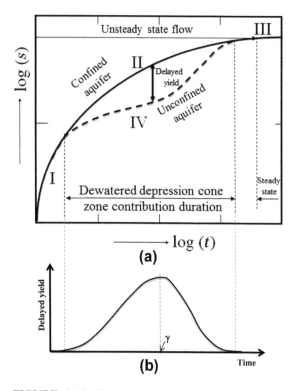

FIGURE 4.13 Time-drawdown relationship, (a) overall response, (b) delayed yield response.

1. During early times, it shows instantaneous aquifer response similar to Theis curve (part I).
2. Comparatively smaller changes take place during medium times, due to the gravity drainage effect (part II).
3. During late times, increase in drawdown exhibits diminishing effect of delayed yield, which results again in the form of the Theis type curve in a delayed manner (part III).

In case of confined aquifer there is no delayed yield effect, and hence, TD curve is continuous with steadily decreasing slope (Chapter 3, Figure 3.22). In unconfined aquifers there is a similarity to confined aquifer at early and late times, but in between the unconfined aquifer type curve has smaller increment rates due to the delayed yield contribution from the depression cone volume.

The change of delayed yield within the cone of depression looks like a Gaussian curve with a single peak (Figure 4.13(b)). It is this curve that causes departure from the confined aquifer response. This brings to one's mind whether it is possible to develop a simple analytical approach for the unconfined aquifer type curve (see Section 4.9). The Gaussian type of response can be written time-wise as a general expression of $e^{-(t-\gamma)^2}$ but should be combined with the depression cone volume, $V_D(t)$, through the specific yield, S_y. Herein, γ is referred to as delayed yield peak parameter. It is suggested to represent the delayed yield volume, $V_{DY}(t)$, as,

$$V_{DY}(t) = S_y V_D(t) e^{-(t-\gamma)^2} \qquad (4.1)$$

4.7 STEADY STATE FLOW TO WELLS

In unconfined aquifers the same set of assumptions is valid as the case of confined aquifer with additional Dupuit–Forchheimer assumptions, which simply imply that the groundwater flow near the well takes place horizontally. Without these assumptions, the groundwater problems in unconfined aquifers are indeterminate theoretically. However, consideration of Dupuit–Forchheimer assumptions provides an opportunity to derive analytical expressions within practically acceptable error limits as good approximations to actual situations.

4.7.1 Radial Flow in Unconfined Aquifer

Compared to the confined aquifer case in Chapter 3, analytical treatment of flow in an unconfined aquifer is more complex because of the saturation thickness variation, and accordingly, the transmissivity decreases within the depression cone as the groundwater flow approaches the well. In reality, the flow lines around the well vicinity are not parallel to each other (Figure 4.14).

In Figure 4.15 the schematic representation of the Dupuit–Forchheimer assumptions is shown.

Since the steady state appears at late times after water withdrawal from the well, the delayed yield effect has already disappeared, and therefore, the pump discharge, Q, is equal to aquifer discharge at any radial distance cylindrical surface. Considerations of the Darcy law and the lateral surface area of such a cylindrical surface gives,

$$Q = 2\pi rhK \frac{dh(r)}{dr}$$

or separation of variables yields,

$$\frac{dr}{r} = \frac{2\pi K}{Q} hdh(r)$$

This expression can be integrated from r_1 to r_2 with corresponding hydraulic heads $h(r_1)$ and $h(r_2)$, respectively. After simple algebraic consideration of the drawdown-hydraulic head relationship at any radial distance as $s(r) = H_o - h(r)$ one can obtain,

$$Q = \pi K \frac{s^2(r_1) - s^2(r_2)}{\ln(r_2/r_1)} \qquad (4.2)$$

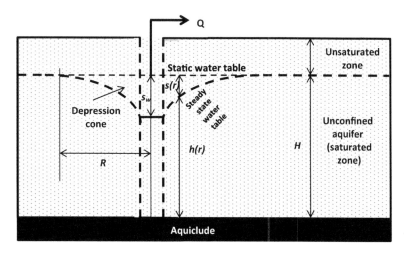

FIGURE 4.14 Radial flows to a fully penetrating well.

This is known in groundwater literature as Dupuit–Forchheimer equation. It is obvious that the relationship between the distance and drawdown is not linear as it was for the confined aquifer (Equation (4.2)) does not give satisfactory results for drawdown measurements at small distances when $r < 1.5H_o$. It is possible to rewrite the last expression mathematically as,

$$Q = \pi K \frac{[s(r_1) - s(r_2)][s(r_1) + s(r_2)]}{\ln(r_2/r_1)} \quad (4.3)$$

Consideration of the average saturation thickness, $m = [s(r_1) + s(r_2)]/2$, one can arrive at the steady state confined aquifer formulation (Eq. (3.15)). Practical use of this expression provides reliable results in cases of small slope water tables. In practice, if the hydraulic head, h_w, in the main well circumference is greater than $0.9H_o$ then the calculation from this expression will differ at the maximum 5%, which is acceptable normally. Consideration of the following points is necessary for successful application of the Dupuit–Forchheimer equation.

1. The drawdown measurements must be taken at large distances (more than $1.5H_o$) and late times (for quasi-steady state).
2. The Dupuit–Forchheimer equation can be rewritten mathematically by considering the average saturation thickness, m, between

Earth's surface

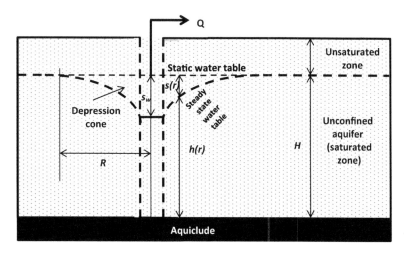

FIGURE 4.15 Dupuit assumption.

two observation wells with no loss of generality as,

$$\left[s(r_1) - \frac{s^2(r_1)}{2m}\right] - \left[s(r_1) - \frac{s^2(r_2)}{2m}\right]$$
$$= \frac{Q}{2\pi T} \ln\left(\frac{r_2}{r_1}\right) \tag{4.4}$$

or succinctly as

$$s'(r_1) - s'(r_2) = \frac{Q}{2\pi T} \ln\left(\frac{r_2}{r_1}\right) \tag{4.5}$$

which is equivalent to the Theim (1906) equation for confined aquifers. Herein, $s'(r_1)$ and $s'(r_2)$ are referred to as the "corrected drawdowns" at distances r_1 and r_2, respectively. Equation (4.5) can be regarded as an equivalent confined aquifer steady state formulation. Even to this correction there is a limitation in the sense that corrected drawdown cannot be less than 10% of the drawdowns observed in the field.

3. If the square of the drawdowns are plotted on a semilogarithmic paper versus the distance on the logarithmic axis, the result should appear as a straight line according to Eq. (4.2). This statement is correct when coupled with the condition in step a. For the domain of distances less than $1.5H_o$, Mansur and Kaufman (1962) presented relationships for predicting $s(r)$. The drawdown $s(r)$ close to a well in an unconfined aquifer can be estimated for $0.3 < r < 1.5H_o$ as,

$$s(r) = \frac{Q\left[0.13\ln\left(\frac{R}{r}\right)\right]\ln\left[10\ln\left(\frac{R}{H_o}\right)\right]}{KH_o}$$

and for $r < 0.3H_o$ as,

In thick unconfined aquifers with small drawdowns $s(t) < 0.10H_o$, the Dupuit–Forchheimer equation is approximated by,

$$s(r) = \frac{Q}{\pi K(H_o + h)} \ln\left(\frac{R}{r}\right) \tag{4.6}$$

By defining the average transmissivity approximately as, $\overline{T} = K(H_o + h)/2 \cong KH_o$, this expression again becomes identical to the Thiem equation given in Eq. (3.15).

EXAMPLE 4.2 UNCONFINED AQUIFER HYDRAULIC CONDUCTIVITY CALCULATION-1

An unconfined aquifer has 20 m saturation thickness and tapped with a constant discharge, $Q = 0.32$ m^3/min through a fully penetrating large diameter well of diameter 3.0 m. Quasi-steady state drawdown is reached after 2 h at two observation wells (OW$_1$, OW$_2$) at radial distances 12.1 and 23.2 m with 2.1 and 1.2 m drawdowns, respectively. What is the hydraulic conductivity of the aquifer and make interpretations.

Solution 4.2

The hydraulic conductivity of the unconfined aquifer can be calculated from Eq. (4.2). The discharge is $0.32 \times 60 \times 24 = 460.8$ m^3/day. The convenient expression of hydraulic conductivity with ordinary logarithm can be written from Eq. (4.2) as,

$$K = \frac{2.3Q}{\pi} \frac{\log(r_2/r_1)}{s^2(r_1) - s^2(r_2)} \tag{4.7}$$

$$s(r) = \frac{Q\left[0.13\ln\left(\frac{R}{r}\right) - 0.0123\ln^2\left(\frac{R}{10r}\right)\right]\ln\left[10\left(\frac{R}{H_o}\right)\right]}{KH_o}$$

The substitution of relevant quantities into this last expression gives,

$$K = \frac{2.3 \times 460.8}{\pi} \times \frac{\log(23.2/12.1)}{2.1^2 - 1.2^2} = 74.21 \text{ m/day}$$

According to the classification in Table 2.6 the aquifer material may have coarse unconsolidated gravel composition.

EXAMPLE 4.3 UNCONFINED AQUIFER HYDRAULIC CONDUCTIVITY CALCULATION-2

In an unconfined aquifer after a long pumping period the quasi-steady state case is reached with discharge 362 m³/day. There are five observation wells at radial distances 8.9, 11.7, 15.1, 19.3, and 27.8 m with corresponding drawdowns as 295, 205, 181, 148, and 136 cm. The radius of influence is 41 m. Hence, calculate the hydraulic conductivity of the aquifer by treating whole data collectively.

Solution 4.3

A close inspection of Eq. (4.2) indicates in general that during a quasi-steady flow in an unconfined aquifer the plot of square difference of drawdowns versus the corresponding ratios of distances should appear as a straight line on a semilogarithmic paper. In case of known radius of influence, R, where drawdown is equal to zero, Eq. (4.2) can be written simply as,

$$Q = \pi K \frac{s^2(r)}{2.3 \log(R/r)} \tag{4.8}$$

which implies in particular that in an unconfined aquifer the square of drawdowns versus R/r on a semilogarithmic paper shows a straight line with slope, $\Delta_{R/r}$

$$\Delta_{R/r} = \frac{2.3 Q}{\pi K} \tag{4.9}$$

From the plot the slope has unit of m², and its substitution into the following equation yields the hydraulic conductivity.

$$K = \frac{2.3 Q}{\pi \Delta_{R/r}}$$

Aforementioned explanations help to determine unconfined aquifer hydraulic conductivity graphically after the completion of the following steps.

1. Plot the ratio of each dimensionless radial distance on the horizontal logarithmic scale axis versus drawdown squares on the vertical linear scale axis (Figure 4.16).
2. Draw the best fitting straight line through the scatter points. This line corresponds to the graphical representation of Dupuit–Forchheimer formula for the homogeneous and isotropic unconfined aquifer under the steady state groundwater flow.
3. Find the slope, $\Delta s_{R/r}$, of the line, which can be read off as equal to 17 m². Hence, the hydraulic conductivity estimate can be calculated from Eq. (4.9) after the substitution values as,

$$K = \frac{2.3 \times 361}{3.14 \times 17} = 15.55 \text{ m}^2/\text{day}$$

EXAMPLE 4.4 DUPUIT FORMULA APPLICATION

In an unconfined aquifer a large diameter well has 60 cm radius and it is pumped with 1.4 m³/min discharge. At 6 and 15 m radial distances the drawdowns are measured as 1.5 m and 0.5 m, respectively. The main well fully penetrates 90 m saturation thickness. What is the hydraulic conductivity value?

Solution 4.4

For the solution Dupuit–Forchheimer formula (Eq. (4.2)) can be used under the light of the Dupuit–Forchheimer assumptions. From Eq. (4.7) the hydraulic conductivity can be calculated as

$$K = \frac{1.4 \times 2.3 \times \log(15/6)}{3.14 \times (1.5^2 - 0.5^2)} = 4.7 \times 10^{-1} \text{ m/min}$$

FIGURE 4.16 Unconfined aquifer square drawdown dimensionless radial distance relationship.

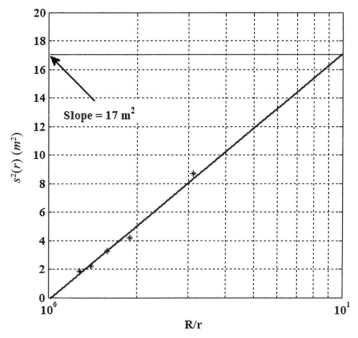

EXAMPLE 4.5 WELL FACE DRAWDOWN CALCULATION

The drawdown at a radial distance of 150 m in an unconfined aquifer is zero after a very long time under 0.5 m³/min continuous water withdrawal. The static water table is 23 m above the impervious layer. The aquifer material hydraulic conductivity is 1.83×10^{-3} m/min and the large diameter well has 0.9 m radius.

1. Determine the drawdown at the well face by considering Dupuit—Forchheimer assumptions.
2. What is the specific capacity of this well?

Solution 4.5

Dupuit—Forchheimer assumptions lead to unconfined steady state radial flow expression as in Eq. (4.2). For calculation in the main well, Eq. (4.2) can be arranged as,

$$s_w^2 = \frac{Q\ln(r_1/r_w)}{\pi K} + s^2(r_1)$$

1. The substitution of the relevant numerical values into this last expression gives,

$$s_w^2 = \frac{0.5 \times 2.3 \log(150/0.9)}{3.14 \times 0.00183} + 0^2 = 1023.9 \text{ m}^2$$

Finally, the drawdown inside the well is 32 m.
2. Specific capacity has been defined in Chapter 2 as the discharge per drawdown inside the main well, and hence, it is $0.5 \times 60 \times 24/32 = 22.5$ m²/day.

EXAMPLE 4.6 TRANSMISSIVITY CALCULATION

A large diameter well (60 cm diameter) penetrates 25 m saturation thickness in an unconfined aquifer. In two observation wells at radial distances 30 and 95 m the drawdowns are measured as 6 and 0.6 m, respectively. The constant pumping duration is 24 h with the discharge 3 m^3/min. Hence, calculate the transmissivity of this aquifer.

Solution 4.6

The hydraulic conductivity of the aquifer can be calculated as in the Example 4.4.

$$K = \frac{3.0 \times 2.3 \times \log(95/30)}{3.14 \times (6^2 - 0.6^2)}$$

$$= 7.11 \times 10^{-2} \text{ m/min}$$

Hence, from the basic definition of the transmissivity one can simply find that $T = 25 \times 7.11 \times 10^{-2} = 1.77 \text{ m}^2/\text{min} = 1.77 \times 60 \times 24 = 2559.6 \text{ m}^2/\text{day}$. According to Table 2.10, the aquifer is highly potential.

4.7.2 Partial Penetration Wells

In case of partial penetration wells the flow lines have curvatures near the water table and within the nonpenetration part of the aquifer. These curvatures increase the difficulty in unconfined aquifer flow analytical derivations even under the steady state conditions (Figure 4.17).

Analytical expression derivations for partial penetrations are almost impossible, but numerical solutions have been achieved by Rushton (2002). Another way of deriving relevant equations for the partially penetrating well calculations is empirical approaches. Logically, the partial penetration will cause more head losses, and hence, bigger drawdowns than the full penetration case under the same conditions. This implies that partial penetration discharge, Q_P, will be smaller than the full penetration pump discharge, Q, under the same conditions. Forchheimer (1898) gave the pump discharge, Q_P, from a partially penetrating well as a percentage of full penetration case by introducing a reduction factor, α,

$$Q_P = \alpha \pi K \frac{h^2(R) - [h^2(r) + t]}{\ln(R/r)} \tag{4.10}$$

in which all of the variables are self-explanatory as in Figure 4.17. The coefficient α is given as follows.

$$\alpha = \left(\frac{L}{H_o}\right)^{1/2} \left(2 - \frac{L}{H_o}\right)^{1/4} \tag{4.11}$$

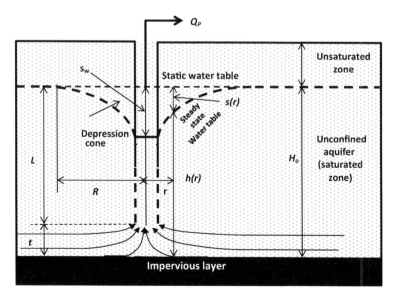

FIGURE 4.17 Partial penetrations and curved flow paths.

For full penetration ($L = H_o$, and hence, $\alpha = 1$, $t = 0$) Eq. (4.10) becomes equivalent to Eq. (4.2). The graphical form of Eq. (4.11) is given in Figure 4.18.

Kozeny (1933) gave an approximate empirical formula for discharge calculation in a partially penetrating well as,

$$Q_P = \frac{2\pi T s_w\, \alpha\left[1 + 7\left(\frac{r_w}{2\alpha m}\right)^{1/2} \cos\left(\frac{\pi \alpha}{2}\right)\right]}{\ln\left(\frac{R}{r_w}\right)} \quad (4.12)$$

where s_w is the steady state drawdown in the main well; α fractional part of the full saturation thickness, m, tapped by main well; r_w is the well radius; and R is the radius of influence. The apparent, T', and true, T, transmissivity values are related as follows.

$$\frac{T}{T'} = \frac{c}{\alpha} \quad (4.13)$$

where c is discharge correction factor given as,

$$c = \left[1 + 7\left(\frac{r_w}{2\alpha m}\right)^{1/2} \cos\left(\frac{\pi \alpha}{2}\right)\right]^{-1} \quad (4.14)$$

Muskat (1937) suggested that the flow of water toward a partially penetrating well in an isotropic unconfined aquifer becomes almost radial at a distance equal to twice the aquifer thickness. However, in anisotropic aquifers Jacob (1963) suggested the distance, r, for radial flow to be calculated as $r = 2m\sqrt{K_h/K_v}$. He also gave a method for correction, δ, steady state drawdown in partially penetrating well screening the top or bottom of a confined aquifer as,

$$\delta = \frac{s_P}{Q_P/2\pi T}$$

$$= \frac{\left(\frac{2}{\pi\alpha}\right) \sum_{n=1}^{\alpha}\left[(\pm 1)^n K_0\left(\frac{n\pi r}{m}\right)\sin(n\pi\alpha)\right]}{n} \quad (4.15)$$

in which s_P, is the drawdown correction as the difference between the observed drawdown

FIGURE 4.18 Reduction factor.

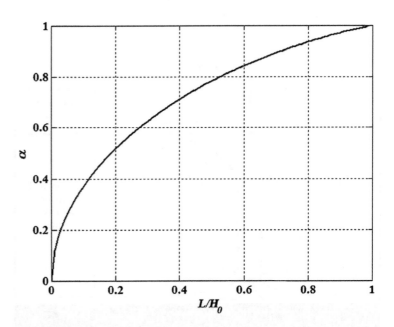

and the drawdown that would have occurred for full penetration. The plus (minus) sign is for the drawdown distribution along the top (bottom) of the aquifer, (Figure 4.19).

The graphical presentations of Eq. (4.15) for top and bottom are given in the same figure. The drawdown correction is given as,

$$s_p = \delta \left(\frac{Q_P}{2\pi T} \right) \qquad (4.16)$$

This correction method is not applicable if the well taps the aquifer somewhere in between the top and bottom. The application procedure involves the following steps.

1. Plot steady state drawdowns in different observation wells versus radial distances from main well. First, preliminary T value is determined either from Dupuit–Forchheimer

steady state formulation or distance-drawdown (DD) CJ method, (Chapter 3, Section 3.9).
2. Calculate values of $\pi r/m$, α and then δ is read for each observation well from Figure 4.19.
3. Calculate the corrected drawdown from Eq. (4.16).
4. Recalculate the value of T by any method utilizing the corrected drawdown values.
5. Compare the transmissivity in steps (a) and (d). If they are different by ±5% relative error (Eq. 3.11) then repeat the whole procedure starting with the recalculated transmissivity value.

Later, by the use of potential theory graphically Nahrgang (1954) related the discharges of fully and partially penetrating wells as,

$$Q_p = \alpha Q \qquad (4.17)$$

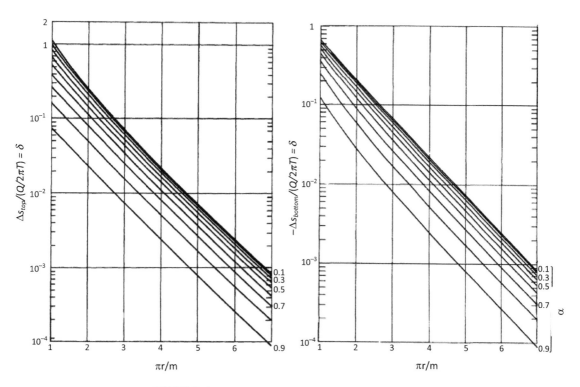

FIGURE 4.19 Drawdown correction (a) top, (b) bottom.

in which α is a factor depending on the two ratios, namely, L/H and h/H where L is the penetration length. Figure 4.20 represents the graphical relationship between h/H and α for a set of given L/H values.

Another alternative is due to purely theoretical consideration of Szechy (1961) who divided the whole flow domain into two parts as follows.

1. If the flow domain along the penetration length is considered as a full penetration then the discharge, Q_1, is expressed as

$$Q_1 = \pi K_h \frac{H^2 - h^2(r)}{\ln\left(\frac{R}{r}\right)} \qquad (4.18)$$

He also suggested that in this formula horizontal hydraulic conductivity, K_h, should be substituted for calculations.?

2. The flow domain beneath the penetration thickness is assumed to have the vertical hydraulic conductivity, K_v, and the discharge, Q_2, contribution is,

$$Q_2 = \frac{2t}{w+1} \pi K_v \frac{H_o^2 - h^2(r)}{\ln\left(\frac{R}{r}\right)} \qquad (4.19)$$

in which w is a factor assuming values in the range -5–10. The total discharge is then,

$$Q_p = Q_1 + Q_2 \qquad (4.20)$$

Finally, Hantush (1964) recommended the use of Eq. (4.20) provided that the pumping time to reach the steady state is short or that the aquifer is relatively thick and the observed drawdowns, s_0, are corrected, s_c, according to the following expression.

$$s_c = s_0 - \frac{s_0^2}{2L} \qquad (4.21)$$

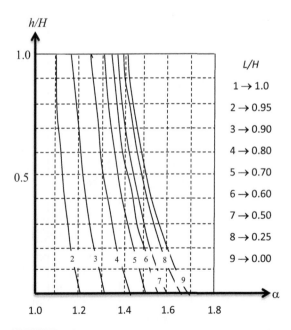

h/H

L/H

1 → 1.0
2 → 0.95
3 → 0.90
4 → 0.80
5 → 0.70
6 → 0.60
7 → 0.50
8 → 0.25
9 → 0.00

FIGURE 4.20 Partially penetrating well dimensionless relationships.

EXAMPLE 4.7 PARTIAL PENETRATION WELL IN UNCONFINED AQUIFER

The first numerical example for the partially penetrating wells is presented by Jacob (1963). Assume that a well with an effective radius of 0.3 m and screened in the top 12 m of a 30 m thick unconfined aquifer is pumped at a constant rate of 3800 m³/day for a period long enough to establish quasi-steady-state condition. The drawdown in the main well is 6.2 m with negligible screen loss. Hence, estimate:

1. The hypothetical drawdown that would occur in a well fully screened in the aquifer,
2. The ratio of true to apparent transmissivity,
3. The hypothetical yield.

Solution 4.7

1. First of all the penetration fraction is $\alpha = 12/30 = 0.40$ and from Eq. (4.14) one can calculate

$$\frac{1}{c} = 1 + \frac{7\cos(3.14 \times 0.40/2)}{\sqrt{2 \times 0.40 \times 30/0.3}} = 1.633$$

By substitution of the relevant values into Eq. (4.12) one can obtain,

$$\frac{s_w \alpha}{c} = 6.2 \times 0.40 \times 1.6333 = 4.048 \text{ m}$$

This expression provides the basis for calculation of the hypothetical drawdown as,

$$s_w = \frac{4.048 \times 1.633}{0.4} = 16.52 \text{ m}$$

2. Ratio of true, T, to apparent, T′, transmissivity can be obtained from Eq. (4.13) as,

$$\frac{T}{T'} = \frac{c}{\alpha} = \frac{0.61}{0.40} = 1.5$$

3. Hypothetical yield, Q, from a fully penetration well can be calculated from Eq. (4.17) as,

$$Q = \frac{Qp}{\alpha} = \frac{3800}{0.4} = 9500 \text{ m}^3/\text{day}$$

4.8 GROUNDWATER RECHARGE

The most significant distinction of unconfined aquifers is their direct recharge intake capabilities and for this reason especially in many regions they are used as subsurface groundwater storages. It is, therefore, useful to be able to model these recharges regularly as radial flow toward wells or subsurface linear flow under the effects of recharge in unconfined aquifers.

4.8.1 Radial Flow in Unconfined Aquifer with Recharge

Unconfined aquifers are the most prone geological formations subject to groundwater recharge. Dupuit—Forchheimer assumptions help to solve unconfined aquifer problems with significant water table slopes around the pumping well (Figure 4.21).

The uniform recharge per unit area is q and the water table does not have significant slope according to Dupuit—Forchheimer assumptions.

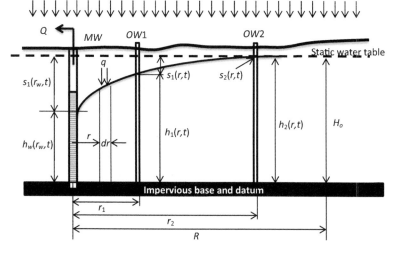

FIGURE 4.21 Unconfined aquifer-well configurations.

The pump discharge, Q, will be equal to the summation of radial aquifer discharge, $Q_a(r)$, at radial distance, r, and the infiltration contribution, $Q_i(r)$, through annulus area of $2\pi r dr$ as $Q_i(r) = 2\pi r dr q$. Hence, water balance concept gives,

$$Q = 2\pi r h(r) K \frac{dh(h)}{dr} + 2\pi r q dr \qquad (4.22)$$

Since the flow within the aquifer is under quasi-steady state, there is no discharge change, $dQ/dr = 0$, and one can write from the previous expression,

$$\frac{d}{dr}\left[r h(r) K \frac{dh(h)}{dr} \right] = -rq \qquad (4.23)$$

Preliminary solution of this expression is possible for small hydraulic head falls, which implies that the saturation thickness is almost constant, and hence, $h(r) \approx H_o$, consequently $T = H_o K$, i.e., a constant. The more the infiltration rate per area the smaller will be the hydraulic head fall in practice. In this case,

$$\frac{d}{dr}\left(r \frac{dh}{dr} \right) = -\frac{qr}{T} \qquad (4.24)$$

The first integration operation leads to,

$$r \frac{dh}{dr} = -\frac{qr^2}{2T} + c_1$$

where c_1 is the integration coefficient and then the second integration after the division of both sides by r gives,

$$h = -\frac{qr^2}{4T} + c_1 \ln r + c_2$$

The boundary conditions are $(r = R, dh/dr = 0)$, since after radial distance, R, there is no groundwater movement. This condition yields $c_1 = qR^2/2T$. Another condition is $(r = R, h = H_o)$, where H_o is the initial saturation thickness. The application of this condition with the

substitution of the first one yields after the necessary arrangements to,

$$H_o - h = -\frac{q}{2T}\left[\frac{R^2 - r^2}{2} + R^2 \ln\left(\frac{r}{R}\right) \right] \qquad (4.25)$$

When the pump discharge comes completely from the recharge, the pump discharge will be equal to the recharge amount over the radius of influence area, which means that $Q = \pi R^2 q$, and the substitution of this quantity into the previous expression with simple algebra gives,

$$s = \frac{Q}{2T}\left[\ln\left(\frac{R}{r}\right) + \frac{R^2 - r^2}{2R^2} \right] \qquad (4.26)$$

The second term within the brackets is due to enhanced groundwater movement toward the well within the aquifer.

EXAMPLE 4.8 WATER WITHDRAWAL FROM A WELL DURING GROUNDWATER RECHARGE

In an area of Quaternary deposits, the groundwater is available in the unconfined aquifer with 24 m saturation thickness. The water table is 5 m below the Earth's surface and the aquifer areal extent is 43 km^2. Aquifer porous medium material specific yield is 0.08. A storm rainfall over the whole area gave rise to direct infiltration duration for 13.5 h with infiltration rate per square meter as 0.09 m/day.

1. Calculate the abstractable groundwater volume over the water table after the storm rainfall end.
2. If the groundwater abstraction starts with direct infiltration event simultaneously, how much water could be abstracted from the groundwater storage only?
3. Compare this amount with the case of without rainfall infiltration event.

Solution 4.8

1. The available volume in the unsaturated zone is $43 \times 10^6 \times 5 = 215 \times 10^6 \, m^3$. The abstractable volume can be found by multiplying this amount by the specific yield, and hence, $215 \times 10^6 \times 0.08 = 17.2 \times 10^6 \, m^3$.

2. The total abstractable groundwater down to the last drop from the saturation layer is $43 \times 10^6 \times 24 \times 0.08 = 3.44 \times 10^6 \, m^3$. Additionally, infiltration continues for 13.5 h at a rate $0.09 \, m/day/m^2$, which yields $43 \times 10^6 \times 0.09 \times 13.5/24 = 2.18 \times 10^6 \, m^3$ of water infiltration. Hence, the total possible volume is $3.44 \times 10^6 + 2.18 \times 10^6 = 5.66 \times 10^6 \, m^3$.

3. The ratio between the case without rainfall and with rainfall is $3.44 \times 10^6/5.66 \times 10^6 = 0.61$, which means that 61% of water comes from the aquifer storage.

4.8.2 Linear Flow in Unconfined Aquifer with Recharge

Linear groundwater flow formulations are necessary for evaluation of the subsurface dam calculations or groundwater flow after surface dams construction for groundwater recharge (Figure 4.9). Subsurface dams are bound to increase in arid and semiarid regions in future, because they enhance groundwater recharge, and hence, protect against evaporation losses.

It is possible to derive expression for linear flow in unconfined aquifer first by considering that there is not significant slope in the hydraulic head. Hence, instead of the radial slice, dr, as in Figure 4.19, the linear slice, dx, along the main channel is considered. In steady or quasi-steady state groundwater flow case, the discharge change within an infinitesimally small distance, dx, is equal to the vertical groundwater recharge. This last statement can be expressed in mathematical symbols logically as,

$$\frac{dQ}{dx} + qdx = 0 \qquad (4.27)$$

where Q is the discharge that passes from the whole aquifer saturation thickness but through unit width and q is the recharge per area. Q can be written for unit channel width as the multiplication of the saturation thickness by Darcy's law,

$$Q = h\left(K\frac{dh}{dx}\right)$$

The substitution of this expression into Eq. (4.27) leads to,

$$K\frac{d}{dx}\left(h\frac{dh}{dx}\right) = -qdx$$

Mathematically, the term in the parenthesis can also be written succinctly as, $(1/2)d(h^2)/dx$. Two subsequent integration operations lead to,

$$h^2 = -\frac{q}{2}\frac{x^2}{K} + c_1\frac{x}{K} + c_2\frac{1}{K}$$

The boundary conditions are $(x = 0, h = h_1)$ and $(x = L, h = h_2)$. The substitution of these conditions into this last expression yields finally,

$$h^2 = -\frac{q}{2K}x^2 + \left[\frac{K}{L}\left(h_2^2 - h_1^2\right) + \frac{q}{x}L\right]\frac{x}{K} + h_1^2$$

$$(4.28)$$

If the recharge does not exist then this expression becomes for $q = 0$ as,

$$h^2 = h_1^2 - \frac{h_1^2 - h_2^2}{L}x$$

4.9 UNSTEADY RADIAL FLOW IN UNCONFINED AQUIFERS

The hydrogeological parameters of interest in confined aquifers are hydraulic conductivity, specific storage, specific yield, and delayed yield.

For determination of these parameters there are varieties of analytical solutions involving different conceptualizations and simplifying assumptions. Accordingly these assumptions limit their applicability. This section presents scientific and engineering thoughts concerning groundwater flow toward a well in unconfined aquifers.

4.9.1 Confined Aquifer Approximation

The basic idea of these models stems from a practical question as how can one apply confined aquifer models to the unconfined aquifers? It is necessary that three-dimensional flows should be approximately planar-radial flow. For the application of available models unconfined aquifer must have approximately similar physical conditions as confined aquifers. The following points are worthy to notice in the applications.

1. In rather thick saturation thickness, m, the maximum drawdown, s_M, is relatively small. In practice, $(s_M/m) \times 100 < 10$ is a good criterion for the application of confined aquifer models to unconfined aquifers. Under these conditions, the three-dimensional flow is approximated as radial flow and gravity drainage contribution is not significant.
2. Based on the theoretical studies of Boulton (1963), the TD record must be taken at observation wells.
3. Boulton (1963) gave the minimum time interval after the start of pumping for unconfined aquifer models to be applicable as,

$$t \geq \frac{5S_y m}{K} \qquad (4.29)$$

in which S_y and K are the specific yield and hydraulic conductivity, respectively. With this condition, CJ methods (TD, DD and composite-drawdown (CD); Chapter 3, Section 3.9) become applicable for unconfined aquifers at large times.
4. For thin saturation thickness aquifers, Jacob (1940) suggested correction of drawdowns

prior to application of confined aquifer models (Chapter 3, Section 3.9). Hence, the equivalent confined aquifer drawdown, s_c, is expressed as,

$$s_c = s_u - \frac{s_u^2}{2m} \qquad (4.30)$$

where s_u is the observed unconfined aquifer drawdown and m is the initial saturation thickness the aquifer. He concluded that if $s_u^2/2m < 3 \times 10^{-3}$ m then the correction in Eq. (4.30) is not necessary, because the condition in (1) is satisfied only in this case.

None of the above mentioned conditions guarantee the full applicability of confined aquifer models. Most often, in practice, there are difficulties as far as the physical plausibility of the final parameter estimates is concerned.

4.9.2 Boulton Model

Unconfined aquifer-well configuration is given in Figure 4.22, where below the static water level, after the pump start, there is a three-dimensional groundwater flow toward the pump well.

Boulton (1963) gave a detailed treatment of groundwater flow problem toward wells in unconfined aquifers with specific initial and boundary conditions. The assumptions listed for the Theis model are the same except that the aquifer is unconfined and shows delayed yield phenomenon due to gravity drainage from dewatered zone. There are three distinct portions in TD plots of unconfined aquifers.

1. Initial portion covers only a short period after the start of pumping as indicated in Figure 4.13 by portion I. The response of the aquifer is the same as for the confined aquifer. There is no significant delayed yield in this portion because the depression cone has not yet expended for the gravity drainage to take place toward the dynamic water table.

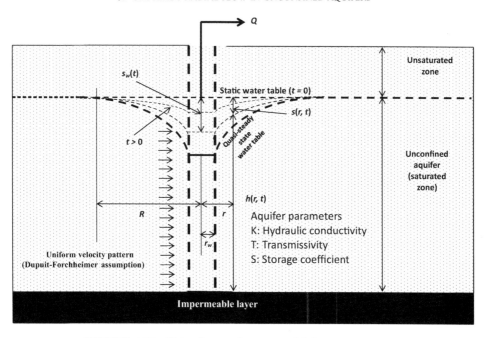

FIGURE 4.22 Unsteady state flows toward fully penetrating well.

2. After this first portion the depression cone expands rather rapidly and accordingly effective gravity drainage takes place vertically toward the saturation zone. There is a decrease in the drawdown rate, which takes place along portion IV in Figure 4.13. The amount of gravity drainage decreases with time.

3. At later stages the increase in the depression cone volume gradually decreases due to the steady or quasi-steady state flow case. The last part of the type curve will return back to the continuation of the confined aquifer type curve as part III in Figure 4.13. The confined aquifer type curve is composed of I−II−III parts collectively whereas unconfined aquifer TD response takes place along I−IV−III trace.

According to given explanations the initial and final portions are expressible in terms of Theis type curves; however, the intermediate portion needs a special attention. Boulton (1963) produced a semiempirical mathematical solution that produces all three portions in an unconfined aquifer. A critical point in his derivation is that the delayed yield is not related to any physical phenomenon. In his approach early and late TD relationships appear according to Theis solution with modified early (late) well function, $W(u_E, r/\beta)$ $[W(u_L, r/\beta)]$ and dimensionless time factor u_E, (u_L) as follows.

$$W(u_E, r/\beta) = \frac{4\pi T_E}{Q} s_E \qquad (4.31)$$

where β is referred to as the drainage factor and defined as,

$$\beta^2 = \frac{T_E}{S_L} \qquad (4.32)$$

where $1/\beta$ is called as Boulton "delay index," which is an empirical constant. For the early portion, the dimensionless time factor is,

$$u_E = \frac{r^2 S_E}{4 t_E T_E} \qquad (4.33)$$

S_E is early time storage coefficient when the unconfined aquifer acts as confined aquifer. The later portion of the TD plot is modeled through the well function,

$$W(u_L, r/\beta) = \frac{4\pi T_L}{Q} s_L \qquad (4.34)$$

and dimensionless time factor as,

$$u_L = \frac{r^2 S_L}{4t_L T_L} \qquad (4.35)$$

in which S_L is the specific yield of the unconfined aquifer. The importance of the vertical flow component is directly related to the magnitude of storage ratio, $\eta = S_L/S_E$.

1. As the value of S_E approaches zero ($\eta \rightarrow \infty$) the duration of the first portion becomes infinitesimally small.
2. As S_E assumes big values ($\eta \rightarrow 0$), the time duration of the first portion increases so that

the aquifer behaves as a confined aquifer with storativity, S_L.

The above mentioned formulas are valid theoretically for $\eta \rightarrow 0$, however, practically for $\eta > 100$. Otherwise the intermediate section is not horizontal and the TD curve for this portion is given with an expression similar to the steady state leaky aquifer formulation. The necessary tables for constructing Boulton type curves are available in any textbook such as (Bear, 1979; Batu, 1998). In Figure 4.23 type curves are presented for a set of parameters.

The type curves are bounded by early and late Theis type curves and their applications require two matching procedures, namely, for early and late field data by taking into consideration connection through the moderate data values similar to Theis case (Chapter 3, Section 3.6). For the early field data matching point $(s_E)_M$, $(t_E)_M$, $(1/u_E)_M$, and $W_M (u_E, r/\beta)$ are read from the type curve and field data sheets. The most

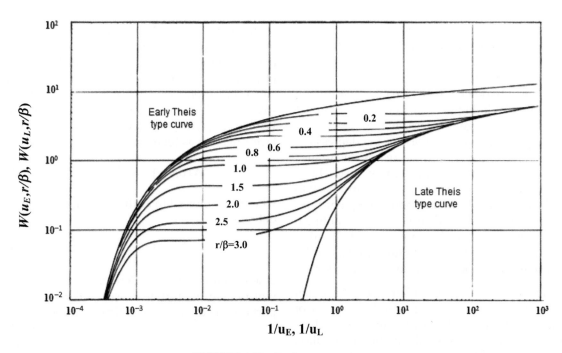

FIGURE 4.23 Boulton type curves.

representative curve label, $(r/\beta)_M$, is also recorded. Substitution of these values into Eqs (4.31) and (4.33) with known Q yields T_E and S_E, respectively. Then the field data sheet is moved to the right and matched with late time type curve with the same $(r/\beta)_M$ label. The coordinates corresponding to the second match point are read as $(s_L)_M$, $(t_L)_M$, $(1/u_L)_M$, and $W_M (u_L, r/\beta)$. Rearrangement of Eqs (4.34) and (4.35) and substitution of the match point coordinates lead to the determination of T_L and S_L, respectively. For the successful application, early and late time calculations should give almost the same value for the transmissivity that is, $T_E \cong T_L$. However, as discussed in Chapter 2 specific yield and storage coefficient are different such that always $S_L > S_E$.

EXAMPLE 4.9 BOULTON METHOD UNCONFINED AQUIFER TEST

A well fully penetrates an unconfined aquifer of 45 m thickness and it is pumped at a constant discharge of 4.9 m³/min. The drawdown registrations in an observation well at 96 m from the pumped well are given in Table 4.1. Calculate the aquifer parameters.

Solution 4.9

First TD data in Table 4.1 are plotted on a double logarithmic paper for the classical matching procedure. The early and late type curve matching procedures are given in Figure 4.24. The calculations are shown succinctly in Table 4.2.

4.9.3 Neuman Model

Neuman (1972, 1973) also reproduced all three portions of the TD curve in an unconfined aquifer without any empirical constant definition. He produced a type curve method very similar to Boulton's model. Although Boulton's method fits field data quite well, it nevertheless fails to provide insight into the physical nature of the delayed yield phenomenon. Neuman's method differs from Boulton by considering well defined physical parameters of the flow phenomenon. This method accounts for the entire delayed yield phenomenon by regarding the following physical points.

1. The water table moves as a material boundary or free surface.
2. The unsaturated flow above the water table has very small influence on the TD response of unconfined aquifer during gravity drainage. As a result, the flow contribution to the water table from dewatered zone is neglected, that is, unsaturated zone is not at all considered.
3. The elastic storage in the aquifer is assumed to play the main role in the groundwater flow toward the well due to the compaction of the aquifer and expansion of water.
4. The groundwater flow to well is axially symmetrical with three-dimensional flow patterns.

Besides, Neuman's method takes into account the aquifer's anisotropy and the partial

TABLE 4.1   Unconfined Aquifer Test Data

Time (min)	Drawdown (m)	Time (min)	Drawdown (m)
1.1	0.004	18	0.068
1.34	0.009	21	0.07
1.7	0.015	26	0.073
2.5	0.03	31	0.075
4	0.047	41	0.078
5	0.054	51	0.081
6	0.061	65	0.085
7.5	0.062	85	0.088
9	0.064	115	0.091
14	0.066	175	0.098

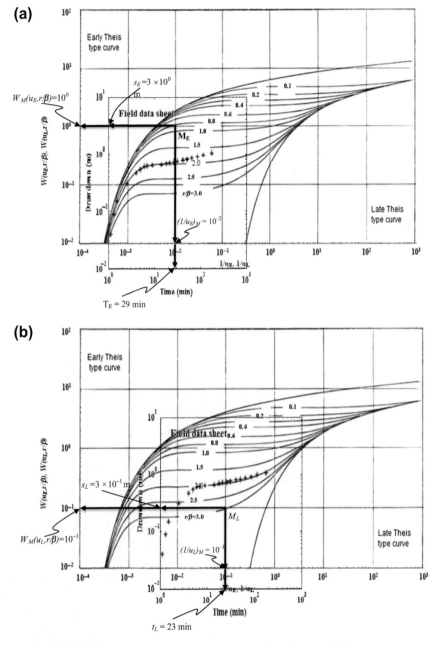

FIGURE 4.24   Boulton type curve matching (a) early times, (b) late times.

TABLE 4.2 Boulton Method Parameter Calculation

	Match Point Coordinates				Parameters Estimate Curve		
	Field Sheet		Type Curve Sheet		Equations (4.31) and (4.34)	Equations (4.33) and (4.35)	r/β
Early times	s_E (m)	t_E (min)	$1/u_E$	$W(u_E, r/β)$	T_E (min/day)	S_E (m)	
	3.00	29.0	1.0	1	187	1.6×10^{-3}	2.0
Late times	s_L (m)	t_L (min)	$1/u_L$	$W(u_L, r/β)$	T_L (min/day)	S_L (m)	
	0.30	23.0	0.1	1	187	1.3×10^{-3}	2.0

penetration of well. His rather complex solution can be written simply for practical purposes, in general, for early times as,

$$W(u_E, u_L, β) = \frac{4\pi T_E}{Q} s_E \qquad (4.36)$$

in which $W(u_E, u_L, β)$ is the unconfined well function and $β = r^2/m^2$. The complete set of type curves is given in Figure 4.25.

For the early portion of Newman type curves the dimensionless time factor is,

$$u_E = \frac{r^2 S_E}{4t_E T_E} \qquad (4.37)$$

where S_E is the elastic storativity responsible for the critical release of water to the well. Similarly, for late time this type curve approaches asymptotically to the Theis curve but with the new definition of parameters as,

$$W(u_L, β) = \frac{4\pi T_L}{Q} s_L \qquad (4.38)$$

and

$$u_L = \frac{r^2 S_L}{4t_L T_L} \qquad (4.39)$$

in which S_L is the specific yield responsible for the delayed release of water to the well. Aquifer parameters can be obtained through a similar type curve matching to the Boulton method as explained above.

EXAMPLE 4.10 NEUMANN TYPE CURVE

Neuman type curve matching is similar to the Boulton method and the same data set in Table 4.1 is adopted for numerical application.

Solution 4.10

The resulting match point coordinates for early and late times are given in Figure 4.26 with the necessary calculations in Table 4.3.

Independently from Neuman, Streltsova (1972,1973) has also obtained type curves for drawdowns in unconfined aquifer. She had a different physical reasoning than Neuman as follows:

1. At any point in the aquifer two types of heads are considered, namely, the free surface of water table and the average head along the depth. The free surface head is greater than the average head. This renders the three-dimensional flow into the two-dimensional axisymmetric flow that depends on the radial distance from the well.
2. The water table is considered as a free surface and there is no delayed yield in the unsaturated zone.
3. The rate of vertical gravity drainage is linearly proportional to the difference between the free surface and average heads.

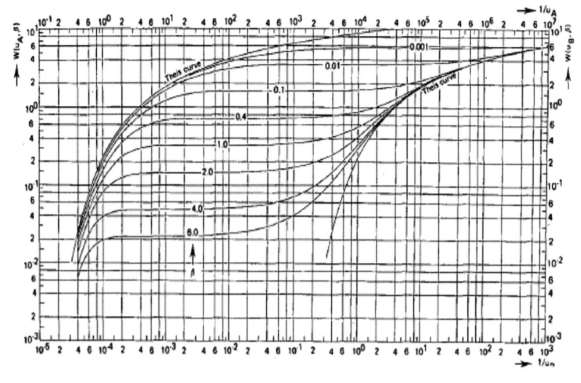

FIGURE 4.25 Newman type curves.

Hence, Neuman and Streltsova had a similar physical look to the problem but her mathematical treatment is different as a result of which estimated aquifer parameters by two methods differ slightly.

The methods proposed by Boulton, Neuman, and Streltsova yield practically identical values of transmissivity, storativity, and specific storage for known aquifer thickness but Boulton's method does not yield a value for vertical permeability, instead yields only the lumped parameter of delayed yield index, β.

4.9.4 Unconfined Aquifer Parameter Estimation Method—Quasi-Steady State Flow

Aquifer potentiality can be estimated on the basis of its transmissivity and storage characteristics (Chapter 2). The valid relationships between discharge and drawdown in an unconfined aquifer under unsteady state flow condition are given by the following expression (Bear, 1979; p. 334).

$$H_o^2 - h^2(r) = \frac{Q}{2\pi K} W(u) \qquad (4.40)$$

where H_o represents the initial saturation thickness, $h(r)$ is the average level of water table at any measurement point, Q is the discharge of the well, and $W(u)$ is the well function as defined by Eq. (3.31). It should be noted that the formulation is valid provided that both H_o and $h(r)$ are measured from the same datum. For approximately $u < 0.05$, Eq. (3.46) becomes valid. Substitution of Eq. (4.47) into Eq. (4.40) yields,

$$H_o^2 - h^2 = \frac{Q}{2\pi K}[-0.5772 - 2.3 \log(u)] \qquad (4.41)$$

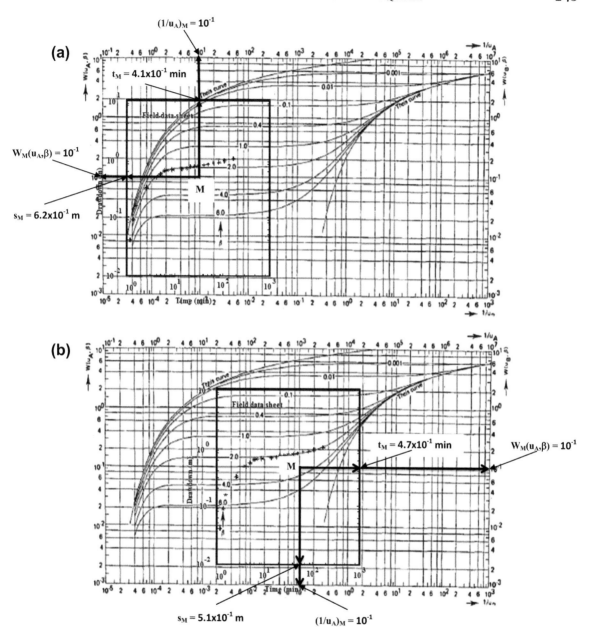

FIGURE 4.26 Neuman type curve matching (a) early times, (b) late times.

TABLE 4.3 Neuman Method Parameter Calculation

	Match Point Coordinates				Parameters Estimate Curve		
	Field Sheet		**Type Curve Sheet**		Equations (4.36), (4.38)	Equations (4.37), (4.39)	β
Early times	s_E (m)	t_A (min)	$1/u_A$	$W(u_A, \beta)$	T_E (min/day)	S_E (m)	
	0.61	0.41	0.1	0.1	137	1.9×10^{-4}	2
Late times	s_L (m)	t_B (min)	$1/u_B$	$W(B_B, \beta)$	T_L (min/day)	S_L (m)	
	0.47	0.51	0.1	0.1	110	1.6×10^{-4}	2

The substitution of this expression between two observation wells leads to,

$$h^2(r_1) - h^2(r_2) = \frac{2.3Q}{2\pi K} \log\left(\frac{u_1}{u_2}\right)$$

On the other hand, the ratio of dimensionless time factors by considering Eq. (3.31) is equal to the inverse ratio of time, that is, $(u_1/u_2) = (t_2/t_1)$ and its substitution in the previous expression gives the hydraulic conductivity as,

$$K = \frac{2.3Q}{2\pi[h^2(r_1) - h^2(r_2)]} \log\left(\frac{t_2}{t_1}\right) \quad (4.42)$$

If both sides of this expression is multiplied by the saturation thickness, H_o, then a new transmissivity equation appears as follows,

$$T = KH_o = \frac{2.3QH_o}{2\pi[h^2(r_1) - h^2(r_2)]} \log\left(\frac{t_2}{t_1}\right) \quad (4.43)$$

On the other hand, storage coefficient has already been defined as the volume of water that an aquifer releases from storage per unit surface area of aquifer per unit decline in the component of hydraulic head normal to that surface (Chapter 2). In order to obtain the storage coefficient value, the right side of Eq. (4.40) can be written as,

$$H_0^2 - h^2(r) = \frac{2.3Q}{2\pi K} \log\left(\frac{0.562}{u}\right)$$

The substitution of Eq. (3.31) into this expression yields,

$$H_o^2 - h^2(r) = \frac{2.3Q}{2\pi K} \log\left(\frac{2.25Tt}{Sr^2}\right)$$

On the other hand, $K = T/H_o$, one can obtain,

$$H_o^2 - h^2(r) = \frac{2.3QH_o}{2\pi T} \log\left(\frac{2.25Tt}{Sr^2}\right) \quad (4.44)$$

The storage coefficient can be found from this last equation after the necessary arrangements as,

$$S = \frac{2.25Tt}{r^2} e^{-\frac{2\pi T[H_o^2 - h^2(r)]}{QH_o}} \quad (4.45)$$

Provided that the pumping discharge is known, the storage coefficient can be evaluated with the drawdown measurement in the abstraction well. The last expression proves useful especially for the unsteady state cases similar to Şen (1987) approach.

On the other hand, from the discussion of the transient drawdown in an unconfined aquifer by Bear (1979), Eq. (4.40) can only be used to represent the transient water level in an unconfined aquifer, if the storativity in Eq. (3.44) is replaced by what may be called the apparent storativity, S', assumed to be a dimensionless constant, which can be expressed in the present notation as,

$$S = \frac{S'H_o}{h} \quad (4.46)$$

Adaptation of this apparent storativity concept renders Eq. (4.45) into the following expression,

$$S' = \frac{2.25Kht}{r^2}e^{-\left[\frac{2\pi K\left[H_o^2-h^2(r)\right]}{Q}\right]} \quad (4.47)$$

The storativity of an unconfined aquifer is therefore not a constant but a parameter that decreases with decreasing water levels, as one would expect from its basic definition for a confined aquifer as $S = mS_o$, where m is the thickness of the aquifer and S_o its specific storativity.

EXAMPLE 4.11 UNCONFINED AQUIFER PARAMETER ESTIMATION—QUASI-STEADY STATE FLOW

The aquifer parameter determinations resulting from various known methods in literature are analyzed for an unconfined aquifer under unsteady state flow condition for the selected numerical model data in Table 4.4. Aquifer saturation thickness is 4 m, hydraulic conductivity is 10 m/day, and storage coefficient is equal to 0.01. Abstraction well radius is 0.15 m and the pump discharge is 30 m³/day. Two observation wells (OW_1 and OW_2) are at radial distances at 5 and 10 m from the main pumping well. These observation wells are important for determination of aquifer parameters by using the methods already available in the literature.

Solution 4.11

For calculations it is necessary to attach a reading from one of the wells with the subsequent

readings from any other wells. In this application, MW second observation piezometric height, $h_1 = 4.00 - 0.621 = 3.329$ m at time 2/24 is attached with the subsequent time piezometric values in the first OW_1 as in Table 4.5.

TABLE 4.4 Calculated Drawdown Values in an Abstraction and Observation Wells in an Unconfined Aquifer

Time (day)	Drawdown (m)		
	MW	OW_1	OW_2
1/24	0.621	0.167	0.088
2/24	0.671	0.209	0.127
3/24	0.700	0.235	0.150
4/24	0.721	0.253	0.167
5/24	0.737	0.267	0.181
6/24	0.750	0.279	0.192
7/24	0.762	0.288	0.201
8/24	0.772	0.297	0.209
9/24	0.780	0.304	0.217
10/24	0.788	0.311	0.223
15/24	0.818	0.338	0.249
20/24	0.840	0.356	0.267
1	0.854	0.368	0.279
2	0.907	0.414	0.323
3	0.938	0.441	0.349
4	0.961	0.460	0.368
5	0.978	0.476	0.383
6	0.993	0.488	0.395
7	1.005	0.498	0.405
8	1.016	0.508	0.414
9	1.025	0.516	0.422
10	1.034	0.523	0.429
15	1.066	0.551	0.456
20	1.090	0.571	0.476

TABLE 4.5   Aquifer Parameters Calculated from the Proposed AnalyticalMethod

t_1 (day)	h_1 (m)	t_2 (day)	h_2 (m)	T_{cal} (m^2/day)	S_{cal}
2/24	3.329	3/24	3.300	92.64	0.023
2/24	3.329	4/24	3.279	92.16	0.025
2/24	3.329	5/24	3.263	92.51	0.025
2/24	3.329	6/24	3.250	92.85	0.023
2/24	3.329	7/24	3.238	92.10	0.025
2/24	3.329	8/24	3.228	91.95	0.025
2/24	3.329	9/24	3.220	92.55	0.025
2/24	3.329	10/24	3.212	92.37	0.025
2/24	3.329	15/24	3.182	92.46	0.025
2/24	3.329	20/24	3.160	92.23	0.025
2/24	3.329	1	3.146	92.16	0.025
2/24	3.329	2	3.093	92.12	0.025
2/24	3.329	3	3.062	92.25	0.025
2/24	3.329	4	3.039	92.10	0.025
2/24	3.329	5	3.022	92.25	0.025
2/24	3.329	6	3.007	92.07	0.025
2/24	3.329	7	2.995	92.14	0.025
2/24	3.329	8	2.984	92.05	0.025
2/24	3.329	9	2.975	92.16	0.025
2/24	3.329	10	2.966	92.02	0.025
2/24	3.329	15	2.934	92.21	0.025
2/24	3.329	20	2.910	91.10	0.025
		Average		92.24	0.025

Calculated transmissivity, T_{cal}, and storage coefficient, S_{cal}, results are presented in the last two columns of Table 4.5 by using the Eqs (4.43) and (4.47), respectively. For instance the first transmissivity value is,

$$T_{cal1} = \frac{2.3 \times 30}{2 \times 3.14 \times (3.329^2 - 3.300^2)} \log\left(\frac{3}{2}\right)$$
$$= 92.68 \text{ m}^2/\text{day}$$

The substitution of this value into Eq. (4.47) with other relevant values leads to storage coefficient estimation as,

$$S_{cal1} = \frac{2.25 \times 92.68}{0.15^2} \exp$$
$$\left[-\frac{2 \times 3.14 \times 10 \times (4.0^2 - 3.3^2)}{30} \right]$$
$$= 0.023$$

The average transmissivity and storage coefficient values are 92.24 m^2/day and 2.5×10^{-2}, respectively.

4.9.5 Quasi-Steady State Flow Storage Calculation Formulation

Similar to what has been explained in Chapter 3, Section 3.12, Şen (1987) derived the storage coefficient calculation formulation in unconfined aquifers from the steady state flow conditions under the light of storage coefficient definition based on the depression cone volume expression as,

$$S = \frac{Qt - \pi r_w^2 s_w(t)}{\frac{1}{2}r_w^2 \sqrt{\frac{2\pi mQ}{T}} \exp\left\{ -\frac{2\pi mT\left[1 - \frac{s_w(t)}{m}\right]^2}{Q} \right\} \int_a^b e^{x^2} dx - \pi r_w^2 s_w(t)} \tag{4.48}$$

Here, r_w is well radius, $s_w(t)$ drawdown in the main well, Q is the pump discharge, which is equal to the discharge from the aquifer into the well in the case of quasi-steady state flow, that is, at large TD cases, m is the aquifer thickness, T is the transmissivity, and x is a dummy variable. Furthermore, in Eq. (4.48) a and b are constants given explicitly as

$$a = \sqrt{\frac{2\pi mT}{Q}} \qquad (4.49)$$

and

$$b = \sqrt{\frac{2\pi mT}{Q}\left[1 - \frac{s_w(t)}{m}\right]} \qquad (4.50)$$

For large times the well contribution may be ignored compared to the overall water volume that originates from the aquifer storage, and hence, it is possible to rewrite Eq. (4.48) as,

$$S = \frac{Qt - \pi r_w^2 s_w(t)}{\frac{1}{2}r_w^2 \sqrt{\frac{2\pi mQ}{T}} \exp\left\{-\frac{2\pi mT\left[1 - \frac{s_w(t)}{m}\right]^2}{Q}\right\} \int_a^b e^{x^2}dx} \qquad (4.51)$$

EXAMPLE 4.12 UNCONFINED AQUIFER STORAGE COEFFICIENT ESTIMATION

Calculate the storage coefficient estimations from each observation well TD data already given in Chapter 3, Table 3.12, for Chaj Doab aquifer.

Solution 4.12

The calculations according to Eq. (4.48) are presented in Table 4.6. In the same table for comparison purposes the late TD, late DD, and CD method applications are also given by Ahmad (1998).

Comparison of the storage coefficients indicates that the proposed approach yields results that are practically in good agreement with conventional methods.

4.10 NATURAL GROUNDWATER RECHARGE

Groundwater recharge depends on several factors such as temporal and spatial variability of hydrometeorological factors especially rainfall and temperature, lithology and subsurface geology, infiltration capacity, and climate factors. Mainly, the spatial and temporal distribution of the rainfall controls the natural groundwater recharge. In arid regions, recharge occurs through the ephemeral streams, which flow through the wadi course but most of the water is absorbed in the unsaturated zone before reaching the aquifer (Şen, 2008). In semiarid regions, the recharge is irregular and occurs only in the periods of heavy rainfalls. In humid regions, recharge is mainly in the winter period. In the summer period, most of the rainfall becomes soil moisture and evaporates. In cold areas the melting of ice suddenly recharges the groundwater.

Groundwater recharge can be defined as water added to the aquifer through the unsaturated zone after infiltration and percolation following any storm rainfall event. Small local floods merely compensate for soil moisture deficits and evapotranspiration particularly during dry season, and therefore, the amount of water that reaches the water table will not be a significant contribution. While comparing direct and indirect recharge, Simmers (1997) concluded the following.

1. Estimates of direct recharge can be more reliable than indirect recharge.
2. With increasing aridity, direct recharge becomes less significant while indirect

TABLE 4.6   Storage Coefficient Values

| Well No | Theis Method | Jacob Method | | | Equation (4.48) |
		Time-Drawdown (TD)	Distance-Drawdown (DD)	Composite-Drawdown (CD)	S
O1	2.7×10^{-5}	1.3×10^{-3}		4.7×10^{-5}	3.0×10^{-5}
O2	3.5×10^{-3}	9.4×10^{-3}		9.8×10^{-3}	1.9×10^{-3}
O3	2.9×10^{-2}	7.0×10^{-3}	2.7×10^{-2}	2.0×10^{-2}	5.1×10^{-3}
O4	3.1×10^{-4}	2.0×10^{-3}		1.7×10^{-3}	1.0×10^{-4}
O5	1.5×10^{-2}	1.2×10^{-2}		9.3×10^{-3}	3.9×10^{-3}
O6	1.6×10^{-1}	4.7×10^{-2}		1.3×10^{-1}	4.5×10^{-3}
Averages	3.5×10^{-2}	1.3×10^{-2}	2.7×10^{-2}	2.8×10^{-2}	2.6×10^{-3}

recharge is more in terms of total recharge to an aquifer.

3. Recharge occurs to some extent even in the most arid regions, although increasing aridity decreases the net downward flux with greater time variability.

4. Successful groundwater recharge estimation depends on identification of the probable flow mechanism and important features influencing recharge for a given locality.

Coupled with the changes in the hydrological cycle and probable inducement of climate change basic elements, the groundwater recharge is also interactively affected due to the following events.

1. Changes in precipitation, evapotranspiration, and runoff are expected to influence recharge. It is possible that increased rainfall intensity may lead to more runoff and less recharge.

2. Sea level rise may lead to increased saline intrusion into coastal and island aquifers, depending on the relative position of sea level.

3. Change in precipitation imply changes in CO_2 concentrations, which may influence carbonate rocks dissolution, and hence,

formation and development of karstic groundwater aquifers.

4. Natural vegetation and crop changes affect the climate and subsequently may influence recharge (Section 4.17).

5. Increased flood events contribute to unconfined aquifers in arid and semiarid zones, and hence, they affect groundwater quality in alluvial aquifers.

6. Changes in soil organic carbon may affect the infiltration properties, and consequently, the groundwater recharge.

These factors must be taken into consideration under the present global climate change issues for protecting the groundwater quantity and quality deteriorations.

EXAMPLE 4.13 WATER BALANCE CALCULATIONS FOR GROUNDWATER RECHARGE

An extensive outcrop of a sedimentary basin has 120 km^2 planar areas over which 80 mm of annual average rainfall falls and 6 m^3/s is abstracted continuously through a set of wells (well field). The total groundwater table level drop is 1.5 m/year. Provided that all other losses are not taken into consideration,

1. What is the amount of groundwater recharge that can take place within one year?
2. What is the percentage of the rainfall that infiltrates?

Solution 4.13

In general, water balance or budget is given in Chapter 2 by Eq. (2.53). For its application the time duration is one year and the area, $A = 12 \, km^2$. The total input during one year is equal to the groundwater recharge, R, whereas the output is the total pumped amount of water, P, which is annually $6 \times 60 \times 60 \times 24 \times 365 = 189.22 \times 10^6 \, m^3/year$. Drop in the groundwater level means negative storage change, that is, $\Delta S \times -1.5 \times 120 \times 10^6 = -180 \times 10^6 \, m^3/year$.

1. The substitution of all these numerical values into the water budget expression yields

$$R - 189,22 \times 10^6 = -180 \times 10^6$$

or

$$R = 189,22 \times 10^6 - 180 \times 10^6 = 9.22 \times 10^6 \, m^3$$

2. The uniform spread of recharge over the study area is $9.22 \times 10^6 / 120 \times 10^6 = 0.0768 \, m/year = 76.8 \, mm/year$. Hence, out of 80 mm/year only 76.8 mm/year appears as infiltration. This means that only $76.8/80 = 0.96$, that is, only 96% of the rainfall contributes to groundwater recharge. In other words, 4% of the rainfall is lost by evaporation, depression, interception, and any other reasons.

1. What is the recharge water amount during March?
2. What is the percentage of the rainfall that is lost?

Solution 4.14

Similar to the previous example the water balance expression will be used by consideration of inputs and outputs with the storage change in the soil moisture. Since there is an increase in the soil moisture the storage change has a positive value as $\Delta S = 25 \, mm$.

1. There are three inputs into the soil, which are the rainfall, R, infiltration, I, and sprinkler water, S_w, that is, $(I + R + S_w)$, whereas the sole output is evaporation, E. Hence, in notations the valid water balance equation takes the following form.

$$(I + R + S_w) - E = \Delta S$$

The substitution of the numerical data into this equation yields,

$$(I + 45 + 0.25 \times 31) - 35 = 25$$

or

$$I = 25 + 35 - (45 + 7.75) = 60 - 52.75$$
$$= 7.25 \, mm$$

Hence, the amount of infiltration water is $50 \times 10^6 \times 7.25 \times 10^{-2} = 3.62 \times 10^6 \, m^3$.

2. The lost rainfall amount in this case is $45 - 7.25 = 37.75 \, mm$. Since 25 mm is for evaporation the direct rainfall is $37.75 - 25 = 12.75 \, mm$. The volume of runoff is $12.75 \times 10^{-2} \times 50 \times 10^6 = 6.375 \times 10^6 \, m^3$.

EXAMPLE 4.14 INFILTRATION-GROUNDWATER RECHARGES CALCULATION

In an outcrop area of 50 km² the average monthly rainfall for March is 45 mm. Evaporation loss is 35 mm during the same month. The soil moisture is observed to increase by 25 mm. During the same month the area is watered by sprinklers at 0.25 mm per day. Under these circumstances,

EXAMPLE 4.15 RECHARGE INTO UNCONFINED AQUIFER

In an unconfined aquifer, effective recharge area is 108.2 km² and during a wet period, the average water table rise is 3.2 m. In the same period, total groundwater abstraction is $1.23 \times 10^6 \, m^3$. The aquifer specific yield value is 0.043. Hence, calculate the total recharge during this season.

Solution 4.15

During a wet season the total natural recharge, T_R, can be calculated by considering two components, which are infiltration into the unsaturated zone, I, and total groundwater withdrawal, Q_w, during this period,

$$T_R = I + Q_w$$

By considering the groundwater level rise, ΔH, effective areal extent of the recharge area, A, and the specific yield, S_y, one can rewrite previous expression as,

$$T_R = A \, \Delta H S_y + Q_w$$

Substitution of relevant quantities into this expression gives wet season total natural recharge as,

$$T_R = 108.2 \times 10^6 \times 3.2 \times 0.043 + 1.23 \times 10^6$$
$$= 16.12 \times 10^6 \text{ m}^3$$

groundwater through deep infiltration (percolation). In the unsaturated zone only molecular adhesion and capillary waters remain (field capacity). Dry soil weight is $(1-0.40) \times 2.65 = 1.59$ t/km^2; field capacity is $0.15 \times 1.59 = 0.239$ t/m^2; at the field capacity the existing water of 450 cm has the height equal to $0.239 \times 450 = 107.5$ cm. Finally, percolation height can be calculated as $180-107.5 = 72.5$ cm.

2. At the wilting (dryness) point the water availability in the soil is $0.06 \times 1.59 = 0.095$ t/m^3. The difference between this and the field capacity is $0.239-0.095 = 0.144$ t/m^3, which corresponds to $0.144 \times 450 = 64.8$ cm water height. Plants can benefit from water until the point of wilting, and accordingly, their usable water height is 64.8 cm. The amount of water that can be used by plants depends on the depth of their roots. The change of water amounts in the unsaturated zone is given schematically in Figure 4.27.

EXAMPLE 4.16 GROUNDWATER RECHARGE

The unsaturated zone over the water table in a geological formation has 4.50 m thickness. The soil is sandy with porosity 0.40 and its specific weight is 2.25 t/m^3. As the percentage of dry soil weight, the soil capacity is 15% and the drying point is 6%.

1. After a storm rainfall calculate the groundwater recharge height due to gravitation as the soil transits from saturation case to soil capacity.
2. Calculate the water height that is useful for the plants.

Solution 4.16

1. Immediately after the storm rainfall the soil is fully saturated, and hence, the existing water height is $0.40 \times 4.50 \times 10^2 = 180$ cm. After a while, some of the water reaches the

EXAMPLE 4.17 GROUNDWATER RECHARGE

The groundwater recharge area in a drainage basin of 50 km^2 is 17 km^2. Daily water abstraction from the groundwater storage is 0.049×10^6 m^3. Annual precipitation, evaporation, and runoff heights are 88, 125, and 33 cm, respectively. What is the change in the groundwater reservoir in one year?

Solution 4.17

Water balance expression is the same as Eq. (2.53). In one year, the input water is composed of runoff volume, V_r, and precipitation volume, V_p, contributions. One can find runoff volume by multiplying annual runoff height by basin area as,

$$V_r = 0.33 \times 50 \times 10^6 = 16.5 \times 10^6 \text{ m}^3$$

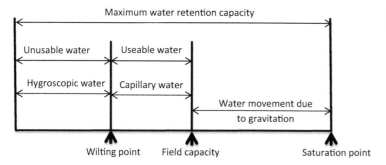

FIGURE 4.27 The change of soil water amount in the unsaturated zone.

The volume of precipitation that falls on the recharge surface is,

$$V_p = 0.88 \times 17 \times 10^6 = 15 \times 10^6 \text{ m}^3$$

The output components in the water balance expression is composed of evaporation, V_e, and water demand, V_d, volumes as,

$$V_e = 1.25 \times 17 \times 10^6 = 21 \times 10^6 \text{ m}^3$$
$$V_d = 365 \times 0.049 \times 10^6 = 18 \times 10^6 \text{ m}^3$$

By substitution of these values into the water balance equation one can calculate the change in the groundwater reservoir volume as,

$$\Delta S = \left(16.5 \times 10^6 + 15 \times 10^6\right) - \left(21 \times 10^6 + 18 \right.$$
$$\left. \times 10^6\right)$$
$$= -7.5 \times 10^6 \text{ m}^3$$

Hence, under these circumstances there is decrease in the groundwater storage.

4.10.1 Water Table Fluctuations and Recharge

In a naturally balanced aquifer, the net recharge hydraulic head change, Δh, is the difference in head between two successive water-level measurements (Figure 4.28).

The difference between the total and net recharge is equal to the sum of evapotranspiration from groundwater, base flow, and net subsurface flow. The water table level fluctuation application methodology is in use since early 1920 (Meinzer, 1923; Meinzer and Stearns, 1929), which have been followed by different researchers (Rasmussen and Andreasen, 1959; Gerhart, 1986; Healy and Cook, 2002). The practicality and simplicity of the methodology are due to the fact that there is no assumption concerning water flow through the unconfined aquifer. In shallow unconfined aquifers sharp groundwater level fluctuations and subsequent infiltration take place and this makes more efficient methodology. On the contrary, in deep unconfined aquifers wetting front covers extensive areas and the time of groundwater table arrival is rather long. The groundwater recharge rate is dependent on the elevation differences, geology, vegetation, and other factors.

Internal and external factors affect groundwater table fluctuations. Long-term changes including climate and human activities (groundwater abstraction, land use, irrigation, agricultural activities, dam construction, etc.) are normal occurrences. Among the practical difficulties are low-intensity rainfall events and their late infiltration contribution to the groundwater table. As a result the rise in height may be less than that expected from a short intense storm of the same total

FIGURE 4.28 Recharge
phenomenon.

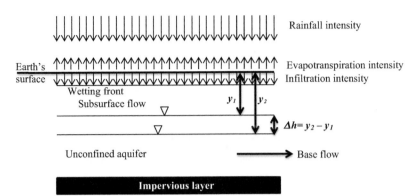

precipitation, and the recharge rate may lead to underestimation.

Groundwater management takes into consideration also pollution protection studies for which groundwater recharge estimation has significant role. The simplest method is based on water table level fluctuation records and aquifer specific yield parameter in unconfined aquifers. Additionally, Darcy law provides a basis for the water table fluctuation record assessments, where hydraulic conductivity is important. The water table level change by time is referred to as the "well hydrograph" (Figure 4.29).

In general, recharge process takes place as a result of the surface water entry into unsaturated zone either directly through water table or indirectly as lateral flow within the aquifer

(see Figure 3.19 and Figure 4.21). Temporal and spatial groundwater recharge methodologies are available in the literature (Simmer, 1988, 1997; Lerner et al., 1990; Scanlon et al., 2002). Temporal models digest groundwater level fluctuation data in a single well by time, but spatial models provide regional groundwater level maps dynamically for recharge volume calculations. Other methods take into consideration interaquifer (like leaky aquifers) flow. In addition to groundwater level fluctuation data they necessitate hydraulic conductivity and hydraulic gradient information for Darcy law application. Such approaches also require more theoretical basis and simplification assumptions such as the Dupuit—Forchheimer principle (Section 4.6.1).

In general, the groundwater balance equation takes into consideration changes in subsurface water storage due to recharge and groundwater flow into the basin minus base-flow including groundwater discharge to streams or springs, evapotranspiration from water table, and groundwater movement outflow. The general form of balance equation is as follows (Schicht and Walton, 1961).

$$G_R = \Delta S_A + Q_B + ET_G + (Q_G - S_F) \quad (4.52)$$

where G_R is groundwater recharge, ΔS_A aquifer storage, Q_B is base-flow, ET_G is

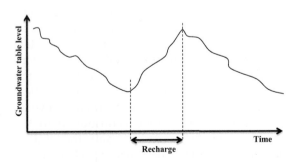

FIGURE 4.29 Well hydrograph.

evapotranspiration from water table, Q_G is flow from the aquifer including pumping, and finally, S_F is subsurface flow from the aquifer. The effect of rainfall on various terms in the groundwater balance expression depends on the nature of time lag between rainfall and subsequent infiltration water arrival. Such a situation can appear for short time durations and the success of this methodology lies within several hours or few days. The shorter the time duration the more reliable is the result from Eq. (4.52). The simplest recharge calculation formulation based on the groundwater level fluctuations, Δh, during a certain time interval, Δt, can be written as follows,

$$R = S_y \frac{\Delta h}{\Delta t} \qquad (4.53)$$

where R corresponds to groundwater recharge through a unit area of the water table and S_y to the specific yield. Furthermore, $\Delta h / \Delta t$ is equal to local slope in the well hydrograph.

In order to avoid overexploitation of an aquifer abstraction rates, natural recharge, and safe yield concepts must be kept in mind. The best and practical way of observing the effects of all these concepts on the aquifer is possible through an observation well water table level multiannual records (well hydrograph). Such a representative well hydrograph is shown for an observation well in the southeastern province of Turkey in Figure 4.30.

It is obvious that there is continuous decline of about 25 m in the water level since 1995. In order to stabilize the decrease, it is necessary to adopt better water resources management. If necessary precautions are not taken, such a decline disturbs aquifer safety and sustainable yields and even may cause aquifer mining. A convenient solution must be sought for keeping the safe yield by an increase in the artificial recharge or a decrease in water abstraction or a combination of the two.

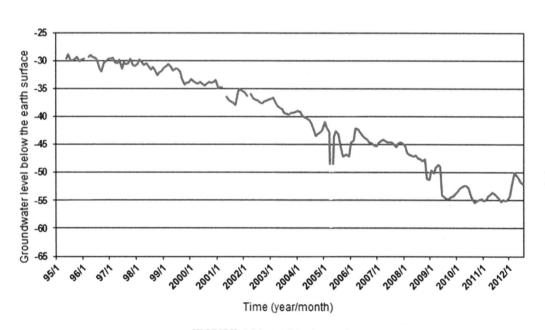

FIGURE 4.30 Well hydrograph.

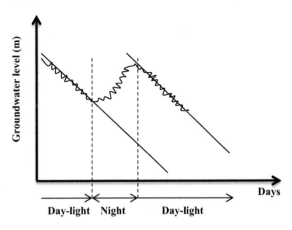

FIGURE 4.31 Diurnal groundwater level fluctuations.

Shallow unconfined groundwater tables show diurnal fluctuations because of evapotranspiration rate difference between day and night. During nights one may take evapotranspiration practically equal to zero with groundwater rise implication (Figure 4.31).

White (1932) assumed that groundwater evapotranspiration is zero between midnight and 4:00 a.m. and he also defined h' as the hourly rate of water table rise during night hours. Finally, he proposed the following formulation for the total amount of groundwater daily evaporation discharge, Q_{DE}, calculation as,

$$Q_{DE} = S_y(24\,h' + s_e) \qquad (4.54)$$

where S_y is the specific yield of the aquifer and s_e is the water-level elevation at midnight at the beginning of the 24-h period minus the water-level elevation at the end of the period. The necessary ingredients of this model are presented in Figure 4.32.

Unsaturated zone intermediates between the atmosphere and saturation zone and also due to some resistance from grains in this zone, the atmospheric pressure may reach water table in a delayed manner. Such a pressure imbalance produces a rise in the water level within the

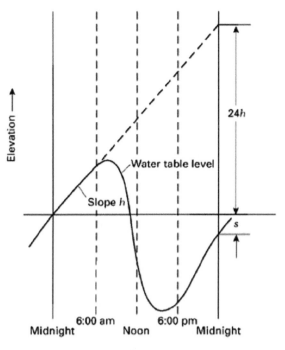

FIGURE 4.32 Diurnal water-level fluctuation indicator diagram (after Troxell, 1936).

well, and hence, water moves from the well to the aquifer. The effect of this atmospheric phenomenon has been explained by Todd (1980) as in Figure 4.33. It shows water-table rise in an observation well due to air entrapped between the water table and an advancing wetting front, where H indicates the initial depth to saturation zone, m_i is thickness of infiltrating saturated front and Δh is water-level rise in well (Todd, 1980).

Atmospheric pressure change, Δp, is directly observable on the water-table fluctuations at rather small amount, Δs, which can be expressed as a percentage of pressure change, $\alpha \Delta p$, where α is the reflection of percentage pressure effect on the water table. Practically, Δs may vary between 1 and 3 cm. Jacob (1940) gave the physical implication of aquifer barometric efficiency and its quantification by a factor, $(1 - \alpha)$.

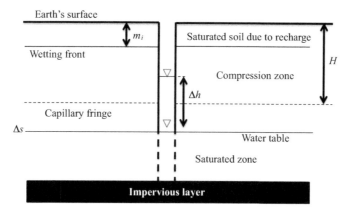

Earth's surface

Wetting front

Capillary fringe

Δs

Saturated soil due to recharge

Compression zone

Δh

Water table

Saturated zone

Impervious layer

FIGURE 4.33 Atmospheric pressure effects on water table.

EXAMPLE 4.18 AQUIFER STORAGE CHANGE

In a semiarid area the hydrologic year is considered from May to October as dry period and from November to April next year as wet period. Based on hydrological analysis average annual groundwater table decline is recorded as 0.71 m. Decline (rise) in dry (wet) period in the ground water table is 4.98 m (2.7 m). Calculate whether there has been net increase or decrease in the aquifer storage during one year. The aquifer area is 98.2 km^2 and the specific yield is 0.027.

Solution 4.18

First let us calculate the amount of groundwater loss volume, V_L, due to decline, $(\Delta H)_D$, during dry period by use of Eq. (4.53) as,

$$V_L = A(\Delta H)_D S_y = 98.2 \times 10^6 \times 4.98 \times 0.027$$
$$= 13.20 \times 10^6 \text{ m}^3$$

Likewise, the amount of groundwater gain volume, V_G, due to water table rise, $(\Delta H)_R$, during dry period can be calculated as,

$$V_G = A(\Delta H)_R S_y = 98.2 \times 10^6 \times 2.70 \times 0.027$$
$$= 7.16 \times 10^6 \text{ m}^3$$

Since $V_L > V_G$ during the year there has been groundwater net volume loss, V_N, as,

$$V_N = V_L - V_G = 13.20 \times 10^6 - 7.16 \times 10^6$$
$$= A \Delta HS_y = 6.04 \times 10^6 \text{ m}^3$$

EXAMPLE 4.19 UNCONFINED AQUIFER WATER BALANCE

In a sedimentary basin unconfined aquifer total inputs during one month are discharges from rainfall recharge, $Q_{RR} = 0.7 \times 10^6 \text{ m}^3$, lateral subsurface inflow, $Q_{LS} = 0.06 \times 10^6 \text{ m}^3$, recharge from nearby rivers, $Q_{RN} = 0.1 \times 10^6 \text{ m}^3$, recharge due to irrigation return, $Q_{IR} = 0.2 \times 10^6 \text{ m}^3$, and sewage return, $Q_{SR} = 0.07 \times 10^6 \text{ m}^3$. The same aquifer total output consists of irrigation water uses, $Q_{IW} = 1.1 \times 10^6 \text{ m}^3$, discharge from springs and qanats, $Q_{SQ} = 0.003 \times 10^6 \text{ m}^3$, lateral subsurface outflow, $Q_{La} = 0.15 \times 10^6 \text{ m}^3$, evaporation from groundwater table, $Q_{EGT} = 0.012 \times 10^6 \text{ m}^3$, domestic and industrial water uses, $Q_{DI} = 0.13 \times 10^6 \text{ m}^3$. Interpret this aquifer from water balance point of view.

Solution 4.19

According to the simplest groundwater equilibrium formulation in Eq. (2.53), first it is necessary to calculate the total input, I_T, and output, O_T, as follows,

$$I_T = 0.7 \times 10^6 + 0.06 \times 10^6 + 0.1 \times 10^6 + 0.2$$
$$\times 10^6 + 0.07 \times 10^6$$
$$= 1.13 \times 10^6 \text{ m}^3$$

and

$$O_T = 1.1 \times 10^6 + 0.003 \times 10^6 + 0.15 \times 10^6$$
$$+ 0.012 \times 10^6 + 0.13 \times 10^6$$
$$= 1.395 \times 10^6 \text{ m}^3$$

Since $I_T < O_T$ the aquifer is not sustainable during this month and additional support must be brought from other sources (for instance, nearby aquifers, surface reservoirs, etc.). The deficit is about $1.395 \times 10^6 - 1.130 \times 10^6 = 0.265 \times 10^6 \text{ m}^3$.

4.11 RECHARGE OUTCROP RELATION (ROR) METHOD

In extremely arid regions of the world, the recharge from rainfall and subsequent runoff to unconfined aquifers is very small, if it exists at all. The recharge calculations in these regions pose special problems, which should be solved by refined techniques suitable for the prevailing conditions, among which are sporadic temporal and areal distribution of rainfall occurrence and amount of high intensity, and barren earth surface with virtually no plant cover. The central Arabian Peninsula is located in an extremely arid zone belt of the world with characteristics of very little and unpredictable amounts as well as irregular rainfall occurrences (Şen, 1983). Rainfall occurs usually during a period from November to May. Basic evaporation rates are high, which lead to excess loss of water if it is available at the surface (Salih and Sendil, 1984).

There are studies for estimating the recharge contribution to the Wasia and Biyadh (WB) aquifers in the Arabian Peninsula (Subyani and Şen, 1991). A brief summary of different studies are presented in Table 4.7. Sogreah (1968) estimated roughly the recharge from the available hydroclimatological data. The average amount of rainfall likely to percolate deep enough is found as 3–5 mm/year.

On the other hand, two distinctive methods are proposed by Dincer et al. (1974) in calculating the recharge amount that goes through sand dunes around Khurais area near Riyadh city (Figure 4.34).

Many studies missed some common points as the data are not enough to give a clear picture of the recharge phenomenon. For example, Sogreah (1968) study depends on very short records of hydroclimatological data (about 5 years). Dincer et al. (1974) measured the recharge rate from isotope technique, which yielded reliable answers about the groundwater mixture and age, but it failed to provide reliable numerical answers for groundwater recharge amounts. S.M.M.P. (1975) study is more precise, especially around unconfined portion. B.R.G.M. (1976) concluded from long period data and model simulation leading to an optimized value, which did not consider the outcrop of geological formations. Besides, it requires digital computer usage for calculations, which might not be practical.

TABLE 4.7   Recharge Rates from Different Studies

Study	Method	Recharge Rate (mm/year)
Sogreah (1968)	Hydroclimatological data	3–5
Dincer et al. (1974)	Isotope	20.0
S.M.M.P. (1975)	Piezometer and transmissivity	5.2
B.R.G.M. (1976)	Hydroclimatological data	6.5
Caro and Eagleson (1981)	Dynamic model	6.0
Subyani and Şen (1991)	Recharge outcrop relation (ROR)	4.0

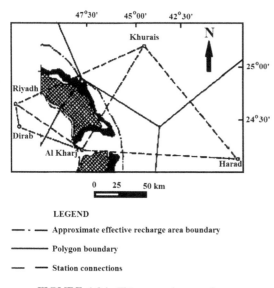

47°30'   45°00'   42°30'

Khurais

N

25°00'

Riyadh

24°30'

Dirab

Al Kharj

Harad

0   25   50 km

LEGEND

— · — Approximate effective recharge area boundary

———— Polygon boundary

— — Station connections

FIGURE 4.34 Thiessen polygon subareas.

EXAMPLE 4.20 ROR FOR WASIA-BIYADH

Subyani and Şen (1991) proposed ROR technique, which combines the geological lithology with water budget. It provides a simple way of calculating mean monthly recharge from daily rainfall amounts. In the study area, there are five climatological stations for rainfall

measurements and they are located at Riyadh, Al-Kharj, Dirab, Khurais, and Harad as shown in Figure 4.34.

Solution 4.20

In order to apply the ROR method, the study area is divided by using the Thiessen polygon technique into representative subareas for each station. Outcrop of WB falls under Riyadh and Al-Kharj subareas only. This means that the recharge of these aquifers is direct, whereas the recharge from Dirab subarea is indirect, that is, prior to recharge there appears runoff. Tables 4.8 and 4.9 show the mean monthly recharge rates (7 months) in Riyadh and Al-Kharj stations.

The mean monthly actual evapotranspiration values for the same daily rainfall period can be calculated by the following expression given by Salih and Şendil (1984).

$$(ET)_{act} = 1.16(ET)_{J-H} - 0.37 \qquad (4.55)$$

where $(ET)_{J-H}$ is the evapotranspiration according to Jensen-Haise method. To calculate the lake ET, the observed ET records are multiplied by pan coefficient, which is equal to about 0.7 for arid zones, (Linsley et al., 1982). The results are presented in Tables 4.8 and 4.9 in column 4.

The study area is devoid of runoff measurements, and therefore, the runoff percentages are

TABLE 4.8 Monthly Recharge Rates for Riyadh Area

Months with Rainy Days (1)	Mean Rainfall (mm) (2)	Mean Actual ET (mm) (3)	Lake ET = 0.7 ET (mm) (4)	Runoff = 0.1 Rainfall (mm) (5)	Recharge Rate (mm) (6)
Jan.	10.2	5.6	3.6	1.0	5.6
Feb.	10.3	5.0	3.5	1.0	5.7
Mar.	12.6	19.7	22.5	15.0	0.0
Apr.	11.5	17.3	12.1	1.1	0.0
May	4.7	10.0	7.0	0.5	0.0
Nov.	2.3	2.3	1.6	0.23	0.5
Dec.	5.0	3.7	2.6	0.5	0.8
				Total	**12.7**

TABLE 4.9   Monthly Recharge Rates for Al-Kharj Area

Months with Rainy Days (1)	Mean Rainfall (mm) (2)	Mean Actual ET (mm) (3)	Lake ET = 0.7 ET (mm) (4)	Runoff = 0.1 Rainfall (mm) (5)	Recharge Rate (mm) (6)
Jan.	5.0	4.1	2.9	0.5	1.6
Feb.	5.0	4.0	2.8	0.5	1.7
Mar.	17.0	23.0	16.1	1.7	0.0
Apr.	10.0	12.0	8.4	1.0	0.0
May	1.8	7.6	5.3	0.2	0.0
Nov.	1.7	2.3	1.6	0.2	0.0
Dec.	3.4	3.8	2.6	0.3	0.5
				Total	**3.8**

calculated empirically from rainfall for arid zones as 10% (Chow, 1964) which are shown in column 5. Finally, the mean monthly recharge is calculated by applying the following water balance expression.

$$\text{Recharge Rate} = \text{Rainfall} - (\text{Runoff} - \text{Lake ET}) \tag{4.56}$$

This leads to the results presented in column 6 of Tables 4.8 and 4.9. In order to calculate the recharge rates by ROR method, estimates of particular outcrop formation-percentages within each subarea are calculated from respective polygons. Subsequently, the multiplication of these percentages with the respective recharge rates gives the annual total recharge received by WB outcrop. Details of the calculations are presented in Table 4.10.

The recharge contribution areas to WB in Riyadh and Al-Kharj are 1924 km^2 and 4530 km^2, respectively. The weighted average ROR recharges in the last column of Table 4.10 is calculated as,

$$(\text{ROR})_w = \frac{r_R A_R + r_K A_K}{A_R + A_K} \tag{4.57}$$

where r_R (r_K) and A_A (A_K) are the recharge rate and area in Riyadh (Al-Kharj). Substitution of the necessary quantities from Table 4.10–4.17 into Eq. (4.57) leads to,

$$(\text{ROR})_w = \frac{9 \times 1920 + 1.8 \times 4530}{1920 + 4530}$$
$$= 4.0 \text{ mm/year}$$

Comparison of this value with the results in Table 4.14 indicates the obvious overestimation of other techniques, which are valid for humid regions only.

TABLE 4.10   Recharge Outcrop Relation (ROR) Calculations

Subarea	Total Area (km^2)	Effective WB Area (km^2)	Percentage of WB Area (%)	Recharge (mm/year)	Recharge to WB Area (mm/year)
Riyadh	2710	1924	71	12.6	9.0
Al-kharj	9250	4350	49	3.8	1.8
			Weighted average (mm/year)		**4.0**

TABLE 4.11 Statistical Summary (Ashafa Station, 1980–1995)

Month	Mean Monthly Rainfall (mm)	Rainfall Chlorine Concentration (mg/l)
Jan.	19.8	7.5
Feb.	13.4	7
Mar.	21.9	8
Apr.	30.6	9.5
May	28	9.9
June	8.6	9.1
July	1.8	NA
Aug.	20.4	8.5
Sept.	12.3	7.8
Oct.	14	8.1
Nov.	17	9
Dec.	15.4	7.3
Sum	203.2	91.7
Mean	17.41	8.34
STD	7.96	0.94
Corr.		0.52

4.12 CHLORIDE MASS BALANCE (CMB) METHOD

Different methods can be used to estimate groundwater recharge, such as empirical approaches, water-balance techniques, tracer techniques, and others depending on data availability and the field situation (Eagelson, 1979; Lerner et al., 1990; Flint et al., 2002; Edmunds et al., 2002). Carter et al. (1994) summarize previous rain-fed groundwater recharge studies in semiarid and arid regions around the world. The groundwater recharge percentages of these studies range from 1% to 30% of the local rainfall. Tracer techniques, such as environmental isotope and chloride mass-balance (CMB), have been commonly used in the overall domain of water resources development and management (Fritz and Fontes, 1980; Wood and Sanford, 1995; Wood and Imes, 1995; Bazuhair and Wood, 1996; Wood, 1999; Shi et al., 2000; Kattan, 2001). In fact, the application of these relatively new techniques has played an important role in solving the envisaged hydrogeological problems that cannot be solved by conventional methods alone (IAEA, 1980, 1983; Clark and Fritz, 1997).

Shallow aquifers such as sedimentary deposits, basalt flows, and fracture crystalline rocks received a good amount of rainfall that can infiltrate directly to shallow groundwater storage. The rate of recharge depends on hydrological phenomena such as rainfall intensity, runoff, temperature and evaporation, and on physical properties of the aquifer such as porosity, permeability fracture spacing, surface weathering, and watershed properties. In general, recharge rate in arid regions does not exceed 10%.

The application of CMB method is simple with no sophisticated instrumental dependence, and it is based on the knowledge of annual precipitation and chloride concentrations in rainfall and groundwater storage. In arid regions, the recharge is sporadic depending on rare rainfall events especially at the upstream portion of drainage basins where most often due to topographic heights, orographic rainfall occurrences take place over limited areas. This leads to small recharge areas and the rainfall reaches the water table through infiltration process. Once the infiltration reaches the groundwater, it will be mixed regionally in the recharge area with the aquifer. The chloride concentration is homogeneously distributed within the aquifer. Rainfall duration is comparatively very short in arid regions, and therefore, after the rainfall occurrence, evapotranspiration takes place from the moist and unconfined soil surfaces due to rather intensive

TABLE 4.12 Hydrogeology Parameter Estimations

Well No.	Storativity, S	Transmissivity, T ($\times 10^3$ m²/day)	Saturation Thickness, D_S (m)	Unsaturated Thickness, D_U (m)
1	0.1099	0.0140	8.06	18.94
2	0.2079	0.3603	10.70	31.10
3	0.2872	0.0149	3.84	32.36
4	0.0324	1.0817	10.21	43.39
5	0.0850	0.0720	32.37	42.63
6	0.0747	0.0924	2.87	15.89
7	0.0641	0.0351	5.85	21.63
8	0.1431	0.0490	2.07	39.25
9	0.1870	0.7914	1.97	45.10
10	0.2207	1.2468	8.09	40.71
11	0.3000	0.1417	10.20	44.87
Aver.	0.1556	0.3546	8.75	34.17
Std. Dev.	0.0914	0.4622	8.12	10.93
Log. Mean	−0.205	−0.2069	1.824	3.473
Log std. dev	0.701	1.634	0.837	0.377
Cor. Coef.	$S{-}T = -0.007$ $S{-}D_s = -0.203$ $S{-}D_U = 0.344$	$T{-}D_S = -0.077$ $T{-}D_U = 0.508$	$D_S{-}D_U = 0.241$	

solar irradiation, which imply increase in the chloride concentration. Generally, in the application of CMB method, a set of assumptions should be taken into consideration in interpretation of the results. The assumptions in the CMB approach for recharge calculations are:

1. In the groundwater storage, there is no chloride source prior to the rainfall. This is the reason why the application of the method is valid for the upstream portions of the drainage basins.
2. There is no additional sources or sinks for chloride concentration in the area of application. This is mostly valid assumption in the upstream portions of drainage basins.

In many geochemistry studies within the drainage basins (wadis) in arid regions, the chloride is found only in negligible amounts, except after a rainfall event.

3. The rainfall either evaporates or infiltrates in the study region without any runoff, which is a rather unrealistic assumption and can be valid just for low intensity rainfall events. However, most often rainfalls are intense, especially, at the upstream portions of the drainage basin system.
4. Long-term rainfall and its chloride concentration amounts have a balanced situation, that is, steady state condition. This implies stable and long-term averages, as the

TABLE 4.13 Hydrogeology Parameter Deviations from Average Values

Well no.	Storativity, S	Transmissivity, T ($\times 10^3$ m²/day)	Saturation Thickness (m)	Unsaturation Thickness (m)
1	−0.0457	−0.3405	−0.6862	−15.23
2	0.0523	0.0058	1.9518	−3.07
3	0.1316	−0.3396	−4.9082	−1.81
4	−0.1232	0.7272	1.4618	9.22
5	−0.0706	−0.2825	23.6218	8.46
6	−0.0809	−0.2621	−5.8782	−18.28
7	−0.0915	−0.3194	−2.8989	−12.54
8	−0.0125	−0.3055	−6.6782	5.08
9	0.0314	0.4369	−6.7782	10.93
10	0.0651	0.8923	−0.6582	6.54
11	0.1444	−0.2128	1.4518	10.70

classical CMB method requires. The standard deviations around the averages are completely ignored in this assumption. A hidden assumption is that the fluctuations around the average rainfall and chloride records must be very small so that they are negligible, and hence, do not appear in the classical CMB equation.

On the basis of aforementioned assumptions, the fundamental equation applicable for recharge calculations is presented by Wood and Sanford (1995) as,

$$q = R(Cl)_{ra} / (Cl)_{gw} \qquad (4.58)$$

in which q is the recharge flux, R is average annual rainfall, $(Cl)_{ra}$ is the weighted average chloride concentration in rainfall, and $(Cl)_{gw}$ is the average chloride concentration in groundwater. It is derived from the physical mass conservation principle with no change in the storage, as simple as input is equal to output, that is, steady state condition (Subyani and Şen, 2006). Unfortunately, there is no consistency in the description of the averages in the classical CMB equations. The following questions are significant and indicate that there are

TABLE 4.14 Alluvium Thickness Distribution in the Wadi

Thickness	Volume ($\times 10^6$ m³)	
	Class Value	Cumulative Value
More than 75 m	50	50
50−75	687	737
25−50	1041	1778
0−25	181	1959

TABLE 4.15 Groundwater Storage Volume

Thickness	Volume ($\times 10^6$ m³)	
	Class value	Cumulative value
More than 75 m	42	42
25−50	680	722
0−25	259	981

TABLE 4.16 Risk Levels for Different Aquifer Parameters

Saturation Layer		Unsaturation Layer		Storage Coefficient		Transmissivity	
Thickness	Risk	Thickness	Risk	Value	Risk	Value	Risk
(m)	(%)	(m)	(%)	(−)	(%)	(−)	(%)
10	0.3349	10	0.999	0.09	0.693	0.18	0.4141
20	0.0936	20	0.8971	0.19	0.2872	0.38	0.25
30	0.0341	30	0.5758	0.29	0.1220	0.58	0.1754
40	0.0147	40	0.2841	0.39	0.0562	0.78	0.1325
50	0.0071	50	0.1227			0.98	0.1049

arbitrariness in the interpretation and use of this equation.

1. Equation (4.58) is valid for short time durations, that is, the smaller the time duration and the areal coverage of the system, the more is the validity of the equation under the steady state condition. Otherwise, the equation must be viewed as a gross simplification.
2. More important than the first point is the use of averages in Eq. (4.58). It is not specified whether it is average, and if average which type, arithmetic or weighted?
3. What about the deviations around the averages? Equation (4.58) does not account for such variations at all.

TABLE 4.17 Alluvium Porosity Values

Material	Porosity (%)
Silts and clays	55.0
Fine sand	45.0
Medium sand	37.5
Coarse sand	30.0
Gravel	25.0
Sand and gravel mixes	15.0

Subyani (2004a) and Subyani and Şen (2006) modified CMB method by including some statistical approaches with effective recharge area and seasonal rainfall. They concluded that the recharge rates in most of the alluvial aquifers in arid regions range from 8% to 10% of effective rainfall.

In practical applications, it is necessary to consider rather long time intervals, such as month or year, and the application area will be at least of several square kilometers. These practical scales ignore temporal and spatial variations in each term. Since, there are many sampling measurements within such practical time and space scales, the question is, which one of these measurements or what type of averages must be inserted into Eq. (4.58)? The simplest view is to consider the arithmetic averages of each term, which leads to a similar expression,

$$\bar{q} = \overline{R}\overline{(Cl)}_{ra} \big/ \overline{(Cl)}_{gw} \qquad (4.59)$$

where overbars indicate arithmetic averages. In practice, hydrogeologists have to sample the recharge phenomena by several measurements. Practically, none of these measurements will be equal, and therefore, there are deviations around the averages. Hence, the basic variables can be considered in addition to their averages with their deviations which are shown by primes in the following equations.

$$q = \bar{q} + q' \tag{4.60}$$

$$R = \bar{R} + R' \tag{4.61}$$

$$(\text{Cl})_{ra} = (\overline{\text{Cl}})_{ra} + (\text{Cl})'_{ra} \tag{4.62}$$

and

$$(\text{Cl})_{gw} = (\overline{\text{Cl}})_{gw} + (\text{Cl})'_{gw} \tag{4.63}$$

In order to simplify the mathematical derivations, it is assumed herein that chlorine concentration deviations within the aquifer are homogeneous due to long term mixture, and therefore, its deviations will be ignored, which means that $\text{Cl}_{gw} = \overline{\text{Cl}}_{gw}$. The substitution of Eqs (4.60)–(4.63) with this simplification into Eq. (4.59) leads to,

$$\bar{q} + q' = (\bar{R} + R') \left[\left((\overline{\text{Cl}})_{ra} + (\text{Cl})'_{ra} \right) \Big/ (\overline{\text{Cl}})_{gw} \right] \tag{4.64}$$

The expansion of the right-hand side parenthesis gives,

$$\bar{q} + q' = \left[(\bar{R}(\overline{\text{Cl}})_{ra} + \bar{R}(\text{Cl})'_{ra} + R'(\overline{\text{Cl}})_{ra} \right.$$
$$\left. + R'(\text{Cl})'_{ra} \right) \Big/ (\overline{\text{Cl}})_{gw} \right] \tag{4.65}$$

If both sides of this expression are averaged keeping in mind that by definition the perturbation terms have zero arithmetic averages, one can simplify it to,

$$\bar{q} = \left[\left(\bar{R}(\overline{\text{Cl}})_{ra} + \overline{R'(\text{Cl})'_{ra}} \right) \Big/ (\overline{\text{Cl}})_{gw} \right] \tag{4.66}$$

The second term in the parenthesis on the right side indicates the average of rainfall deviation multiplied by chloride concentration deviation. The arithmetic average of multiplication of two variables is defined as the covariance. By definition, covariance is equal to the

multiplication of the correlation coefficient, by the standard deviations. Hence, it is possible to write the final equation as follows,

$$\bar{q} = \left[(\bar{R}(\overline{\text{Cl}})_r + \hat{\rho}_{RCl_r} \hat{\sigma}_R \hat{\sigma}_{Cl_r}) \Big/ (\overline{\text{Cl}})_{gw} \right] \tag{4.67}$$

where $\hat{\rho}_{RCl_r}$ is the correlation coefficient between the rainfall and its chloride concentration; $\hat{\sigma}_R$ and $\hat{\sigma}_{Cl_r}$ are the standard deviations of rainfall and its chloride concentration measurements. This last expression is the refined CMB equation, which reduces to Eq. (4.59) under the following two circumstances.

1. Either the correlation coefficient is equal to zero between the rainfall and its chloride concentration, which implies independence of chloride concentration from the rainfall amount, or
2. When one of the standard deviations is equal to zero. In this case, the variable with zero standard deviation implies homogeneity, that is, temporal or spatial constancy or both.

In practical applications the correlation coefficient between the rainfall and its chloride concentration appears as a positive value. This is physically plausible, because the more the rainfall, the bigger is the chloride concentration. This is the main reason why in the classical CMB method, weighted average for the rainfall and its chloride concentration are used. However, Eq. (4.70) is expected to yield comparatively bigger results than Eq. (4.59). It is further possible to write the classical CMB equation within the refined expression by considering Eqs (4.62) and (4.70) as,

$$\bar{q} = \bar{q}_c + \hat{\rho}_{RCl_r} \hat{\sigma}_R \hat{\sigma}_{Cl_r} \Big/ (\overline{\text{Cl}})_{gw} \tag{4.68}$$

where $\bar{q}_c$ is the average recharge rate according to classical CMB method. It is obvious that the additional term appears extensively on the statistical parameters including standard deviations and correlation coefficient.

EXAMPLE 4.21 CHLORIDE MASS BALANCE METHOD

In order to show the performance of the refined CMB method, data from wadi Yalamlam in the western part of the Arabian Peninsula is considered by Subyani (2004) (Figure 4.35).

This basin drains a wide catchment area of about 1600 km^2. The catchment area is characterized by high amount of annual rainfall of more than 200 mm. After opening on the plain, the drainage basin loses its defined course and becomes wide spans of sheet wash, while further downstream; it is integrated as part of the Red Sea coast. The hydrology and precipitation features of

the region are identified extensively by Şen (1983) and Al-Yamani and Şen (1993).

Solution 4.21

In order to perform the application of refined CMB as presented by Eq. (4.67), mean monthly rainfall and rainfall chloride concentration data are collected from the Ash-Shafah station, which lies at the upstream of Yalamlam drainage basin. Groundwater samples for average chloride concentration are taken from the upstream wells. The necessary data for the application of refined CMB method are given in Table 4.11.

The average concentrations of chloride in rainfall and in the groundwater during the same period

FIGURE 4.35 Wadi Yalamlam (Subyani, 2004).

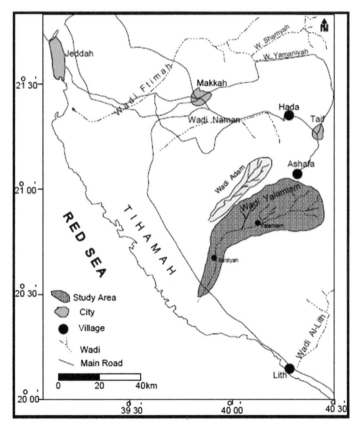

are $\overline{(Cl)}_r = 8.34$ mg/l and $\overline{(Cl)}_{gw} = 91.7$ mg/l, respectively. Monthly rainfall and groundwater chloride concentration standard deviations are $\hat{\sigma}_R = 7.96$ mm and $\hat{\sigma}_{gw} = 0.94$ mg/l, respectively. The average monthly rainfall is 17.41 mm. The estimated correlation between monthly rainfall and chloride concentration is $\hat{\rho}_{RCl_r} = 0.52$, which indicates that there is a significant correlation between the two variables, which is not taken into consideration in the classical CMB approach. Substitution of all the relevant values into Eq. (4.70) gives 1.7 mm/month, which in the case of classical calculations using Eq. (4.62) yields 1.6 mm/month. It is possible to conclude that the refined and classic approaches result in 10% and 9% of recharge rates, respectively. As expected, the refined approach recharge estimation from Eq. (4.70) is a little more than the recharge estimation from Eq. (4.62). However, the difference between the two approaches will increase with the increase especially in the standard deviations of monthly rainfall, and/or rainfall concentration, or in both.

4.13 ARTIFICIAL GROUNDWATER RECHARGE

Groundwater resources are completely dependent on occasional rainfall occurrences and afterward recharge amounts. It is, therefore, necessary to enhance the recharge as much as possible so that unsaturated zone over the groundwater table can be filled with maximum amount of infiltration water. Although there are many practices, the most effective one is the dam construction across alluvium deposits (Section 4.5.4, Figure 4.9). Surface dams are convenient for upstream parts where the sedimentary layer thickness is not big, slope is high, and alluvium consists of mostly gravel and sand. Such dams also protect the downstream lands against possible flood threats. The temperature is higher in middle or downstream watershed parts than upstream, and construction of surface dams at these localities may lead to excessive evaporation with little benefit from the water stored behind the dam.

For replenishment of groundwater resources through artificial recharge, it is necessary to modify the natural local hydrological cycle, especially by enhancing the infiltration component in order to give more room for groundwater storage. It can be accomplished either directly by increasing the infiltration from the rainfall or indirectly from the surface runoff by convenient hydraulic structures such as recharge ponds and very large diameter recharge (injection) wells and spreading of runoff over extensive areas.

If flood water is not exploited it is lost by evaporation or discharge to seas or deserts. Proper arrangements enhance recharge into unconfined aquifers. There are different ways and means for artificial recharge of such aquifers.

1. Infiltration basins aid water to penetrate through sand and gravel layers to reach water table. The basin may have surface dam reservoir or artificial depressions with boreholes to lead surface water into unconfined aquifers.
2. Water traps are used to increase infiltration along rainwater flood courses. These traps are surface dams with different heights ranging from 3 to 10 m along the runoff channels.
3. Cutwaters are applicable in places where there is no river or ephemeral stream. It is similar to ditches of different dimensions, where water is impounded and then recharged into the aquifer.
4. Through the injection wells surface waters are pumped directly to the underlying unconfined aquifer. These wells are considered as the best ways for artificial recharge in the most effective manner. It depends on the availability of surface water volumes either as rain water or water diverted from river or treated or untreated sewage waters.

EXAMPLE 4.22 GROUNDWATER RECHARGES APPLICATION IN WADI HANIFAH, RIYADH

The groundwater recharge application is achieved at Al-Ghat Dam, which is situated around the Riyadh City region (about 220 km north of Riyadh) (Prince Sultan Research Center, 2009). It is an earthen dam of 250 m length with 11 m height. The storage capacity of the reservoir is about 1.0×10^6 m^3 and the dam is built for groundwater recharge purpose. The catchment area is 43 km^2 with 82 km total stream length. After the impoundment of surface water behind the small dam vertical pipe feeders are located for water injection into the subsurface Quaternary deposits (see Figure 4.9). Figure 4.36 shows the location of injection pipes for speeding up the groundwater recharge into the unconfined aquifer and then onward subsurface flow toward downstream.

Groundwater recharge possibility from the impounded water behind the dam provides additional groundwater source into already available aquifer domain and such structures are expected to increase in arid regions for groundwater augmentation especially with the effect of climate change impacts. In the Mediterranean region annual average precipitation predictions have decreasing trends by 10–20% (IPCC, 2007).

The Arabian Peninsula including Riyadh region is found to have increasing rainfall trends by 10– 30% (Şen et al., 2012a, b).

If a storm rainfall of 20 mm occurs over the Al-Ghat drainage basin, is it sufficient to fill in the surface dam? If not sufficient then what amount of rainfall can fill the dam completely?

Solution 4.22

One can simply calculate the runoff volume through the rational approach (Chapter 1, Eq. 1.16) as follows.

1. The rain water volume over the catchment area is $43 \times 10^6 \times 20 \times 10^{-3} = 0.86 \times 10^6$ m^3/s.
2. A certain portion of this volume will be lost either by evaporation, depression fillings, or vegetative interceptions. In practice about 60% of the available rainfall water will be lost, and hence, the runoff coefficient is equal to 0.40. The surface water available for the impoundment behind the Al-Ghat dam can be calculated as $0.40 \times 0.86 \times 10^6 = 0.32 \times 10^6$ m^3.
3. Since this amount is smaller than the dam capacity (1×10^6 m^3) the dam will be full only by one-third of its capacity.
4. The empty part of the dam will be $1 \times 10^6 -0.32 \times 10^6 = 0.68 \times 10^6$ m^3.
5. The question now is what amount of a single storm rainfall will fill this volume? If the

FIGURE 4.36 Ghat site (a) before the storm rainfall, (b) after the storm rainfall.

unknown rainfall height is R_x then $0.4 \times 43 \times 10^6 \times R_x = 0.68 \times 10^6$, and hence, $R_x = 0.0395$ m $= 39.5$ mm just for the surface runoff. Since this is amount after 0.6% loss, $39.5/0.6 = 65.83$ mm of rainfall is expected to fill the whole dam reservoir. Total rainfall amount is $20 + 65.83 = 85.83$ mm.

4.14 AQUIFER PROPERTIES AND GROUNDWATER AVAILABILITY CALCULATIONS

For groundwater availability calculations the two most important hydrogeological parameters are the storativity and transmissivity. In order to explain the methodology, let us consider that a given study area has 11 wells that tap an unconfined aquifer as in Figure 4.37.

Saturation and unsaturated alluvium deposits consist of sand and gravel mixtures. The necessary parameters can be obtained through efficient assessment and interpretation of aquifer tests that are performed in the field. Among the field measurements are the discharge, well radius, and observation well radial distance, static and dynamic water levels. It is possible to match the most suitable type curve or use the CJ straight-line methods (Chapter 3, Section 3.9) for the calculation of aquifer parameters. The aquifer parameters together with saturation and unsaturated thicknesses obtained from the geophysical prospecting are given in Table 4.12.

A first glance on this table indicates that none of the four variables are homogeneous, which necessitates the use of probabilistic and

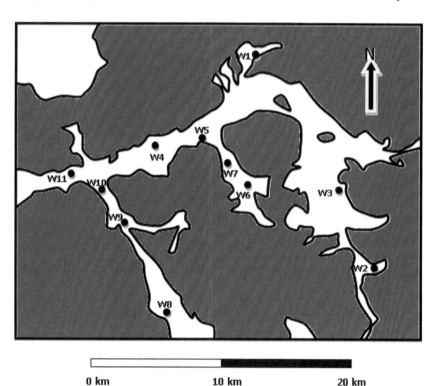

FIGURE 4.37 Aquifer test well locations.

statistical evaluation techniques for strategic groundwater storage risk assessments. Intervariable dependence of these variables is also depicted by the correlation coefficient as shown in the last row. There is a negligible correlation between storage coefficient and transmissivity. The transmissivity is in a weak dependence with regional saturation thickness. However, the saturation thickness is in a stronger, but reverse relationship with the storage coefficient, due to negative correlation coefficient sign. In order to appreciate the heterogeneous composition of the aquifer, error terms are given as deviations from respective arithmetic mean values in Table 4.13.

The alluvium consists of coarse sands with some rare intercalations of gravel, while in the upstream mainly coarse materials exist as gravel and boulders. The volume of alluvial deposits in the basin is given in Table 4.14.

The groundwater volumes in the saturated zone are summarized in Table 4.15 for different thickness classes by Italconsult (1969).

4.14.1 Deterministic Groundwater Storage Calculation

In general, the groundwater resources evaluation requires three quantities, namely, available, abstractable, and storable water volume calculations. Available water volume is equal to the multiplication of the saturation volume by the specific yield (effective porosity). Its calculation needs information about the storage domain volume dimensions, which are thickness, width, and length. The saturation and unsaturated volumes can be calculated as trapezoidal prism between two successive cross sections by use of the cross-sectional areas and the distance between the sections.

The saturation volume plays the major role in any groundwater abstraction calculations whereas the unsaturated volume is important for groundwater recharge possibilities. In calculating the present abstractable water volume, it is necessary to multiply the saturation volume by the storage coefficient. The storable groundwater volume can be calculated with the multiplication of unsaturation volume by the specific yield (effective porosity).

An unsaturated domain next to the earth's surface must be left for accommodation of capillary zone, reduction of evaporation, and groundwater losses. There is also hydrochemical reason for avoiding salinization, and hence, chlorine increment by leaching to groundwater storage. The maximum groundwater table level can be considered at least 1.5 m below the earth surface. For storable water volume calculation, the difference between the earth surface and existing static water table levels corresponds to the unsaturated zone thickness from which 1.5 m must be subtracted for storable (recharge) water volume calculations. The result is effective unsaturated thickness for recharge water storage. The effective porosity values in the alluvium channel and in its tributaries are necessary for the calculation of storable water volume in the Quaternary alluvium deposit.

4.14.2 Risk Considerations

Haphazard variations in the hydrogeological parameters can be taken into consideration by two new definitions. The first is the specific groundwater storage capacity, which is defined as the amount of available groundwater per unit water table area. Similarly, the next definition is the specific groundwater recharge capacity that gives the possible amount of rechargeable water also for per unit area. It is necessary to represent uncertainty by some probability distribution function, (pdf). Such a function indicates the relationship of concerned uncertain variable, say x, with its relative frequency of occurrence, that is, probability, P(x). For many hydrogeological variables this relationship is best represented by logarithmic normal pdf (Benjamin and Cornell, 1970;

Şen, 1999). After the identification of such a functional relationship it is possible to calculate the probability of exceedance, which is equivalent to risk concept. In fact, risk is the probability of exceedance. The risk calculation of different groundwater storage and recharge related variables can be achieved on the basis of the following information and completion of five steps.

1. Obtain a series of aquifer parameter estimations at a set of locations as in Figure 4.39.
2. Calculate the logarithmic mean, α, and standard deviation, β, of the hydrogeological variable.
3. Calculate the cumulative pdf values according to the logarithmic normal pdf expression as,

$$f(x) = \frac{1}{x\beta\sqrt{2\pi}} \int_0^x \exp\left[-\frac{1}{2}\frac{(\ln x - \alpha)^2}{\beta^2}\right] dx$$

(4.69)

4. Plot of x on the horizontal axis versus f(x) on the vertical axis gives the logarithmic normal cumulative pdf.
5. Similar plot, but x versus 1−f(x) provides the necessary risk graphs for the variable concerned.

Application of these steps for the hydrogeological variables (storativity, transmissivity, and saturation and unsaturated thickness) lead to risk curves in Figures 4.38−4.41. In all these figures logarithmic normal pdf is used with different means and standard deviations.

These figures help to estimate the corresponding variable value for any given risk level or vice versa. In order to facilitate the risk assessment of each hydrogeological parameter, Table 4.16 is prepared through the use of previous five steps. However, if there is a need for values that are not included in this table, then Eq. (4.58) can be used with the parameters given in Figures 4.38−4.41.

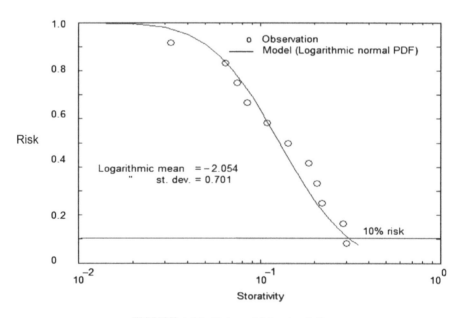

FIGURE 4.38 Risk model for storativity.

FIGURE 4.39 Risk model for transmissivity.

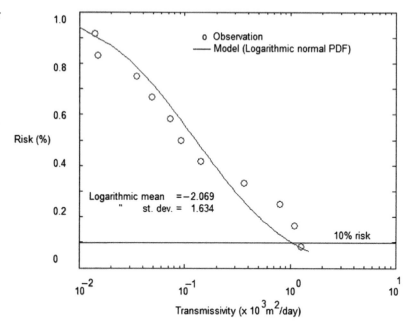

FIGURE 4.40 Risk model for saturation thickness.

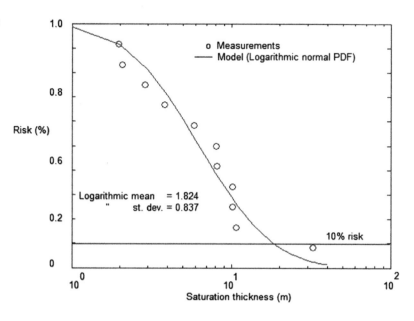

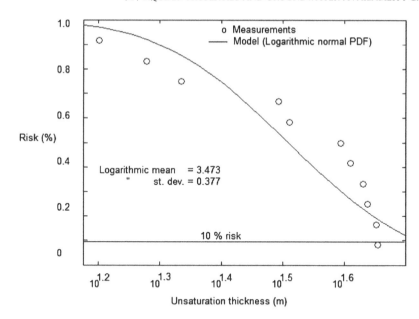

FIGURE 4.41 Risk model for unsaturation thickness.

4.14.3 Parameter and Water Storage Volume Risk Calculation

As already presented by Al-Sefry et al. (2004) the possible groundwater recharge storage, G_{Rs}, volume can be calculated as,

$$G_{Rs} = Adn \qquad (4.70)$$

where A is the unsaturation surface area, d is the possible recharge layer depth, and n is the porosity. The larger the area, the more complex is the subsurface geology. Therefore, for refining the estimation procedure, it is possible to define the specific groundwater recharge capacity, g_{Rs}, as the volume of groundwater recharge per unit area, $g_{Rs} = G_{Rs}/A$. Hence, from Eq. (4.59) it can be written as,

$$g_{Rs} = dn \qquad (4.71)$$

Abstractable groundwater volume from an aquifer can be calculated after the multiplication of groundwater table (piezometric) area by g_{Rs} Eq. (4.70). If the risk attachments to various g_{Rs} values are known then it is simple to obtain the abstractable groundwater volumes for any given risk level. For this purpose, the risk model for g_{Rs} can be developed similar to previous risk graphs by employing the logarithmic normal pdf model (Figure 4.42).

One of the most significant parameters for groundwater recharge calculations is the porosity of the unsaturated layer material (Chapter 2, Section 2.7.1). Porosities of specific materials are listed by considerations from Davis (1969) and for arid areas by Red Sea Mining Co. (1986) as in Table 4.17.

It is not well-known how different types of materials are mixed in any alluvium layers, and therefore, they are treated as uncertain percentages and Figure 4.43 indicates the general porosity values with a suitable logarithmic normal pdf model.

4.14.4 Groundwater Recharge Volume Risk Calculation

The mixture and distribution of different porosity materials within the saturation layer

FIGURE 4.42 Specific groundwater capacity risk model.

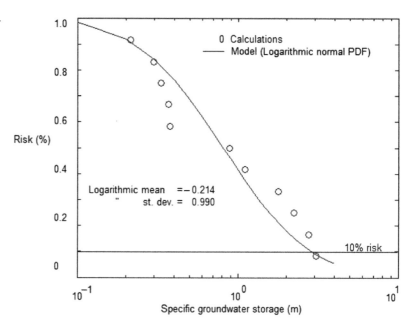

FIGURE 4.43 General porosity
risk levels in nature.

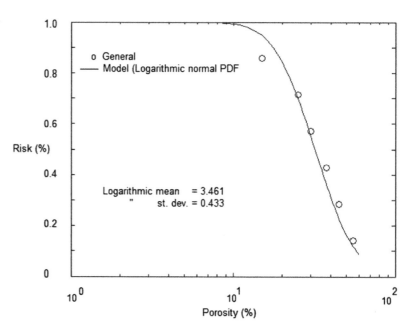

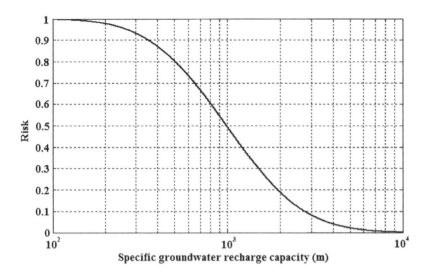

FIGURE 4.44 Specific groundwater recharge risk graph.

are independent from the saturation thickness. It is possible to calculate the logarithmic mean and standard deviation of the specific groundwater recharge capacity as the summation of porosity and unsaturated thickness logarithmic means and standard deviations. Hence, the logarithmic mean and standard deviation of the specific groundwater recharge capacity are (from Figures 4.42–4.44) $3.473 + 3.461 = 6.934$ and $0.377 + 0.433 = 0.80$, respectively. With these values at hand, the risk model of specific groundwater recharge capacity can be developed as a logarithmic normal pdf and its representation is given in Figure 4.44.

4.15 CLIMATE CHANGE AND GROUNDWATER RECHARGE

Human induced climate change in the coming decades may further affect groundwater resources in several ways, including changes in groundwater recharge, changes in annual and seasonal distribution of precipitation and temperature, more severe and longer lasting droughts, frequent floods and flash floods, changes in evapotranspiration as a result of vegetation cover, and possible increase for groundwater demands. Shallow aquifers that supply much of the water in streams, lakes and wetlands are likely to be the part of groundwater system most sensitive to climate change. However, little attention has been directed at determining the possible effects of climate change on shallow aquifers and their interaction with surface water (Alley, 2001).

Impacts of climate change on groundwater are poorly understood and the relationship between climate variables and groundwater is more complicated than that with surface water. Groundwater is the major freshwater source especially for arid and semiarid regions, but unfortunately there has been very little attention or study on the potential climate change effects on these freshwater resources. Most of the works are concentrated on humid regions. Aquifers in arid and semiarid regions are replenished by floods at possible recharge outcrop areas through fractured and fissured rocks, solution cavities in dolomite or limestone geological setups as well as through main stream channels of Quaternary alluvium deposits (Section 4.11). At convenient places along the main channel engineering infrastructures such as levees, dikes, successive small-scale groundwater recharge dams may be constructed for groundwater recharge augmentation (Section 4.5.4, Figure 4.8).

Groundwater and climate are linked in many ways. Groundwater hydrologists and hydrogeologists need to be more attuned to the effects of climate. The links between groundwater and climate should receive greater attention as part of the drought monitoring and assessments, studies of climate change and variability, analysis of water quality, and assessment of the availability and sustainability of groundwater resources and their management. Groundwater level responses to climate variation have a time delay. Without proper consideration to variations in aquifer recharge and sound pumping strategies, the water resources of aquifer could be severely impacted under a warmer climate. Many aquifer regions are very vulnerable to climate-change impacts because of the following reasons.

1. The regions that are largely dependent on the aquifer to meet municipal, agricultural, industrial/military, and recreational water demands, with limited large-scale alternative water supplies are expected to be subject to large climatic variability.
2. There is a strong linkage between climatic inputs, precipitation, and regional hydrology, through the conversion of rainfall to runoff and runoff to aquifer recharge by drainage basin surface and especially stream bed seepage.
3. The historical climatic record shows large variability in precipitation and the occurrence of occasional multiyear droughts (North et al., 1995) which can reduce natural aquifer recharge to negligible level, (Loaiciga et al., 2000).

Due to natural replenishment of groundwater resource by infiltration after each precipitation event and subsequent percolation of water through geologic materials, a decline in precipitation or an increase in evapotranspiration would result in a decline in recharge, possibly resulting in groundwater table level declines.

Changes in any key climatic variables could significantly alter recharge rates for major aquifer systems and thus the sustainable yield of groundwater in the region. Chen et al. (2002) showed that that climate trends have good correlations with groundwater level fluctuations. They also mentioned that both precipitation and annual mean temperature display a strong correlation with annual groundwater levels in this aquifer. In areas where the aquifer is found at shallow depth, temperature has a greater influence than precipitation on groundwater levels. Results suggest that a trend of increasing temperatures, predicted by global climate models for some regions, may reduce net recharge and affect groundwater levels (Chen et al., 2004; Şen et al., 2012a).

Climate change impact predictions and development of adaptation strategies are essential for ensuring a sustainable groundwater supply in many regions. However, potential impacts are difficult to accurately assess, as the influence of climate change on groundwater levels cannot be detected immediately.

The potential impacts of climate change on groundwater supplies in the carbonate aquifers could include the following points (Chen et al., 2004).

1. Less groundwater recharge available due to more evaporative loss of surface water and less precipitation in the winter and spring.
2. Longer residence time due to changes in hydraulic properties in the aquifer after prolonged droughts.
3. Groundwater quality could deteriorate as a result of saline water intrusion. In response to climate change fresh groundwater heads may drop relative to saline waters.

Consideration of climate variations can be a key factor in ensuring the proper management of groundwater resources. Climate is of utmost importance to groundwater at seasonal to decadal scales. Groundwater systems tend to respond more slowly to variability in climate conditions than do surface water systems. As surface water storage becomes more limited, use of groundwater storage to modulate the effects of droughts increases in importance, as do potential enhancements by artificial recharge (Alley, 2001).

Unconfined or shallow aquifers that are in contact with present day hydrological cycle are affected by climate change more significantly. Deep and especially confined aquifers are not in contact with the present day hydrological cycle, and therefore, their effect from climate change is virtually negligible (Loáiciga et al., 1996).

Unsaturated zone has a unique capability in helping to assess impacts of climate change on groundwater resources. The potential impacts of climate change can be assessed by focusing on porous, fractured, and karstic (carbonate rock, dolomite, limestone) aquifer systems. Especially, fractured and karstic aquifers are the most responsive to changes in recharge as typically they have low specific yields (i.e., they have drainable porosities) in comparison with porous flow systems. Karstic rocks are soluble and the aquifers might show exacerbated water table lowering if predicted increases in atmospheric CO_2 content along with temperature rise induce rapid enlargement of fracture apertures and enlargement in the solution cavities. Dissolution of carbonate rocks (karstic media) might become more vigorous by time and accordingly the hardness of groundwater sources is expected to increase, leading to possibly unacceptable water quality.

Relative sea-level rise adversely affects groundwater aquifers and freshwater coastal ecosystems. Rising sea level causes an increase in the intrusion of salt water into coastal aquifers. Other impacts of sea-level rise are likely to include changes in salinity distribution in estuaries, altered coastal circulation patterns, destruction of transportation infrastructure in low-lying areas, and increased pressure on coastal levee systems. Higher sea levels associated with thermal expansion of the oceans and increased melting of glaciers will push salt water further inland in rivers, deltas, and coastal aquifers. It is well understood that such advances would adversely affect the quality and quantity of freshwater supplies in many coastal areas.

Application of the climate change scenarios and software is necessary to forecast changes in aquifer geometries, hydraulic parameters (permeability, storage coefficient), flows, water balances, and water quality.

4.16 UPCONING

In nature fresh water layers are always above saline water storages and they are exploited easily especilly at the coastal unconfined aquifers. These fresh waters are withdrawn by partially penetrating wells where well bottom remains in the fresh water layer, but if more than enough water is withdrawn then the stagnant lower saline water rises above due to the upward potential gradient that is caused by pumps as in Figure 4.45.

According to Ghyben–Herzberg theory the interface rise is $[\rho_f/(\rho_s - \rho_f)]xs$, where ρ_f and ρ_s are the density of fresh and saline water, respectively and s is the drawdown of the water table. The first upconing calculation approximations can be achieved according to Ghyben–Herzbery theory. It has been suggested by McWhorter and Sunada (1977) and Motz (1992, 1994) that the interface remains stable provided that it does not rise more than one third of the distance from the undisturbed (initial) interface to the bottom of the borehole. In case of the steady state flow the well discharge is proportional to the saturation depth between the interface and water table multiplied by the hydraulic gradient.

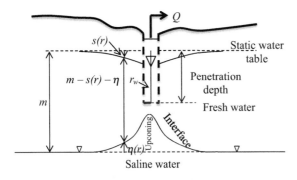

FIGURE 4.45 Upconing.

Considering the Dupuit's assumptions one can write the water balance equation as,

$$Q = 2\pi rK[m - s(r) - \eta(r)]\frac{ds}{dr} \qquad (4.72)$$

Integration of this expression is possible provided that the upconing amount can be expressed in terms of drawdown. Herein, it is assumed that $\eta(r) = cs(r)$, where c is referred to as the upconing proportionality constant. Its substitution, into Eq. (4.72) and after the simple integration operation by taking into consideration the measurements $s(r_1)$ and $s(r_2)$ at distances r_1 and r_2, results in,

$$Q = \frac{2\pi\,T[1 - (1 + c)/m]}{\ln(r_2/r_1)} \qquad (4.73)$$

The ratio $(1 + c)/m$ is always less than 1, and hence, $C = 1 - (1 + c)/m$ is also less than 1. In case of low upconing even though the aquifer is unconfined the steady state confined aquifer formulation (Eq. (3.16)) can be used, because the saline water plays the role of lower confining layer. The more the salinity the better is the role of lower confining layer. Furthermore, the same confined aquifer formulation is valid when the drawdown at the main well is small. After all that have been explained, it is obvious that

at coastal aquifers the unconfined aquifer formulation (Eq. (4.2)) is not valid.

References

Ahmad, N., 1998. Evaluation of Groundwater Resources in the Upper Middle Part of Chaj Doab Area, Pakistan (Unpublished Ph. D. thesis). Technical University of Istanbul, Istanbul, Turkey; p. 308.

Al-Sefry, S., Şen, Z., Al-Ghamdi, S. A., Al-Ashi, W., and Al-Baradi, W. Strategic ground water storage of wadi Fatimah — Makkah region, Saudi Arabia, 2004: Hydrogeology Project Team, Final Report. Saudi Geological Survey.

Al-Yamani, M.S., Şen, Z., 1993. Determination of hydraulic conductivity from grain-size distribution curves. Groundwater 31, 551—555.

Alley, W.M., 2001. Ground water and climate. Ground Water 39, 161.

Amin, M.I., Qazi, R., Downing, T.E., 1983. Efficiency of infiltration galleries as a source of water in arid lands. Int. Symp. Water Res. Saudi Arabia 1, A-107.

Batu, V., 1998. Aquifer Hydraulics: A Comprehensive Guide to Hydrogeologic Data Analysis. John Wiley and Sons, Inc., New York, 727 p.

Bazuhair, A.S., Wood, W.W., 1996. Chloride mass-balance method for estimating ground water recharge in arid areas: examples from western Saudi Arabia. J. Hydrol. 186, 153—159.

Bear, J., 1979. Hydraulics of Groundwater. McGraw-Hill, New York.

Benjamin, J.R., Cornell, C.A., 1970. Probability, Statistics, and Decision for Civil Engineers. McGraw Hill, New York, NY.

Boulton, N.S., 1963. Analysis of data from nonequilibrium pumping test allowing for delayed yield from storage. Proc. Inst. Civil Eng. 26. London.

B.R.G.M. Hydrogeological investigation of Al-Wasia aquifer in the eastern province of Saudi Arabia, regional study, 1976: Final Report. Ministry of Agriculture and Water Riyadh, Saudi Arabia.

Caro, R., Eagleson, P.S., 1981. Estimation aquifer recharge due to rainfall. J. Hydrol 53, 185—211.

Carter, R.C., Morgulis, E.D., Dottridge, J., Agbo, J.U., 1994. Groundwater modeling with limited data: a case study in a semi-arid dune field of northeast Nigeria. Quart. J. Eng. Geol 27, S85—S94.

Chen, Z.H., Grasby, S.E., Osadetz, K.G., 2002. Predicting average annual groundwater levels from climatic variables: an empirical model. J. Hydrology 260 (1—4), 102—117.

Chen, Z., Grasby, S., Osadetz, K., 2004. Relation between climate variability and groundwater levels in the upper carbonate aquifer, Southern Manitoba. Can. J. Hydrol 290, 43—62.

Chow, V.T., 1964. Handbook of Applied Hydrology. McGraw-Hill Book Co.

Clark, I.D., Fritz, P., 1997. Environmental Isotopes in Hydrogeology. Lewis Publishers, New York.

Davis, S.N., 1969. Porosity and permeability of natural materials. In: DeWeist, R.J.H. (Ed.), Flow through Porous Media. Academic Press, New York, pp. 54–89.

Dincer, T., Al-Mugren, A., Zimmerman, V., 1974. Study of infiltration and recharge through the sand dunes in arid zones with special reference to the stable isotopes and thermonuclear tritium. J. Hydrol 23, 79–109.

Dupuit, J., 1863. Etude theoretique et pratique sur le movement des eaux dans les cannaux decouverts et a tranvers les terrains permeables. Dunod, Paris.

Eagleson, P.S., 1979. The annual water balance. J. Hydraul., ASCE 105 (HY8), 923–941.

Edmunds, W.M., Fellman, E., Goni, I.B., Prudhomme, C., 2002. Spatial and temporal distribution of groundwater recharge in northern Nigiria. Hydrogen J. 10, 205–215.

Flint, A., Flint, L., Kwicklis, E., Bodvarsson, G., 2002. Estimation recharge at Yucca mountain, Nevada. USA, comparison methods. Hydrogen J. 10, 180–204.

Forchheimer, P.H., 1898. Grundwasserspiegel bei brunnenanlagen. Z. Osterreichhissheingenieur Architecten Ver 44, 629–635.

Forchheimer, P.H., 1901. Wasserbewegung durch Boden. Zitschrifft Ver. Dtsch. Ing (49), 1736–1749. No.50: 1781–1788.

Fritz, P., Fontes, J.C., 1980. Handbook of Environmental Isotope Geochemistry. Elsevier, Amsterdam.

Gerhart, J.M., 1986. Ground-water Recharge and its Effects on Nitrate Concentration beneath a Manured Field Site in Pennsylvania, vol. 24. No. 4: 483–489.

Hantush, M.S., 1964. Hydraulics of Wells. In: Chow, V.T. (Ed.), Advances in Hydroscience. Academic Press, New York, pp. 282–432.

Healy, R.W., Cook, P.G., 2002. Using groundwater levels to estimate recharge. Hydrogeol. J. 10, 91–109.

IAEA, 1980. Arid zone hydrology: investigations with isotope techniques. In: Proceeding of an Advisory Group Meeting. IAEA, Vienna.

IAEA, 1983. Paleoclimate and paleowaters: a collection of environmental isotope studies. In: Proceeding of an Advisory Group Meeting. IAEA, Vienna.

International Panel on Climate Change (IPCC), 2007. Fourth Assessment Report: Climate Change. Entered on 13 October 2013. http://www.ipcc.ch/publications_and_data/publications_and_data_reports.shtml#.Ulr3iCdrPIU.

Italconsult, 1969. Water and Agricultural Development Studies for Area IV (Eastern Province, Saudi Arabia: Unpublished report to Ministry of Agriculture and Water, Riyadh, Saudi Arabia).

Jacob, C.E., 1940. On the flow of water in an elastic artesian aquifer. Am. Geophys. Union Trans 72 (Part II), 574–586.

Jacob, C., 1963. Recovery Method for Determining the Coefficient of Transmissibility. Water Supply Paper 15361. U.S. Geological Survey, Washington D.C.

Kattan, Z., 2001. Use of hydrochemistry and environmental isotopes for evaluation of groundwater in the Paleogene limestone aquifer of the Ras Al-Ain area (Syrian Jazirah). J. Environ. Geol. 41, 128–144.

Kozeny, J., 1933. Theorie und Berehnung der Brunnen. Wasserkraft Wasserwirtschaft 39, 101.

Lerner, D.N., Issar, A.S., Simmers, I., 1990. International Association of Hydrogeologists. UNESCO International Hydrogeological Program. Hannover, Germany. Groundwater Recharge- a Guide to Understanding and Estimating Natural Recharge, vol. 8, pp. 345.

Linsley, R.K., Kohler, M.A., Paulhus, J.L.H., 1982. Hydrology for Engineers. McGraw-Hill Book Co., 508 pp.

Loaiciga, H.A., Maidment, H.A., D.R., Valdes, J.B., 2000. Climate-change impacts in a regional karst aquifer, Texas, USA. J. Hydrol 227, 173–194.

Loáiciga, H.A., Valdes, J.B., Vogel, R., Garvey, J., Schwarz, H., 1996. Global warming and the hydrological cycle. J. Hydrol 174, 83–127.

Mansur, C.I., Kaufman, R.I., 1962. Dewatering Foundation Engineering. In: Leonards, G.A. (Ed.). McGraw-Hill, New York, pp. 241–350.

McWhorter, D.B., Sunada, D.K., 1977. Ground Water Hydrology and Hydraulics. Water Resources Publications, Highland Ranch, Colorado.

Meinzer, O.E., 1923. The Occurrence of Groundwater in the United States, with a Discussion of Principles. U.S. Geological Survey Water Supply Paper 489.

Meinzer, O.E., Stearns, N.D., 1929. A Study of Ground Water in the Pomperaug, Connecticut: With Special Reference to Intake and Discharge. USGS. Water Supply Paper, 597-B: 73–146.

Milanovic, P., 1981. Karst Hydrogeology. Water Resources Publications, Littleton, Colorado, USA.

Motz, L.H., 1992. Salt-water upconing in an aquifer overlain by a leaky confining bed. J. Ground Water 30 (2), 192–198.

Motz, L.H., 1994. Predicting salt-water upconing due to wellfield pumping. In: Soveri, J., Suokko, T. (Eds.), Future Groundwater Resources at Risk (Proceedings of the Helsinki Conference).

Muskat, M., 1937. The Flow of Homogeneous Fluids through Porous Media. McGraw-Hill.

Nahrgang, G., 1954. Zur Theorie des vollkommenen und unvollkommenen Brunnens, 43 p.

Neuman, S.P., 1972. Theory of flow in unconfined aquifers considering delayed response of the water table. Water Resour. Res 8 (4), 1031–1045.

Neuman, S.P., 1973. Saturated-unsaturated seepage by finite elements. J. Hydraul. Div., ASCE, HY12, 2233–2250.

North, et al., 1995. Multi-Year Drought Climate Change.

Rasmussen, W.C., Andreasen, G.E., 1959. Hydrologic budget of the Beaverdam Creek Basin. U.S. Geol. Surv, 106. Water Suppl Paper 1472.

Rushton, K.R., 2002. Will reduction in groundwater abstraction improve low river flow?. In: Hiscock, K.M., Rivett, M.O., Davison, R.M. (Eds.), Sustainable Groundwater Development, vol. 193. Geological Society, London, Special Publications, pp. 193–204.

Salih, A.M.A., Sendil, U., 1984. Evapotranspiration under extremely arid climates. J. Irrig. Drain. Eng. 110 (3), 289–303.

Scanlon, B.R., Healy, R.W., Cook, P.G., 2002. Choosing appropriate techniques for quantifying groundwater recharge. Hydrogeol. J. 10, 18–39.

Schicht, R.J., Walton, W.C., 1961. Hydrologic budgets for three small watersheds in Illinois. State Water Surv. Rep. Invest 40, 40.

Şen, Z., 1983. Hydrology of Saudi Arabia. In: Proceeding of Water Resources in Saudi Arabia, pp. A68–A94 (Riyadh, Saudi Arabia).

Şen, Z., 1987. Storage coefficient determination from quasi-steady state flow. Nordic Hydrol. 18, 101–110.

Şen, Z., 1995. Applied Hydrogeology for Engineers and Scientists. Lewis Publishers, Boca Raton, 444 pp.

Şen, Z., 1999. Simple probabilistic and statistical risk calculations in an aquifer. Ground Water 37 (5), 748–754.

Şen, Z., 2008. Wadi Hydrology. Taylor and Francis Group, CRC Press, Boca Raton, 347 pp.

Şen, Z., Al Alsheikh, A.S., Alamoud, A.M., Al-Hamid, A.A., El-Sebaay, A.S., Abu-Risheh, A.W., 2012a. Quadrangle Downscaling Model and Water Harvesting in Arid Regions: Riyadh Case.

Şen, Z., Al Alsheikh, A., Al-Dakheel, A.M., Alamoud, A.I., Alhamid, A.A., El-Sebaay, A.S., Abu-Risheh, A.W., 2012b. Climate change and Water Harvesting possibilities in arid regions. Int. J. Glob. Warming. 3 (4), 2011.

Shi, J.A., Wang, Q., Chen, G.J., Wang, G.Y., Zhang, Z.N., 2000. Isotopic geochemistry of the groundwater system in arid and semiarid areas and its significance: a case study in Shiyang River basin, Gansu province, northern China. J. Environ. Geol. 40 (4–5), 557–565.

Simmer, I., 1988. Estimation of Natural Groundwater Recharge. In: NATO Advanced Study Institute Series C, vol. 222.

Simmers, I., 1997. Recharge of Phreatic Aquifers in (Semi-) Arid Areas. In: IAH Int. Contrib. Hydrogeol, vol. 19. AA Balkema, Rotterdam, 277 pp.

SMMP (1975). Riyadh - Additional Water Resources Study - Wasia Hydrogeology. Ministry of Agriculture and Water

p. Riyadh Additional Water Resources Study - Appendix A3 - Wasia Hydrogeology. npubl. Rep., Ministry of Agriculture and Water, Riyadh. 230p.

Sogreah, 1968. Water and Agricultural Development Survey for Area IV Ministry of Agriculture and Water (Riyadh, Saudi Arabia).

Streltsova, T.D., 1972. Unsteady radial flow in an unconfined aquifer. Water Resour. Res 8 (8), 1059–1066.

Streltsova, T.D., 1973. Flow near a pumped well in an unconfined aquifer under nonsteady conditions. Water Resour. Res 9 (1), 227–235.

Subyani, A., 2004a. Study evaluation of groundwater resources in wadi yalamlam and wadi Adam basins, makkah al-Mukarramah. Int. Conf. Water Resour. Arid Environ, 1–18.

Subyani, A.M., 2004b. Use of chloride mass-balance and environmental isotopes for evaluation of groundwater recharge in the alluvial aquifer, Wadi Tharad, Western Saudi Arabia. J. Environ. Geol. 46, 741–749.

Subyani, A., Şen, Z., 1991. Study of recharge outcrop relation of the Wasia aquifer in Central Saudi Arabia. J. King Abdulaziz Univ. (Earth Sci.) 4, 137–147.

Subyani, A.M., Şen, Z., 2006. Refined Chloride Mass-Balance method and its application. Hydrol. Process 20, 4373–4380.

Subyani, A.M., Bayumi, T.H., Matsah, M.I. and Al-Garni, M.A. Quantitative and Qualitative Evaluation of Water Resources in Wadi Malakan and Wadi Adam Basins, Makkah Area, Project No. 205/422, 2004: Final Report. King Abdulaziz University, Jeddah; p. 161.

Szechy, C., 1961. Foundation Failures. Concrete Publications Limited, London.

Theim, G., 1906. Hydrologische Methoden. J.M. Gebhart, Leipzig, 56.

Todd, D.K., 1980. Groundwater hydrology. John Wiley and Sons, 535.

Troxell, H.C., 1936. The diurnal fluctuation in the groundwater and flow of the Santa Ana river and its meaning. Eos Trans. AGU 17 (4), 496–504.

White, W.N., 1932. A Method of Estimating Ground-water Supplies Based on Discharge by Plants and Evaporation from Soil–Results of Investigations in Escalante Valley, Utah. USGS Water Supply Paper, 659-A: 1–150.

Wood, W.W., Imes, J.L., 1995. How wet is wet? Precipitation constraints on late quaternary climate in the Southern Arabian Peninsula. J. Hydrol. 164, 263–268.

Wood, W.W., 1999. Use and misuse of the chloride-mass balance method in estimating ground water recharge. Groundwater 37, 2–3.

Wood, W.W., Sanford, W.E., 1995. Chemical and isotopic methods for quantifying groundwater recharge in a regional, semiarid environment. Groundwater 33, 458–468.

Copyright © 2015 Elsevier Inc. All rights reserved.

5.1 HYDROCHEMISTRY

Groundwater contains a wide variety of dissolved inorganic constituents as a result of chemical interactions with geological materials and to lesser extent contributions from the atmosphere. The study of hydrochemistry is of prime importance in deciding about the quality of groundwater supply. Hydrochemistry helps to evaluate hydrogeochemical processes responsible for temporal and spatial changes in the chemistry of groundwater. Information about changes in groundwater chemistry and hydrogeochemical processes is available in literature (Miller, 1991; Kimblin, 1995; Mayo and Loucks, 1995; Hudson and Golding, 1997). Groundwater chemistry is affected from many factors such as movement through the rocks, recycling by irrigation practices, natural or artificial recharge and discharge. As water flows, it assumes a diagnostic chemical composition as a result of interaction with lithological and stratigraphical frameworks. The term hydrochemical facies is used to describe the spatial variations in aquifer groundwater quality variations. The facies are a function of the lithology, solution kinetics, and flow patterns of the aquifer (Back, 1960).

Hydrochemical results provide a basis for the characterization of groundwater within each hydrogeological unit. From this information, it is possible to draw inferences concerning the processes that may occur within the aquifers, the effect of groundwater abstraction on the aquifer, host rock, and the suitability of ground waters for use.

Groundwater flow triggers various hydrochemical, mechanical and solution activities in the subsurface. As a result of these processes there appears temporal and spatial changes in the chemical composition, porosity, hydraulic conductivity, pore fluid pressure, fissures, fractures, and solution cavities.

5.2 IONIC CONSTITUENTS

Subsurface water except in the vicinity of recharge areas is rarely found without dissolution of minerals from the host geological environment through which it makes its way according to hydraulic gradients. Water quality determination is the primary prerequisite for public, agricultural, and industrial supply purposes. It is a function of physical, chemical, and biological characteristics. Water has the highest capability to dissolve salts as compared to any other solvents found in nature. Its dissolving power stems from the molecular structure as shown in Figure 5.1.

Water is formed by the combination of two hydrogen and a single oxygen atoms. The oxygen atom is bonded to the hydrogen atoms unsymmetrically with a bond angle of 105°. This arrangement gives rise to an unbalanced electrical charge that imparts a polar characteristic to the molecule. In liquid form, water is a weak

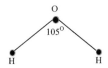

FIGURE 5.1 **Structure of water molecule.**

ionic compound, which reacts according to the following chemical reaction.

$$H_2O \Leftrightarrow H^+ + OH^-$$

Herein, + and − signs imply charge of ionic species. The natural composition of groundwater is a combination of atmospheric activities, especially rainfall, lithospheric weathering of rocks and soils, subsurface chemical reactions, and mixtures of various groundwater sources.

Carbon dioxide (CO_2) helps to increase the solubility of minerals. The origin of hydrochemistry evolution process starts in the atmosphere by rainwater, which dissolves mineral particles from pollutant sources or marine spray in coastal areas. Supply of oxygen from the atmosphere produces organic matter oxidization in the soil, and finally, CO_2 is generated. Weak carbonic acid, H_2CO_3, is the most important acid production agent in the soil.

After rainfall and infiltration processes, CO_2 laden water dissolves calcium (Ca^{2+}), magnesium (Mg^{2+}), sodium (Na^+), and potassium (K^+) along its flow path in subsurface geological formations. Consumption of weak carbonic acid (H_2CO_3) in the subsurface enhances CO_2 in the soil, air in the voids, and accelerates oxidation of organic matter providing respiration to root zones. Groundwater in recharge areas has lower dissolved solid composition than in discharge zones.

In general, the groundwater quality is suitable for any use in the upstream reaches of a drainage basin, but it deteriorates toward the middle and downstream parts. Generally, in sedimentary basins, such as the Quaternary alluvium deposits, naturally the groundwater movement is from the upstream toward downstream with some lateral effects. After the rainfall occurrence, in the upstream the subsequent infiltration water is rich in bicarbonates (HCO_3^-). Bicarbonate type of groundwater becomes sulfate (SO_4^{2-}) type in the middle stream, and finally, turns into chlorine (Cl^-) water types at downstream. The chemical evolution of groundwater from upstream to downstream occurs along the following chain modification.

$$HCO_3^- \Rightarrow HCO_3^- + SO_4^{2-} \Rightarrow SO_4^{2-} + HCO_3^- \Rightarrow$$
$$SO_4^{2-} + Cl^- \Rightarrow Cl^- + SO_4^{2-} \Rightarrow Cl^- + Ca^{2+} \Rightarrow$$
$$Ca^{2+} + Mg^{2+} \Rightarrow Ca^{2+} + Na^+ \Rightarrow Na^+Ca^{2+} \Rightarrow Na^+$$

Groundwater starting from H_2CO_3 composition evolves during its travel through the saturation zone toward seawater composition as the age increases according to residence time. This chain of hydrochemical reaction is possible, provided that there are mineral availability and solubility within the geological formations.

Igneous rock passing groundwater has slight mineral matter because such rocks have very difficultly soluble constituents. Such groundwater has silica from the silicate minerals. Sedimentary rock groundwater has many solvable minerals and most of the constituents come from these rocks. They are rich in Ca^{2+} and Na^+ as cations and among the dominant anions are HCO_3^- and SO_4^{2-} with little Cl^- that may inborn from the seawater intrusion and connate water. Limestone terrains provide bicarbonate and calcium ions.

5.3 MAJOR IONS

The major inorganic species which constitute the overall quality groundwater are Ca^{2+}, Mg^{2+}, Na^+, and K^+, major cations, and Cl^{-1}, SO_4^{2-}, CO_3^{2-}, and HCO_3^{-1}, major anions. The total concentration of these eight major ions normally comprises more than 90% of the total dissolved solids in groundwater, regardless of whether the water is dilute or has salinity greater than seawater. These inorganic constituents occur in ionic forms in water solutions.

5.3.1 Major Cations (Ca^{2+}, Mg^{2+}, Na^+, K^+)

Cation concentrations are not reliable tools to read the history of groundwater. Large variations in cation concentrations occur due to exchange processes, which give rise to reversals in cation concentrations. Ca^{2+} and Mg^{2+} are similar in many respects and are dissolved freely from many rocks and soils (Salomons and Forstner, 1984). Ca^{2+} is an abundant constituent in waters associated with calcite, dolomite, and gypsum. Sources of Mg^{2+} include dolomite, magnetite, and micas. Concentrations of Ca^{2+} and Mg^{2+} found in natural waters vary over a wide range depending on the history of the water sample with respect to geological formations and the geochemical characteristics. In general, concentrations of Ca^{2+} in freshwaters are somewhat greater than those of Mg^{2+}, because of the greater abundance of Ca^{2+} in the Earth's crust. Mg^{2+} concentrations may often be greater than the Ca^{2+} concentrations especially in oceanic or estuarine waters. The Ca^{2+} and Mg^{2+}, so called earth alkaline metals, enter into ground water by the dissolution of carbonate minerals in water containing CO_2. Traces to about 30 ppm are characteristic in magmatic aquifer; up to 100 ppm are common in the sedimentary terrain; up to 300 ppm occur in carbonate aquifers; and up to 600 ppm are found in aquifers containing gypsum or in contact with gypsum layers. Ca^{2+} and Mg^{2+} are susceptible to ion exchange with Na^+ and vice versa.

Na^+ is the most important and abundant of the alkali metals in natural waters. Nearly all Na^+ compounds are readily soluble, and since it does not take part in significant precipitation reactions, Na^+ generally remains in solution. The principal source of Na^+ in water is the evaporate sediments, although sewage, industrial wastes, and oilfield drainage also contribute significant amounts. All waters contain some Na^+ ranging from a few ppm where leaching has removed most soluble rock minerals up to 100,000 ppm provided that water is in contact with evaporate beds. In seawater, Na^+ is the most abundant cation; its concentration is about 10,000 ppm. Groundwater pollution by sodium chloride (NaCl) is the unavoidable byproduct

of human activity, such as return flows from irrigation and disposal of industrial and urban wastes. The process usually takes the form of a very slow "salt creep" and may remain masked for a long time by the noise inherent in salinity data. Intensive groundwater abstraction creates water level depressions with little or no outflow and thus accentuates the process.

K^+ is also in the alkali-metals group and it is similar in many respects to Na^+, although it differs significantly in several factors, which relate to its presence in water and its effects on other uses. While Na^+ is readily available for solution, rocks containing significant amounts of K^+ are generally resistant to weathering, and evaporates containing large amounts of K^+ are uncommon. In addition, several processes tend selectively to return K^+ to the solid phase by refixing it in new minerals, which resist attacks. Thus K^+ is normally present in water in smaller amounts than Na^+. Concentrations of K^+ in most natural waters are generally less than 10 ppm (Chow, 1964).

5.3.2 Major Anions (SO_4^{2-}, HCO_3^{2-}, Cl^-, CO_3^-)

Different anion concentration measurements in a groundwater sample at a particular time provide the inference about the distance traveled by the groundwater from the source areas and also give indication about the contact time with soluble minerals. Availability of the mineral types in the way of the groundwater may also be revealed by the chemical analyses of the anions.

Chlorides (Cl^-) are present in all natural waters originating from many sedimentary rocks and particularly the evaporates. The dominant source of Cl^- is the ocean, with approximately 20,000 ppm. Chloride salts precipitate from solution only at concentrations exceeding that of seawater. Because of this conservative property (i.e., the tendency to remain in solution) Cl^- is a reliable indicator of groundwater chemical processes. Decreasing Cl^- concentrations

TABLE 5.1   Ionic Constituent Equivalent Weights

Ion	Atomic Weight (g/mol)	Charge	Equivalent Weight
CATION			
Calcium (Ca^{2+})	40.08	+2	20.04
Magnesium (Mg^{2+})	24.32	+2	12.16
Sodium (Na^+)	23.00	+1	23.00
Potassium (K^+)	39.10	+1	39.10
ANIONS			
Carbonate (CO_3^{2-})	60.01	−2	30.00
Bicarbonate (HCO_3^-)	61.01	−1	61.01
Sulfate (SO_4^{2-})	96.06	−2	48.03
Chloride (Cl^-)	35.46	−1	35.46
Nitrate (NO_3^-)	62.01	−1	62.01

show fresh water entrance into aquifer. Increasing concentrations can be caused by several processes such as mixing with more saline water, solution of solid salt from the rock, and concentration of Cl^- by evapotranspiration from a shallow unconfined aquifer (water table). In a well-flushed aquifer the small concentrations of Cl^- that stem from airborne salts may serve as data for estimating natural replenishment. In mixed waters Cl^- faithfully reflects the mixing ratio of the end members (Chow, 1964 and, Mandel and Shiftan, 1981). Chloride concentration mass balance studies help to determine the groundwater recharge amounts at the upstream of drainage basins (Chapter 4).

In most natural waters, SO_4^{2-} is found in smaller concentrations than Cl^-. The major source of SO_4^{2-} in fresh water is evaporites, mainly gypsum and anhydrite. Other sources are airborne SO_4^{2-} compounds originating from the sea and from dust, gaseous sulfur oxides produced by the combustion of fossil fuel and washed down by rainfall, decaying organic matter, volcanic exhalations, and the weathering products of some magmatic rocks. Concentrations in excess of about 30 ppm in groundwater suggest contact with gypsum-bearing rocks. The maximum concentration in fresh groundwater is 1360 ppm and higher concentrations may be reached in saline waters because of the other ions' presence. SO_4^{2-} is not appreciably affected by adsorption or ion exchange processes. Table 5.1 indicates major ionic constituents and their atomic weights, charges, and equivalent weights.

5.4 CHEMICAL UNITS AND BALANCE

In general, water sample analyses in the laboratory are given in milligram per liter (mg/l) for the major ions; μg/l for the trace elements and some rare elements are expressed in ng/l (nanograms/liter). It is known that 1 mg is 0.001 g and 1 l of water is very close to 1000 g, and mg/l is equivalent to parts per million (ppm), while μg/l is equivalent to parts per billion (ppb). Pure water at standard temperature and pressure has a density of 1 kg/l, therefore, $mg/l = mg/kg = ppm$. Groundwater in the nature has different densities according to salinity, and therefore, an accurate conversion between mg/l and ppm is not possible.

For chemical balance, the ion concentrations are expressed in milliequivalents per liter (meq/l), which is similar to molality, except that the charge on the ion is taken into consideration.

In laboratory analysis, concentrations are usually measured in parts per million but conversion to meq/l takes into account the equivalent weights and charges of the individual ions. As the chemical reactions consume and produce these ions in the natural environment according to their equivalent weights, therefore, this conversion to meq/l is a good approximation and helps in determining the underlying hydrochemical process. The conversion of concentration from ppm to meq/l is possible by the following relations.

$$ppm \cong \frac{\text{grams of solid}}{10^6 \text{gram of solution}} \tag{5.1}$$

$$meq/l \cong \frac{ppm}{\text{equivalentweight}} \tag{5.2}$$

where

$$\text{equivalentweight} \cong \frac{\text{atomic weight}}{\text{charge number on the ion}} \tag{5.3}$$

After expressing the anions and cations in meq/l units, the following two important points provide useful information in practical applications.

1. The sum of cations is equal to the sum of anions within practically acceptable error limits of preferably ±5% or at the maximum ±10%. If the error is outside these ranges then there is either an analytical error, calculation error, or presence of undetermined constituents.

2. The summation of total anions and cations gives the total dissolved solid (TDS) content in the groundwater sample.

Chemical balance can be achieved in terms of meq/l. To ascertain the reliability of the chemical data, the ionic charge balance percentage error, E, criteria is used for each analysis, which is defined as,

$$E \cong 100 \times \frac{\sum_{i=1}^{n}(C_i - A_i)}{\sum_{i=1}^{n}(C_i + A_i)} \tag{5.4}$$

where all the concentrations of cations, C_i, and anions, A_i, are in meq/l.

EXAMPLE 5.1 MOLALITY CALCULATIONS

If a calcium solution is about 39.1 mg/l then calculate the molality of calcium knowing that its atomic weight is 40.08 g/mole (see Table 5.1).

Solution 5.1

According to Eq. (5.2) molality is defined as the division of mg/l by gram per molality, then the molality of this solution is $39.1/40.08 = 0.976$ millimoles per liter (mM/l)

EXAMPLE 5.2 MILLIEQUIVALENTS PER LITER CALCULATION

In a groundwater sampling analysis the amount of calcium (Ca^{+2}) is found about 123.2 ppm. What is its meq/l value?

Solution 5.2

It is known that $ppm = mg/l$. Given 123.2 mg/l can be converted to mM/l by use of Eq. (5.2) leading to $123.2/40.08 = 3.0738$ mM/l. Since calcium is bivalent, the result is $3.0738 \times 2 = 6.1476$ meq/l.

EXAMPLE 5.3
HYDROCHEMICAL BALANCES

The chemical analysis results of a water sample are given in Table 5.2. Convert them to meq/l and check whether there is practically acceptable chemical balance.

Solution 5.3

In order to convert the given mg/l values into meq/l the corresponding atomic weights must be considered from Table 5.1 together with the monovalent or bivalent cases. According to definitions in Eqs (5.1) and (5.2) one can write that

$$meq/l = \frac{(ppm) \times (ion\ charge\ number)}{atomic\ weight}$$

Hence, the final conversion calculations are presented in Table 5.3, where the total anions and cations are 77.562 meq/l and 73.284 meq/l, respectively.

The relative error is $100 \times (77.562 - 73.284)/77.562 = 5.52 > 5$ and hence, the equilibrium cannot be accepted as 5% level. However, it can be accepted at 10%, but the authors of this book would recommend repetition of the chemical analysis.

EXAMPLE 5.4: CHEMICAL BALANCE

In Table 5.4 there are eight groundwater sample hydrochemical analysis results. Notice that there are four cations and three anions in all the samples.

1. Convert each ion into meq/l.
2. Control chemical balance.

Solution 5.4

1. According to Eqs (5.1) and (5.2) by knowing atomic weights and charges from Table 5.1 for each ion the following meq/l table (Table 5.5) can be prepared. The columns 2−8 include the meq/l values for each ion.
2. In order to check for the chemical balance total cation and anion amounts are calculated in meq/l and they are in columns 9 and 10,

TABLE 5.2 Chemical Constituents

Cations (mg/l)				Anions (mg/l)			
Ca^{+2}	Mg^{+1}	Na^+	K^+	Cl^-	SO_4^{-2}	NO_3^-	HCO_3
211.894	126.623	1292.272	15.237	2279.328	256.395	91.523	324.786

TABLE 5.3 Conversion Calculation Results

Ions	Cations				Anions			
	Ca^{+2}	Mg^{+2}	Na^+	K^+	Cl^-	SO_4^{-2}	NO_3^-	HCO_3^-
mg/l	211.894	126.623	1292.272	15.237	2279.328	256.395	91.523	324.786
Atomic weight	40.08	24.32	23.00	39.10	35.56	96.06	62.01	61.01
Charge	2.0	2.0	1.0	1.0	−1.0	−2.0	1.0	−1.0
Meq/l	10.574	10.413	56.186	0.390	64.098	5.338	1.476	5.324

TABLE 5.4 Ion Concentrations

Sample Number	mg/l (ppm)						
	Na$^+$	K$^+$	Ca^{++}	Mg^{++}	Cl$^-$	SO$_4^-$	HCO$_3^-$
1	134.87	11.33	49.20	104.26	195.03	263.20	375.2
2	160.04	3.13	46.80	49.94	141.84	173.87	375.2
3	1786.14	8.60	741.40	856.50	3404.25	4091.25	305.0
4	1387.06	6.64	533.00	476.12	2836.90	2098.94	183.0
5	1126.09	7.89	577.00	508.43	2482.26	2013.92	527.6
6	85.91	3.13	44.00	43.62	65.60	100.86	366.0
7	343.19	4.70	36.00	66.10	277.65	429.90	375.1

TABLE 5.5 Chemical Balance and Control

Sample Number (1)	meq/l							Total Cation (9)	Total Anion (10)	Relative Error (%) (11)
	Na$^+$ (2)	K$^+$ (3)	Ca^{++} (4)	Mg^{++} (5)	Cl$^-$ (6)	SO$_4^{--}$ (7)	HCO$_3^-$ (8)			
1	58.64	0.29	24.55	85.74	55.00	54.80	61.50	169.22	171.30	1.21
2	69.58	0.080	23.35	41.07	40.00	36.20	61.50	134.08	137.70	2.63
3	776.58	0.22	369.96	704.36	960.03	851.81	49.99	1851.1	1861.8	0.57
4	603.07	0.170	265.97	391.55	800.03	437.01	29.99	1260.8	1267.0	0.49
5	489.60	0.202	287.92	418.12	700.02	419.30	86.48	1195.8	1205.8	0.83
6	37.35	0.080	21.96	35.87	18.50	21.00	60.00	95.26	99.50	4.26
7	149.21	0.120	17.96	54.36	78.30	89.51	61.48	221.65	229.29	3.33

respectively. The relative error according to Eq. (3.11) yields the values in the last column of this table. It is obvious that each sample is in practical chemical balance, because all the relative errors are less than 5%.

5.5 GROUNDWATER SAMPLING AND ANALYSIS

Reconnaissance field trips help to gather linguistic information about many aspects of hydrogeology in addition to preliminary measurements. They provide guidance regarding selection of appropriate sampling methods and analytical variables. In order to achieve an effective sampling procedure, it is necessary to consider site-specific geology, hydrology, and environmental aspects. Anthropogenic potential influences are also searched in the area as for their impacts to groundwater quality. In cases of large open areas or surface waters shallow contamination can be carried down into a deeper aquifer and even airborne chemicals may cause trace contamination of groundwater samples, if they enter an open borehole. Groundwater samples are taken from a set of convenient wells, springs, or oases. The groundwater quality determinations and regional quality assessments are possible through water sampling from available wells and springs. Laboratory analysis might yield significant clues about the groundwater quality variations, mixtures, and movements.

The number of wells as well as the extend of inspection region, hydrogeological conditions, possible network complexities, and required analytical aspects provide a basis for an effective sampling plan prior to field excursion. Water samples can be taken according to the following alternatives.

1. If the purpose is to assess the groundwater quality variation spatially then a set of representative wells must be selected for water sampling.
2. For dynamic groundwater quality change one may take water sampling as in the previous step but on monthly, seasonal, or annual periods.
3. Short term local and temporal groundwater quality variations can be traced during a pumping test.
4. Some of the quality variables can be measured easily in the field (pH, electrical conductivity (EC)) and others in the laboratory.

It is recommended that groundwater monitoring should be carried out on a regular basis depending on the purpose of water supply. The water sampling must not be just for mechanical collection, but at each well location after each sample measurement (practically EC) can be compared with the previous ones so as to identify linguistically similar and dissimilar groups, which may help to reduce the number of samplings. Such comparisons are also helpful in the identification of possible mistakes, biases, mishaps, and contamination may be traced in the field.

Prior to an effective field survey and sampling, short-term and long-term goals and information needs must be understood on the basis of available knowledge. A high level of confidence for hydrologic, hydrogeology and hydrochemical sampling must be appreciated and in a way effective sampling procedure must be visualized in the office. Monitoring program must be designed such that in the possible shortest duration the maximum source of information could be obtained. In the meantime, sampling cost minimization must be cared for. Poorly designed monitoring programs cannot help for efficient, sufficient, and objective data collections. In any field sampling there are possible random and human errors. The systematic errors may be due to calibration lack of the used instruments. The instruments must be checked and calibrated prior to any field trip for better and accurate measurements. For an effective and sustainable monitoring program, monitoring well locations, numbers, and constructions are among the most significant decision variables. The monitoring wells must be arranged in such a manner that they capture possible contamination traces.

5.6 COMPOSITE QUALITY INDICATORS

Different ions including salinity, infiltration and permeability, specific ion toxicity, trace element toxicity and hazards due to other factors effect on sensitive crops (FAO, 1985). There are various combinations of the major ions that express collectively the water quality indicators, which can be calculated for each well (groundwater quality sampling points) and then according to given guidelines one can make quality classifications.

5.6.1 Electrical Conductivity (EC)

Depending on the total ion concentration water samples convey electric current, which is referred to as the electrical conductivity (EC). Its measurement in any groundwater sample provides useful means and information in assessing the total dissolved solids content. It is related to the reciprocal of resistance, and hence, it increases with water salinity. The basic unit of conductivity is the mho (from right to left ohms) or siemens. Conductivity is measured in micromhos per centimeter (μmhos/cm) or equivalently micro-siemens per centimeter (μs/cm). Specific EC is defined as the conductance of a milliliter of water at a standard temperature, 25 °C; an increase of 1 °C increases conductance by approximately 2%. The warmer the water, the higher is the conductivity. For this reason, conductivity is reported as conductivity at 25 °C.

The water conductivity is the ability of water to pass an electrical current and it helps to determine the quality of the water. It also provides an indirect way of water salinity measurement. Higher EC readings imply inorganic dissolved solids such as Cl^-, SO_4^-, Na^+, and Ca^{2+}.

Waters that have been in contact with granitic rock environments tend to have lower EC, because granite is composed of more inert materials. Runoff waters that run through clay soils tend to have higher EC, because of the presence of materials that ionize when washed into the water. Groundwater inflows can have the same effects depending on the bedrock they flow through.

Unsuitable waters for human consumption or irrigation activities have high content of EC, which is generally expressed in mg/l. A general salinity scale is given in Table 5.6 (Wilcox, 1955)

TABLE 5.6   EC Classification

EC Range	Quality Classification
<250	Excellent
250–750	Good
750–2000	Permissible
2000–3000	Doubtful
>3000	Unsuitable

EC, electrical conductivity.

5.6.2 Total Dissolved Solid (TDS)

The sum of dissolved constituents' concentrations in a water sample is known as the total dissolved solid (TDS), which can be estimated by adding up the concentrations of all the analyzed constituents in meq/l. It can also be determined indirectly by measuring the electrical conductivity of the water (Section 5.7.1).

Ions are measured collectively through the TDS irrespective of any ion specification. TDS is composed of solid residue made up mostly by inorganic elements and small amounts of organic material. It is obtained after evaporating a certain amount of groundwater volume. According to WHO specification TDS up to 500 mg/l is the highest desirable and up to 1500 mg/l is the permissible maximum. The classification of groundwater according to TDS is given in Table 5.7 (USSL, 1954).

TABLE 5.7   TDS Classification

TDS Range	Quality Classification
<200	Fresh water
200–500	Brackish water
500–1500	Saline water
>1500	Brine water

TDS, total dissolved solid.

EXAMPLE 5.5 TOTAL DISSOLVED SOLID CALCULATION

Table 5.5 includes the meq/l amounts of major ions, where the total anions and cations are also given separately. What is the TDS value for each sample?

Solution 5.5

A total dissolved solid is equal to the summation of anion and cation concentrations. From columns of Table 5.5 the TDS values are calculated as the summation of them and they are given in column 11 of Table 5.8. In the last column of this table are the classifications according to Table 5.7.

5.6.3 Total Hardness (TH)

For domestic and industrial water quality classifications hardness is a very important property. It is dependent mainly on the presence of alkaline earths, Ca^{2+} and Mg^{2+} (WMO, 1977).

Hardness of water is usually expressed as total hardness (TH) given by,

TABLE 5.9 Hardness Classification

Hardness as $CaCO_3$ (ppm)	Water Class
0–75	Soft
75–150	Moderately hard
150–300	Hard
>300	Very hard

$$TH = Ca^{2+} \times \frac{CaCO_3}{Ca^{2+}} + Mg^{2+} \times \frac{CaCO_3}{Mg^{2+}} \quad (5.5)$$

This expression can be written succinctly as,

$$TH = 2.5\,Ca^{++} + 4.1\,Mg^{++} \quad (5.6)$$

where TH and Ca^{2+} and Mg^{2+} concentrations are all in mg/l (Todd, 1980). Waters on the basis of hardness can be classified according to Table 5.9.

Water in boilers causes excessive scale formation (carbonate mineral precipitation) if the hardness is above 60–80 ppm (Freeze and Cherry, 1979).

TABLE 5.8 TDS Values and Classifications

Sample Number (1)	Na$^+$ (2)	K$^+$ (3)	Ca^{++} (4)	Mg^{++} (5)	Cl$^-$ (6)	SO$_4^{--}$ (7)	HCO$_3^-$ (8)	Total Cation (9)	Total Anion (10)	TDS (11)	Classification (12)
						meq/l					
1	58.64	0.29	24.55	85.74	55.00	54.80	61.50	169.22	171.30	340.52	Brackish
2	69.58	0.080	23.35	41.07	40.00	36.20	61.50	134.08	137.70	271.78	Brackish
3	776.58	0.22	369.96	704.36	960.03	851.81	49.99	1851.1	1861.8	3712.90	Brine
4	603.07	0.170	265.97	391.55	800.03	437.01	29.99	1260.8	1267.0	2527.80	Brine
5	489.60	0.202	287.92	418.12	700.02	419.30	86.48	1195.8	1205.8	2401.60	Brine
6	37.35	0.080	21.96	35.87	18.50	21.00	60.00	95.26	99.50	194.76	Fresh
7	149.21	0.120	17.96	54.36	78.30	89.51	61.48	221.65	229.29	450.94	Brackish

TDS, total dissolved solid.

TABLE 5.10　Groundwater Major Ion Concentrations in ppm

	mg/l (ppm)								
	Cations				Anions			TH	Classification
Sample Number (1)	Na^+ (2)	K^+ (3)	Ca^{++} (4)	Mg^{++} (5)	Cl^- (6)	SO_4^{--} (7)	HCO_3^- (8)	(ppm) (9)	(10)
1	134.87	11.33	49.20	104.26	195.03	263.20	375.2	550.5	Very hard
2	160.04	3.13	46.80	49.94	141.84	173.87	375.2	321.8	Very hard
3	1786.14	8.60	741.40	856.50	3404.25	4091.25	305.0	5365.1	Very hard
4	1387.06	6.64	533.00	476.12	2836.90	2098.94	183.0	3284.6	Very hard
5	1126.09	7.89	577.00	508.43	2482.26	2013.92	527.6	3527.1	Very hard
6	85.91	3.13	44.00	43.62	65.60	100.86	366.0	288.8	Hard
7	343.19	4.70	36.00	66.10	277.65	429.90	375.1	361.0	Very hard

TH, total hardness.

EXAMPLE 5.6: TOTAL HARDNESS (TH)

Calculate total hardness values for all the given water sample ionic concentration values (mg/l) in column 9 of Table 5.10.

Solution 5.6

It is equivalent to the consequence of divalent metallic cations (Ca^{++}, Mg^{++}) in the groundwater. Their chemical analyses results are necessary in mg/l (= ppm). TH formulation is given by Eq. (5.6) which leads to the calculations in the 9-th column of Table 5.10. The groundwater classifications according to Table 5.9 are presented in the last column of the same table.

5.6.4 pH

The pH indicates the strength of the water to react with the acidic or alkaline material present in the water. It is defined as the negative logarithm of its hydrogen ion activity (Boyd, 2000). It indicates the acidity or alkalinity of water (Langmuir, 1997). It is a measure of the balance between the concentration of hydrogen (H^+) ions and hydroxyl (OH^-) ions in water. The

pH of water provides vital information in many types of geochemical equilibrium or solubility calculations (Hem, 1970). The limit of pH value for drinking water is specified as 6.5 to 8.5 (WHO, 2004) and the classification is presented in Table 5.11.

5.6.5 Sodium Adsorption Ratio (SAR)

It is a measure of groundwater suitability for use in agricultural irrigation activities. Na^+ concentration can reduce the soil permeability and soil structure (Todd, 1980). Sodium adsorption ratio (SAR) is a measure of alkali/sodium

TABLE 5.11　pH Classification

Variation Range	Specification
3–3.5	High acid water
3.5–5.5	Acid water
5.5–6.8	Low acid water
6.8–7.2	Neutral water
7.2–8.5	Low water
>8.5	Alkaline water

hazard to crops and it can be estimated by the following formula.

$$SAR = \frac{Na^+}{\sqrt{\frac{Ca^{2+}+Mg^{2+}}{2}}} \qquad (5.7)$$

Herein, Ca^{2+} and Mg^{2+} should have meq/l unit. It reflects the degree of Na^+ absorption by the soil. At its high levels, the soil needs soil amendments for preventing from long-term damages. It is well-known that the Na^+ in the water can displace the Ca^{2+} and Mg^{2+} in the same soil leading to decrease in the soil ability to form stable aggregates and loss of soil structure. In practice, this leads to subsequent decrease in infiltration and permeability of the soil. Such a phenomenon damages the crop yield in the area. The groundwater classifications on the basis of SAR are given in Table 5.12.

TABLE 5.12 Irrigation Water Classification

SAR (epm)	Classification
<10	Excellent
10–18	Good
18–26	Permissible
>26	Unsuitable

SAR, Sodium adsorption ratio.
Richards (1954).

EXAMPLE 5.7 SODIUM ADSORPTION RATIO (SAR)

What are the sodium adsorption ratios for the groundwater samples analyses given in the following table? Classify these samples according to the classification given by Todd (1980).

Solution 5.7

Sodium adsorption ratio is dependent on two cations, namely, Na^+ and Mg^{++} and its calculation can be achieved by use of Eq. (5.7) leading to SARs in the 5-th column of Table 5.13. According to the classification criteria in Table 5.12 all groundwater samples are unsuitable for agricultural and irrigation purposes.

5.6.6 Sodium Content (SC)

Sodium ion concentration is significant because it causes reduction of the soil permeability by filling the void space partially. Soils with sodium and carbonate are referred to as alkali in comparison to saline soils in which chloride and sulfate are the dominant anions. Sodium content (SC) is expressed in % and its formulation is given as,

TABLE 5.13 SAR Calculations

Sample Number (1)	meq/l			SAR (5)	Classification (6)
	Na+ (2)	Ca++ (3)	Mg++ (4)		
1	58.64	24.55	85.74	78.966	Unsuitable
2	69.58	23.35	41.07	122.600	Unsuitable
3	776.58	369.96	704.36	335.069	Unsuitable
4	603.07	265.97	391.55	332.605	Unsuitable
5	489.60	287.92	418.12	260.580	Unsuitable
6	37.35	21.96	35.87	69.459	Unsuitable
7	149.21	17.96	54.36	248.133	Unsuitable

SAR, Sodium adsorption ratio.

TABLE 5.14   Irrigation Water Quality Classification

SC Range	Quality Classification
<20	Excellent
20–40	Good
40–60	Permissible
60–80	Doubtful
>80	Unsuitable

SC, sodium content.

$$SC\ (\%) = \frac{100\ (Na^+ + K^+)}{Ca^{2+} + Mg^{2+} + Na^+ + K^+} \quad (5.8)$$

SC classification is given by Wilcox (1955) as in Table 5.14.

EXAMPLE 5.8 SODIUM CONTENT (SC)

In Table 5.15 the anionic concentrations are given in meq/l. Calculate the SC for each of the samples.

Solution 5.8

The sodium content is dependent on certain combinations and ratios of anions only as given in Eq. (5.8). The application of this expression requires that all the cations are expressed in meq/l. The 6-th column in Table 5.15 includes SC values that are calculated according to Eq. (5.8). In the last column of the same table the classifications are given according to Table 5.14.

5.6.7 Residual Sodium Carbonate (RSC)

The bicarbonate content of groundwater is expressed by the residual sodium carbonate (RSC), where high concentrations of bicarbonate lead to an increases in pH value that may cause dissolution of organic matter. As a result of RSC increase, Ca^{2+} and M^{2+} precipitate leading to an increase in Na^+ content in the soil. The high concentration of HCO_3^- ion in irrigation water leads to toxicity and affects the mineral nutrition of plants. According to Richard (1954) classification, water with RSC greater than 2.5 meq/l is considered unsuitable for irrigation. The RSC can be calculated according to the following expression.

$$RSC = (CO_3^- + HCO_3^-) - (Ca^{2+} + Mg^{2+}) \quad (5.9)$$

RSC classifications are given in Table 5.16.

TABLE 5.15   Cation Ion Concentrations

Sample Number (1)	meq/l				SC (6)	Classification (7)
	Na^+ (2)	K^+ (3)	Ca^{++} (4)	Mg^{++} (5)		
1	58.64	0.29	24.55	85.74	34.82	Good
2	69.58	0.080	23.35	41.07	51.95	Permissible
3	776.58	0.22	369.96	704.36	41.97	Permissible
4	603.07	0.170	265.97	391.55	47.85	Permissible
5	489.60	0.202	287.92	418.12	40.96	Permissible
6	37.35	0.080	21.96	35.87	39.29	Good
7	149.21	0.120	17.96	54.36	67.37	Doubtful

SC, sodium content.

TABLE 5.16   RSC Classification

RSC (meq/l)	Specification
<1.25	Generally safe for drinking
25.25–2.50	Marginal as an irrigation source
>2.5	Usually unsuitable for irrigation without amendment

RSC, residual sodium carbonate.

EXAMPLE 5.9 RESIDUAL SODIUM CARBONATE (RSC)

In Table 5.17 major anion and cation concentrations are given in meq/l. Calculate RSC and provide the corresponding classifications.

Solution 5.9

The relevant expression for this calculation is given in Eq. (5.9), where all the ion concentrations should be in meq/l. The resulting values are

shown in the 9-th column of Table 5.17. They are all less than zero. The classifications are decided by considerations from Table 5.16 from which it is clear that all the samples imply "Generally safe for drinking."

5.6.8 Permeability Index (PI)

This index is proposed to classify water for its suitability in irrigation activities and it is defined in terms of Na^+, Mg^{2+}, and HCO_3^- as follows (Doneen, 1964).

$$PI = 100 \times \frac{Na^+ + \sqrt{HCO_3^-}}{Na^+ + Ca^{2+} + Mg^{2+}} \quad (5.10)$$

In this expression all the ions must be expressed in meq/l. A classification based on PI was proposed by World Health Organization (WHO) for assessing suitability of groundwater for irrigation purpose. According to this

TABLE 5.17   Ion Concentrations

Sample Number (1)	meq/l								Classification (10)
	Na^+ (2)	K^+ (3)	Ca^{++} (4)	Mg^{++} (5)	Cl^- (6)	SO_4^{--} (7)	HCO_3^- (8)	RSC (9)	
1	58.64	0.29	24.55	85.74	55.00	54.80	61.50	−82.88	Generally safe for drinking
2	69.58	0.080	23.35	41.07	40.00	36.20	61.50	−49.15	Generally safe for drinking
3	776.58	0.22	369.96	704.36	960.03	851.81	49.99	−1430.95	Generally safe for drinking
4	603.07	0.170	265.97	391.55	800.03	437.01	29.99	−964.63	Generally safe for drinking
5	489.60	0.202	287.92	418.12	700.02	419.30	86.48	−821.24	Generally safe for drinking
6	37.35	0.080	21.96	35.87	18.50	21.00	60.00	−13.22	Generally safe for drinking
7	149.21	0.120	17.96	54.36	78.30	89.51	61.48	−142.09	Generally safe for drinking

classification PI values less than 25% falls into Class-I and Class-II includes PI values that range between 25% and 75%.

EXAMPLE 5.10 PERMEABILITY INDEX (PI)

Table 5.18 provides the necessary ion concentrations for PI calculation. What is the PI and state the quality classification of this water?

Solution 4.10

Permeability index is concerned with four ions, namely, Na^+, Ca^{++}, Mg^{++}, and HCO_3^- and it indicates the suitability of water for irrigation purposes. The substitution of the necessary concentration values into Eq. (5.10) gives the PI values in the 6-th column of Table 5.18. The last column shows that all the samples fall within Class-II classification.

5.6.9 Chloride Classification

Fertilizer, human, and animal waste sources can result in significant concentrations of Cl^- in groundwater because it is readily transported through the soil (Stallard and Edmond, 1981). Increasing Cl^- concentrations toward the coastline are good indicators of seawater intrusion (Chapter 4). In the process of seawater intrusion, mixing between saline and freshwater and water—rock interaction may influence groundwater salinity (Appelo and Postma, 1999; Vengosh et al., 2002).

Stuyfzand (1989) proposed the Cl^- classification in meq/l values with fuzzy words such as "extremely fresh," "very fresh," "fresh," "fresh Brackish," "brackish," "brackish-salt," "salt," and "hypersaline" (Table 5.19).

TABLE 5.19 Chloride Classifications

Chloride Range	Quality Classification
<0.14	Extremely fresh
0.14—0.84	Very fresh
0.84—4.23	Fresh
4.23—8.46	Fresh brackish
8.46—28.21	Brackish
28.21—282.1	Brackish-salt
282.1—564.1	Salt
>564.1	Hyperheline

TABLE 5.18 Ion Concentration Values

Sample Number (1)	meq/l				PI (%) (6)	Classification (7)
	Na^+ (2)	Ca^{++} (3)	Mg^{++} (4)	HCO_3^- (5)		
1	58.64	24.55	85.74	61.50	39.35	Class-II
2	69.58	23.35	41.07	61.50	57.77	Class-II
3	776.58	369.96	704.36	49.99	42.34	Class-II
4	603.07	265.97	391.55	29.99	48.27	Class-II
5	489.60	287.92	418.12	86.48	41.73	Class-II
6	37.35	21.96	35.87	60.00	47.38	Class-II
7	149.21	17.96	54.36	61.48	70.89	Class-II

5.7 COMPOSITE VARIABLE RELATIONSHIPS

These are functions of ion compositions, which express the groundwater quality as a universal value in terms of especially EC, TDS, and SAR. The easiest of these variables to measure directly in the field is EC. As ion concentrations increase EC also increases.

5.7.1 EC−TDS Relationship

EC and TDS are direct proportionality to ionic concentrations, and therefore, there is a definite relationship between them. EC is a more flexible variable as it can be directly measured in the field. TDS, being a measure of total dissolved chemical species in water, is more useful variable in the hydrochemical equilibrium studies. For quick and economic determination of groundwater quality, it is possible to estimate TDS from EC values measured in the field for a considerable number of groundwater samples. Once this empirical relation is developed between EC and TDS for a particular region, it then helps in future to determine TDS by just measuring EC from any well in the same area and also to observe quickly whether the sample in hand belongs to the same population or not. The empirical relationship between EC and TDS is also available from literature such as Hem (1970) and Şen and Al-Dakheel (1986). Quick field determination of TDS values save time, effort, and cost. EC and TDS increase with an increase in ion concentration expressed as,

$$TDS = \alpha \, EC \qquad (5.11)$$

where TDS is in mg/l, EC is in micro-mhos/cm at 25 °C; and α is a conversion constant. Hem (1970) reported the range of α as 0.54−0.96 which represents nearly all types of natural waters.

This empirical relation has been formulated by taking into consideration the physical reasoning such that if the EC measurement is zero, the water will have no dissolved chemical contents. Theoretically, it is true but in practice there is no natural water found totally free of dissolved solids.

EXAMPLE 5.11 ELECTRICAL CONDUCTIVITY CLASSIFICATION

Measured EC of the same water sample is 7631.82 μS/cm. Compare the TDS value with the EC. Classify it under the light of classification in Table 5.7.

Solution 5.11

According to Eq. (5.11) the TDS value can be calculated provided that the coefficient, C, is known. In practice, its value changes from between 0.54 and 0.96 with an average value of 0.75. It is necessary to convert 7631.82 μS/cm in one of the TDS units. One μS/cm is 0.640 ppm, and therefore, $0.64 \times 7{,}631.82 \, \mu S/cm = 4884.36$ ppm. Hence, estimation of TDS based on EC with minimum, average and maximum constant values are 2637.60 ppm, 3663.56 ppm, and 4689.00 ppm, respectively. Under the light of Table 5.7, the water quality is of brine type.

5.7.2 EC−SAR Relationship

EC together with SAR determine whether groundwater can be used for agricultural purposes. SAR (alkali hazard) and EC (salinity hazard) data are plotted in a USSL (1954) diagram to determine the suitability of water for irrigation (Figure 5.2). The USSL diagram has 16 subareas specified with EC on the horizontal and SAR on the vertical axes. The specification of each subarea is given in Table 5.20.

For instance, the water samples in categories of C2-S1, C1-S1, C3-S1, C3-S2, C2-S2, C3-S3, and C3-S4 indicate a wide range of salinity and alkalinity. However, samples in C2-S1 and C1-S1 indicate that the water can be used for

FIGURE 5.2 **SAR–EC water classification diagrams for irrigation purposes.** SAR, sodium adsorption ratio; EC, electrical conductivity.

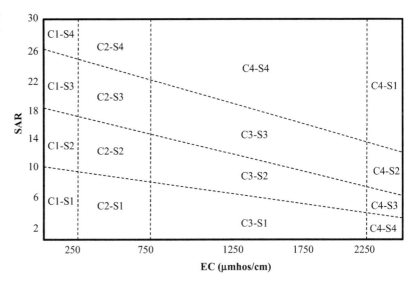

TABLE 5.20   EC and SAR Joint Relationships

SAR		C1—Low	C1—Medium	C3—High	C4—Very high
		EC			
	S1—Low	Low—low	Low—medium	Low—high	Low—Very high
	S2—medium	Medium—low	Medium—medium	Medium—high	Medium—Very high
	S3—High	High—low	High—medium	High—high	High—Very high
	S4—Very high	Very high—low	Very high—medium	Very high—high	Very high—Very high

EC, electrical conductivity; SAR, sodium adsorption ratio.

irrigation in almost all soils, with caution. There is a little danger of harming the soil by Na^+ exchange (Karanth, 1989).

The relationship between the EC and SAR are expressed through fuzzy words (Section 5.15).

5.8 WATER QUALITY GRAPHICAL REPRESENTATIONS

Graphical representations of different ions in hydrochemistry have been developed from time to time and such diagrams make groundwater quality variations' understanding easier, efficient, quicker, and more informative. In hydrogeology studies the assessment of groundwater quality constitutes a significant part of the overall work, because depending on the water quality groundwater activities (drilling wells, management, etc.) can be directed in a more efficient manner. After analysis of water samples from any study area chemically in the laboratory, it is necessary to decide about the quality, suitability, and convenience of groundwater for

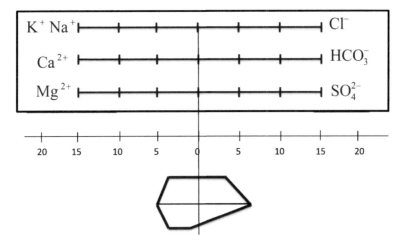

FIGURE 5.3 **Stiff diagram representation of groundwater quality.**

different objectives. For this purpose, different collective diagrammatic representations of the quality variables are suggested in the literature.

5.8.1 Stiff Diagram

Stiff (1951) proposed pattern diagram for representing chemical analysis results on three parallel axes. To the left (right) are cation (anion) concentrations in meq/l. The resulting points are then connected by straight-lines, which appears in the form of irregular polygonal shape. The same ion composition water samples have the same shapes, theoretically. However, in practical applications two or more shapes that are similar to each other within ±5% or at the maximum ±10% are regarded as the same. The basics of Stiff diagram with a representative sample are given in Figure 5.3.

5.8.2 Circular Diagram

The area of the circle is proportional to the total ionic concentration of the groundwater sample. The ionic sectors show the fraction of different ions expressed in meq/l. A representative circular diagram is given in Figure 5.4.

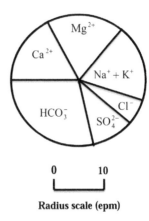

FIGURE 5.4 **Circular diagram representation of groundwater quality.**

5.8.3 Schoeller Diagram

This is a semilogarithmic diagram for groundwater quality concentration representations (Schoeller, 1967). On the horizontal axis anions and cations are shown in sequence with their equivalent per million amounts on the vertical axis logarithmically (Figure 5.5). The reason for logarithmic scale is for rather uniform distribution appearance of ionic concentrations.

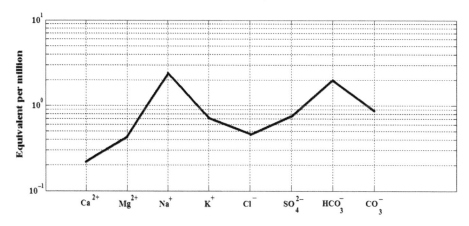

FIGURE 5.5   **Shoeller diagram of semilogarithmic scale for groundwater quality.**

Logarithmic scale reduces the high ion concentrations and enlarges small concentrations, and hence, they appear collectively and distinctively from each other.

In general, the main ionic concentrations are plotted on eight equally spaced logarithmic scales and the points are then joint in sequence. It is possible not only to see the ion concentrations, but also the differences between any two pairs of ion concentrations. In case of more than one groundwater sample plot, if the straight line between two samples are parallel to each other then one can conclude that these ionic concentrations are in ratio between the samples. These diagrams can be used also for depicting the degree of calcium bicarbonate and calcium sulfate saturations in the groundwater sample. Figure 5.6 indicates such a graph with seven major ion concentrations.

5.8.4 Piper Diagram

There are a number of chemical variables which define together the overall worth of water from the stand point of its use. Therefore, the joint presentation of multivariate chemical associations has been a problem. Researchers have been trying to achieve the combined effect of major constituents responsible for groundwater quality on a single diagram. In this pursuit, Hill (1940) succeeded first time in transforming the combination of ions into a practical form by making the use of triangular coordinates. Piper (1944) further refined this work and introduced an explicit form of triangular coordinates which are known after his name as Piper diagram (Figure 5.7).

One triangle can hold at the maximum three variables as it has three axes, and therefore, on the whole association of six different variables is possible. In plotting three cations (Ca^{2+}, Mg^{2+} and $Na^+ + K^+$) expressed as percentages of total cations in meq/l and hence, a single point is obtained on the cations triangle. The other triangle is for anions (SO_4^{2+}, Cl^-, $HCO_3^- + CO_3^-$) in the same way (Figure 5.7). These two points each within the triangles are projected into a central diamond (rhombohydral rectangular) shaped area by parallels to the upper edges of the central area.

Percentage of individual ions is evaluated from each respective group of cations and anions separately. The percentages of cations (anions) are plotted on the left (right) triangle. Every axis of each triangle is divided between 0 and 100 units, which represent the percentage of the particular ion or group of lumped ions drawn on that axis. In each chemical analysis, a

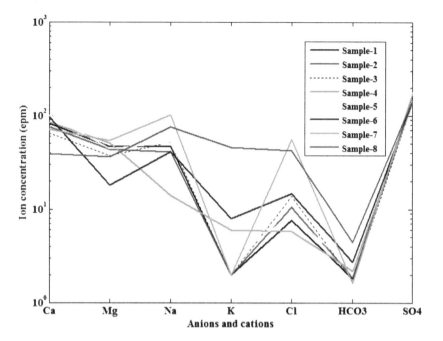

FIGURE 5.6 Shoeller diagram.

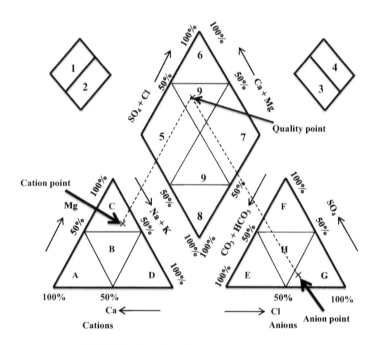

FIGURE 5.7 Piper diagram interpretations.

TABLE 5.21   Trilinear Portion Interpretations

	Equilateral Triangles	Diamond
Cations	A—Calcium type	1—(Ca + Mg) exceed alkalies (Na + K)
	B—No dominant type	2—Alkalies exceed alkaline earths
	C—Magnesium type	3—Weak acids (CO$_3$ + HCO$_3$) exceed strong acids (SO$_4$ + Cl)
	D—Sodium and potassium type	4—Strong acids exceeds weak acids
Anions	E—Bicarbonate type	5—Magnesium bicarbonate type
	F—Sulfate type	6—Calcium-chloride type
	G—Chloride type	7—Sodium-chloride type
	H—No dominant type	8—Sodium bircarbonate type
		9—Mixed type (no cation—anion exceed 50%)

point is constructed on both triangles separately. In this way, all the chemical analyses are plotted on the cation and anion triangles. It is possible to single out the dominant and nondominant cations and anions from their locations in the respective triangles. Each triangle is divided into four subregions. Regions B and H are reserved for nondominant cation and anion, respectively. Other regions show the dominance of cations and anions accordingly. After differentiating this individual dominance or nondominance status of cations and anions, there remains evaluation of the mutual cation and anion combinations, which are useful in assessing the overall groundwater quality and the underlying hydrogeochemical process. For this purpose, the cation and anion points in the base triangles are further transferred to above rhombohydral rectangular field as quality point (see Figure 5.7). A point is constructed in the diamond-shaped rectangle where the projected images of respective cation and anion cross each other. When all the chemical analyses are transferred to this new rectangular field, one gets a modified field, which describes interrelationships between cations and anions. For the sake of analysis the rhombohydral rectangle is divided first into four smaller rhombohydral fields as shown at the upper left and right sides of Figure 5.7.

The descriptions in Table 5.21 are valid for trilinear diagram groundwater classification at different portions of this diagram.

As far as individual anions and cations are concerned, it gives significant information about their dominance and nondominance, but Piper diagram has drawbacks in evaluation of their combined effect. Şen and Dakheel (1986) pointed out some of these shortcomings as follows:

1. The Piper diagram can represent only three cations and anions. Joining them to the nearest chemical partner, for example, Na$^+$ and K$^+$ are taken as one variable; CO$_3^-$ is combined with HCO$_3^{2-}$, and F and NO$_3^-$ with Cl$^-$, reduces their individual effect.
2. Water samples showing different water chemistry with the same percentages in the base triangles may appear as a single point after projection onto the rhombic rectangle and it is not possible to distinguish them as different water qualities.
3. If the number of data points is large then the utility of the diagrams becomes rather vague.

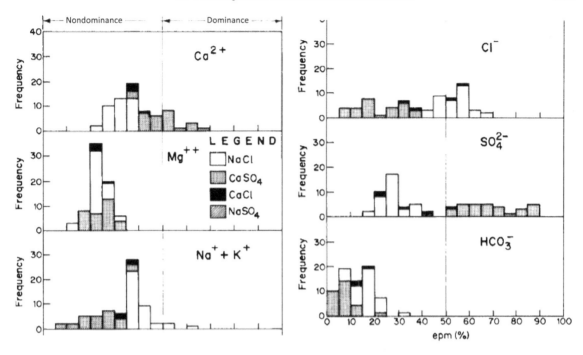

FIGURE 5.8 Ion concentration percentage frequencies.

This is because differentiation between different water categories is inefficient as most of the points in the cation and anion triangles and the rhombohydral field occupy the same places, and hence, their individual effects cannot be identified with ease.

5.8.5 Ion-Concentration-Frequency Diagrams

These graphs are proposed as supplements for the conventional Piper diagrams by Şen and Dakheel (1986). The aforementioned shortcomings can be avoided by ion concentration frequency diagrams as shown in Figure 5.8.

Any number of ions can be exhibited individually through these graphs. In addition, it yields the frequency (number) of water samples falling within a certain percentage of meq/l values. In

the preparation of these diagrams the following steps are necessary.

1. Each groundwater sample is classified into a chemical type according to maximum major cation and anion concentrations. In general, there are pairwise combinations of the chemical water type according to the following matrix.

		Anions			
		SO_4^{2-}	HCO_3^-	CO_3^-	Cl^-
Cations	Ca^{2+}	$CaSO_4$	$CaHCO_3$	$Ca(CO_3)_2$	$CaCl_2$
	Mg^{2+}	$MgSO_4$	$MgHCO_3$	$Mg(CO_3)_2$	$MgCl_2$
	Na^+	Na_2SO_4	Na_2HCO_3	$NaCO_3$	$NaCl$
	K^+	K_2SO_4	K_2HCO_3	KCO_3	KCl

2. Decide about the type of water based on the maximum anion and cation couple. For instance, if any sample has maximum Mg^{2+} (CO_3^-) concentration among the cations (anions) then the type of water is magnesium carbonate $Mg (CO_3)_2$. Hence, each water sample will have similar chemical type attachment.

3. For each ion find the number of chemical types and prepare the following matrix, which includes the numbers. In this matrix the number of $CaSO_4$ is shown by N_{CaSO_4}. These numbers are also referred to as the frequency of the water types.

		Anions			
		(SO_4^{2-})	$(HCO_3^{2-})-$	(CO_3^-)	$Cl-$
Cations	Ca^{2+}	N_{CaSO_4}	N_{CaHCO_3}	$N_{Ca(CO_3)_2}$	N_{CaCl_2}
	Mg^{2+}	N_{MgSO_4}	N_{MgHCO_3}	$N_{Mg(CO_3)_2}$	N_{MgCl_2}
	Na^+	$N_{Na_2SO_4}$	$N_{Na_2HCO_3}$	N_{NaCO_3}	N_{NaCl}
	K^+	$N_{K_2SO_4}$	$N_{K_2HCO_3}$	N_{KCO_3}	N_{KCl}

The summation of all the chemical water types is equal to the number of data, N, collected from the field.

4. Find the percentage of each cation (anion) among the total cation (anion) similar to the Piper diagram percentage calculations. In this manner each ion percentage will have water type attached with it.

5. Plot the statistical frequency diagram of each ion based on the percentage values (see Figure 5.8). Adopt convenient class numbers depending on the number of ion data.

EXAMPLE 5.12 TYPE OF GROUNDWATER

Specify the types of groundwater for each sample given in Table 5.22.

Solution 4.12

The groundwater quality types can be identified by coupling the maximum cations with the maximum anions available in the table. Hence, four water samples are of sodium chloride (NaCl) type; two has sodium bicarbonate (NaHCO$_3$) type; one is of magnesium bicarbonate [Mg (CO$_3$)$_2$] and another one is of sodium sulfate (NaSO$_4$) type.

TABLE 5.22   Chemical Ion Concentrations

Sample Number	meq/l							Water Type
	Na^+	K^+	Ca^{++}	Mg^{++}	Cl^-	SO_4^{--}	HCO_3^-	
1	58.64	0.29	24.55	85.74	55.00	54.80	61.50	Mg (HCO$_3$)$_2$
2	69.58	0.080	23.35	41.07	40.00	36.20	61.50	NaHCO$_3$
3	776.58	0.22	369.96	704.36	960.03	851.81	49.99	NaCl
4	603.07	0.170	265.97	391.55	800.03	437.01	29.99	NaCl
5	489.60	0.202	287.92	418.12	700.02	419.30	86.48	NaCl
6	37.35	0.080	21.96	35.87	18.50	21.00	60.00	NaHCO$_3$
7	149.21	0.120	17.96	54.36	78.30	89.51	61.48	NaSO$_4$

5.8.6 Durov Diagram

Groundwater quality interpretations in hydrogeological studies are valuable, well established and have become very sophisticated in terms of hydrochemical evolution (Appelo and Postma, 1999). While various process tools are invaluable, there is frequently the need to handle large numbers of hydrochemical analyses and to understand initial relationships between different groundwater types and their geographical distributions prior to embarking upon detailed quality evolution analysis. The hydrochemical interpretations are based on various diagrams such as trilinear diagrams (Piper, 1944) and to a lesser extent Durov diagram (Durov, 1948). Al Bassam et al. (1997) introduced a code to rapidly process data for representation using Durov procedure.

In the primary Durov (1948) diagram, the base triangles representing cations and anions separately are drawn with their bases perpendicular to each other and then points from both triangles are projected in a lower square field where new points are plotted at the cross points (Figure 5.9(a)). Durov diagram poses the same difficulties as discussed for Piper diagram. The recognition of different quality waters is not possible clearly, because in the square field some samples with different water chemistry may fall on the same point. The only advantage of this diagram over Piper diagram is that it is possible to draw two more variables (TDS and pH) as in the extended Durov diagram (Figure 5.9(b)).

5.8.7 Multirectangular Diagram (MRD)

As the groundwater has more than six ions (usually eight ions or more), the Piper diagram provides insufficient space to accommodate all these chemical constituents (Ahmad, 1998). In order to overcome the difficulties of classical diagrams for groundwater quality data interpretation, another scheme is developed and proposed by Ahmad et al. (2003), which covers not only the information about the cations and anions separately, but also, gives very clear hydrochemical facies classification. The new diagram is named as multirectangular diagram (MRD) and is shown in Figure 5.10.

All major cations (Ca^{2+}, Mg^{2+}, Na^+, K^+) are plotted on the horizontal axis and major anions (HCO_3, CO_3, SO_4, Cl) on the vertical axis. In Figure 5.10 only 3×3 MRD is shown, however it can be expanded to n × m dimensions depending on the number of cations, n, and anions, m. In this scheme, meq/l percentages of total cations and anions are considered for each ion. The MRD procedure covers all the shortcomings of Piper and Durov trilinear diagrams. In MRD, initially plotted points are not further projected to any new field such as rhombohydral rectangular or rectangular areas as in the Piper and Durov diagrams, respectively. Each major field is further divided into four subareas in order to decide about dominance or nondominance of cations and anions (see Figure 5.10). On the whole, the diagram in this figure has 36 subareas and each subarea delivers complete information about the respective cation and anion pair, which falls in that area. Twelve subareas are available for each cation and anion, respectively. For instance, the first subareas like A1 in each major rectangle show that both cation and anion are nondominant type (Figure 5.10). If the point lies in the second subarea, (like A2) then anion is the dominant but cation is nondominant. The third subarea, A3, shows the dominance of both cation and anion while the fourth subarea, A4, is indicative of dominance of cation but nondominance of anion. Therefore, subareas one, two, and four give mixed type of water but the third subarea in each major square shows the dominant type of water. After this classification of waters, hydrochemical facies maps can be prepared which show prevalent types of waters in the study area.

It is also possible to say something about the history of groundwater from the present diagram. For example, waters falling in square A1 (see Figure 5.10), are indicative of fresh waters

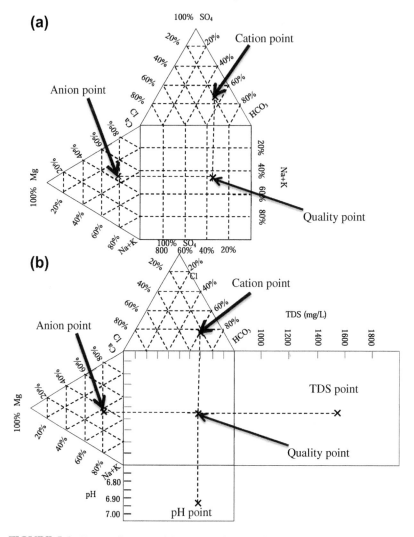

FIGURE 5.9 **Durov diagrams.** (a) Primary, (b) extended. TDS, total dissolved solid.

in recharge areas. The subarea A2 shows that ion exchange has occurred in which Ca^{2+} in has been replaced most probably by Na^+ from the sediments/rocks through which water has moved. The subarea A3 gives the indication of calcite dissolution. Occurrence of waters in subarea A4 is not common but may be possible in coastal areas where more saline seawater intrudes the fresh aquifers and reverse ion exchange process

takes place in which Ca^{2+} in the solution increases at the cost of Na^+. If the chemical ionic pair falls in square D, the dissolution of dolomite minerals is expected. Again the four subareas in D show the extent to which dolomite minerals have been dissolved. Square G is very important as in most of the sedimentary areas of the world; the waters are of $NaHCO_3$ type. Na^+ enters the groundwater system by ion exchange process

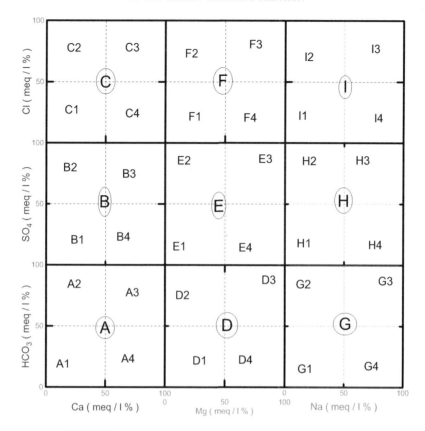

FIGURE 5.10 Innovative multirectangular diagrams (MRD).

with Ca^{++} and dissolution of halite mineral (NaCl) (Freeze and Cherry, 1979; Mandel and Shiftan, 1981). If the waters lie in square G then there is a strong indication of ion exchange and dissolution geochemical processes. Squares H and I favor dissolution of gypsum and halite as the water gets older along its flow path. Further detailed information is given by Ahmad et al. (2003).

5.9 THE GHYBEN–HERZBERG RELATION

More than 100 years ago it has been observed that saline water occurred in the subsurface at a depth approximately 40 times the height of fresh water below sea level (Drabbe and Baydon-Ghyben, 1889; Herzberg, 1901). This situation is considered under the natural equilibrium in the form of hydrostatic state due to difference between the saline and fresh water densities. The representative natural balance in the subsurface between the two different densities is shown in Figure 5.11.

Furthermore, in Figure 5.12 the natural situations are shown before the pumping and the dynamic case after pumping.

The hydrostatic balance between the two water types is similar to U tube balance and by considering the pressures on each side it is possible to write by considerations from Figure 5.11 that,

$$\rho_S g z = \rho_F g(z + h_F) \qquad (5.12)$$

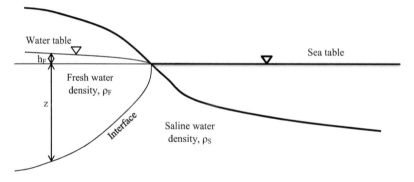

FIGURE 5.11   Ghyben–Herzberg relations in nature.

(a)   **(b)**

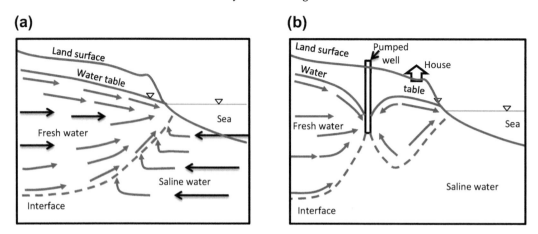

FIGURE 5.12   Ghyben–Herzberg relations. (a) Before pumping, (b) after pumping.

where ρ_S and ρ_F are the saline and fresh water densities, respectively and g is the acceleration due to gravity. One can arrange the last expression to calculate the z variable as,

$$z = \frac{\rho_F}{\rho_S - \rho_F} h_F \tag{5.13}$$

Frequently, ρ_S has the value equal to 1.025 g/cm^3, but fresh water density is 1 g/cm^3. Substitution of these values into Eq. (5.13) gives,

$$z = 40 \, h_F \tag{5.14}$$

It is obvious from Figure 5.11 that h_F is the elevation of the water table above sea level and z is the depth to the interface (freshwater/saline

water) below sea level in the case of hydrostatic balance.

In these equations only density is considered, and therefore, in case of no flow occurrence, when the interface has exactly or almost horizontal situation, the Ghyben–Herzberg equation does not hold correctly. Furthermore, at coastal areas with significant vertical flow the same equation does not yield reliable results. The above arguments are valid for coastal unconfined aquifers. It is possible to extend similar expressions into the case of confined aquifers provided that water table surface is considered as piezometric levels.

Hubbert (1940) expanded the use of Ghyben–Herzberg equation by considering the groundwater movement to cases where subsurface saline water is in motion with heads either above or below the sea level. This corresponds to nonequilibrium case and more general formulation is given as,

$$z = \frac{\rho_F}{\rho_S - \rho_F} h_F - \frac{\rho_F}{\rho_S - \rho_F} h_S \qquad (5.15)$$

here h_F and ρ_F are the fresh water elevation and density, respectively, which terminate at a depth z. Furthermore, h_S is the elevation of water level in the saline water well with density ρ_S, which also terminates at the same depth. When h_S is equal to zero then the saline water is in equilibrium with stagnant fresh water.

5.10 ARTIFICIAL GROUNDWATER MIXTURE

Groundwater in any part of the world is not homogeneous in quality. Its occurrence, distribution, and movement lead to natural mixture of different water types (e.g., meteoric, connate, juvenile). The natural mixture determines the groundwater quality in nature depending on the geological setup and the hydrological cycle activities. For instance, during metamorphism, the hydrous clay minerals are converted to less hydrous forms and water is expelled from the rock. It appears that as this metamorphic water migrates toward the Earth's surface, it generally mixes with meteoric water. Consequently, the groundwater quality might vary spatially to a significant extent within the same aquifer at different well locations and depths. In addition, due to the lowering of groundwater levels by means of pumping, some additional temporal changes might also occur. Furthermore, in some aquifers the groundwater quality deteriorates with depth. As a result, depending on the well location, penetration depth and pumping discharges, and periods, groundwater quality may show significant differences from one well to another. In order to protect or to improve the water quality many authors presented different suggestions, (Fair et al., 1971; Ayers and Wescot, 1976). A comprehensive review and interpretation of the chemical characteristics of natural water are presented by Hem (1970). These studies provide practical suggestions only after detailed laboratory studies based either on individual water sample analysis or continuous groundwater monitoring.

The artificial mixture is not an alternative to any one of the aforementioned studies but it is an effective and supplementary technique, which increases the availability of groundwater especially for agricultural purposes based on simple field measurements. The conceptual model of the groundwater withdrawal and artificial mixture can be idealized as in Figure 5.13.

Two wells, with different groundwater quality, pump water from the same or different aquifer with respective discharges Q_S (saline) and Q_P (potable). The mixture discharge is denoted by, Q. During a specific time period, t,

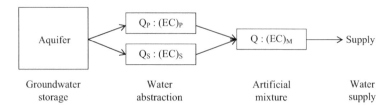

FIGURE 5.13 **Two well artificial water mixture model.** EC, electrical conductivity.

the pumped waters are placed in a common reservoir prior to water supply so that an artificial mixture takes place. In Figure 5.13 $(EC)_S$, $(EC)_P$, and $(EC)_M$ show the electrical conductivity values of saline, potable, and artificially mixed waters. The question is what volume ratios should one mix water from each well so as to achieve a predetermined water quality at $(EC)_M$ value for water supply?

5.10.1 Experimental Model

Water samples collected from two different sources with great difference in their EC values are mixed proportionally. One of the sources has the greatest salinity among the oceans of the world (Red Sea) ($EC = 59,000\,\mu mhos/cm$). The other source is groundwater from an aquifer ($EC = 1300\,\mu mhos/cm$). Two sets of laboratory experiments are carried out, namely, addition of saline water to fixed volume of potable water and vice versa. After each addition, the mixture volume and its $(EC)_M$ values are measured. Detailed information about the artificial groundwater mixture is presented by Şen et al. (2004). The graphical representations of the complete results are presented in Figures 5.14 and 5.15.

Figure 5.14 represents the final result of fresh water mixture to a set of fixed ocean (saline)

water volumes. It is obvious that the mixture EC decreases steadily until it reaches almost the groundwater $(EC)_P$.

In this figure a set of potable water volume is added steadily onto ocean water, and hence, the mixture EC increases until it reaches the ocean water $(EC)_S$.

In order to arrive at a universally usable curve between the volumes and the EC values, all of the curves in Figures 5.14 and 5.15 are rendered into a dimensionless form by defining two set of ratios.

Figure 5.16 shows the combination and standardization of the previous two figures such that the relationship between the EC ratio of the potable and saline waters on the vertical axis with the ratio of their volumes on the horizontal axis in logarithmic scale. The volume ratio, V_r, is defined as,

$$V_r = \frac{V_S}{V_P} \qquad (5.16)$$

where V_S and V_P are the saline and potable water volumes, respectively. The electrical conductivity ratio, $(EC)_r$, is a dimensionless value and defined as,

$$(EC)_r = 100\frac{(EC)_M - (EC)_P}{(EC)_S - (EC)_P} \qquad (5.17)$$

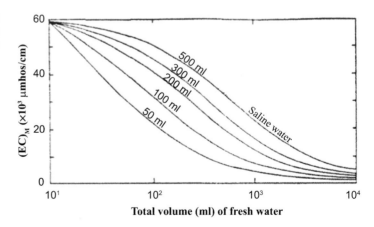

FIGURE 5.14 Relationship between mixture of electrical conductivity $(EC)_M$ and volume of fresh water added to given volume of saline water.

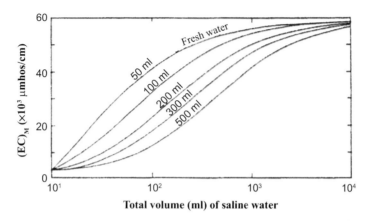

FIGURE 5.15 Relationship between mixture of electrical conductivity (EC)$_M$ and volume of saline water added to given volume of fresh water.

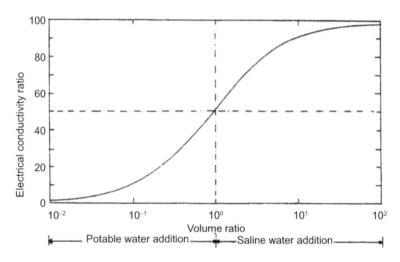

FIGURE 5.16 Standard mixture curve.

It is obvious that the (EC)$_r$ assumes any value between 0 and 100. When equal amounts of potable and saline water are mixed, ($V_S = V_P$), $V_r = 1.0$, then correspondingly, (EC)$_r = 50$. However, excess of potable water, $V_P > V_S$, i.e., ($V_r < 1.0$) results in (EC)$_r$ value between 0 and 50. Otherwise, if $V_S > V_P$ then $V_r > 1.0$ then, $50 < (EC)_r < 100$. After different sets of experiments in the laboratory it was observed that whatever the electrical conductivity values of the two sources are, the same dimensionless standard curve is obtained as in Figure 5.16.

5.10.2 Theoretical Model

In order to find an analytical expression for the standard curve, a mass balance equation was used to describe the mixture of different water qualities. The solution is assumed as an ideal case with each ion behaving independently. The magnitude of interactions depends, among other factors, on the ionic concentration and on the ion electric charge. The effects of these two major factors can be combined together through the ionic strength, I. The relationship between I and TDS, was estimated by Langelier (1936) as,

$$I = 2.5 \times 10^{-2} \, TDS \qquad (5.18)$$

Ionic strength can be estimated with a reasonable degree of accuracy by including the concentration of the major ions and cations only (Şen and Dakheel, 1986). When two different quality waters are mixed the ionic strength of individual water changes and attains a new value in the mixture. Let the volumes of potable, saline and mixture waters be denoted by V_P, V_S, and V_M, respectively. The mass balance equation by considering TDS values of constituents and the mixture can be written as,

$$V_S(TDS)_S + V_P(TDS)_P = V_M(TDS)_M$$
$$= (V_S + V_P)(TDS)_M$$

Substitution of Eq. (5.18) for TDS values leads to,

$$V_S I_S + V_P I_P = (V_S + V_P)I_M \qquad (5.19)$$

In order to render this expression into a more practical form the ionic strength can be expressed in terms of the electrical conductivity (Şen and Al-Dakheel, 1985),

$$I = C \times (EC) \qquad (5.20)$$

in which C is a proportionality factor depending on the type of water quality. For groundwater in the Kingdom of Saudi Arabia, Şen and Dakheel (1986) gave the following relationships.

$$I = 1.47 \times 10^{-5} EC$$

(NaCl type and EC < 2000 μmhos/cm)

$$I = 1.87 \times 10^{-5} EC$$

(mixed type)

$$I = 2.25 \times 10^{-5} EC$$

(CaSO$_4$ type and EC > 2000 μmhos/cm)

The substitution of Eq. (5.20) into Eq. (5.19) yields,

$$V_S C_S(EC)_S + V_P C_P(EC)_P$$
$$= (V_S + V_P)C_M(EC)_M \qquad (5.21)$$

in which C's are the water quality factors. If the source waters are of the same type, say, NaCl type, then $C_S = C_P = C_M$ hence Eq. (5.21) becomes,

$$V_S(EC)_S + V_P(EC)_P = (V_S + V_P)(EC)_M \quad (5.22)$$

Division of both sides by V_P gives,

$$V_r(EC)_S + (EC)_P = (1 + V_r)(EC)_M \qquad (5.23)$$

Solution of $(EC)_P$ from this expression and its substitution into Eq. (5.17) leads to,

$$(EC)_r = 100 \frac{V_r}{(1 + V_r)} \qquad (5.24)$$

In fact, this expression gives the experimentally obtained standard curve in Figure 5.16.

EXAMPLE 5.13 GROUNDWATER MIXTURES

An application of the standard mixture curve will be shown for Wadi As-Safra in the northwestern part of the Kingdom of Saudi Arabia (see Figure 5.17).

Solution 5.13

In Wadi As-Safra region more groundwater is needed to irrigate additional agricultural areas. The most distinctive characteristic of this wadi is that within the same unconfined aquifer there are areas of extremely different water qualities. This is due to the hydraulic connection of the aquifer with adjacent aquifers through several lateral faults along the wadi course. The groundwater from adjacent aquifers is comparatively far better

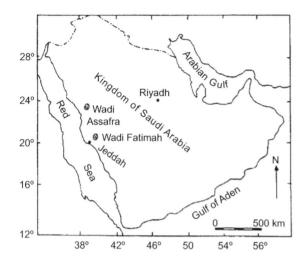

FIGURE 5.17 **Location of the study area.**

than the Wadi As-Safra waters. Two of the wells in the adjacent aquifers have EC values as $(EC)_S = 9000\,\mu mhos/cm$ and $(EC)_P = 300\,\mu mhos/cm$. They are about 1000 m apart from each other. From the classification in Table 5.6 one of the wells in Wadi As-Safra is close to excellent type since it has an EC of 300 μmhos/cm whereas the other one is not usable at all. By means of artificial mixture it is possible to obtain any EC value between 300 and 9000 μmhos/cm depending on the volume mixture ratio.

A good groundwater quality of $(EC)_M = 750\,\mu mhos/cm$ was chosen as the representative artificially mixed water. In order to attain this value, saline water (well with 9000 μmhos/cm) must be mixed with fresh water. If they are mixed at equal volumes then $(EC)_M$ would be $(9000 + 300)/2 = 4650\,\mu mhos/cm$. This is less than the volume ratio that should be considered as $V_r = V_S/V_P$. The application of the aforementioned electrical conductivity ratio Eq. (5.17) yields $(EC)_r = 5.17$. The corresponding volume ratio from Figure 5.16 is $V_r = 4.7 \times 10^{-2}$ which implies that $V_S = 4.7 \times 10^{-2}V_P$. Thus, the volume of water available for use in the good category can be increased by 4.7% by mixing saline and potable water from different sources.

In order to obtain permissible water quality for agricultural uses from the artificial mixture it is sufficient to substitute $(EC)_M = 2000\,\mu mhos/cm$ in the $(EC)_r$ equation (Eq. (5.17)), which leads to the dimensionless value 19.54. With this value at hand, Figure 5.16 gives the volume ratio value as 2×10^{-1}, which is greater than the previous ratio value for good water quality mixture. This volume ratio implies that $V_S = 2 \times 10^{-1}V_P$. This statement is tantamount to saying that water available for agricultural uses would be increased by 20% by mixing potable and saline water to yield a mixture with $(EC)_M = 2.000\,\mu mhos/cm$.

Finally, the mixture of the same groundwater sources to obtain $(EC)_M = 3000\,\mu mhos/cm$ after similar calculations, one sees that $V_S = 3.9 \times 10^{-1}V_P$. Thus, the volume of water available for use in the permissible category can be increased by 39% by mixing saline and potable waters. The complete relationships between the saline and fresh waters on the basis of groundwater quality have been shown in Figure 5.18.

The discharges or volumes are without units in this figure which means that any unit can be employed. Any point within the same quality region gives the volume ratio as well as the quality. According to this ratio, the two volume amounts from each well can be determined rather arbitrarily. For instance, if the potable water well is pumped at 500 l/min then the saline water well must pump between 0 and 23 l/min for good quality after artificial mixture; between 23 and 100 l/min for moderate water quality; and finally between 100 and 195 l/min for permissible water quality for the artificial mixture.

5.11 ENVIRONMENTAL ISOTOPES

Various researchers indicated that the environmental isotopes ($\delta^{18}O$ and δD) can be used to know the attitude of recharge (Musgrove and Banner, 1993; Scholl et al., 1996). In isotope hydrology, the depletion of $\delta^{18}O$ and δD with altitude has been used to show the groundwater

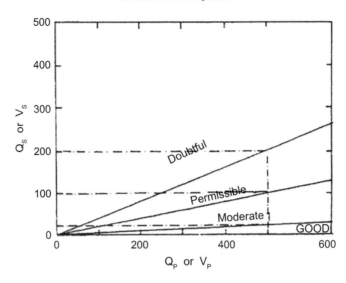

FIGURE 5.18 **Quality and discharge relationship.**

from different sources as well as the identification of groundwater flow system. Since deuterium is not affected by water–rock interaction its use has been preferred over oxygen-18 by many researchers (Lyles and Hess, 1988; Noack, 1988; Kirk and Campana, 1990; Thomas et al., 1986). As stated by Clark and Fritz (1997) the altitude effect of $\delta^{18}O$ ranges generally between 0.1% and 0.5% for each 100 m gained. This figure is valid for humid regions. However, in arid and semiarid regions the isotopic composition is affected by other factors such as the rainfall amount, variation and seasonality and air temperature. Therefore, Al-Yamani (2001) has shown that per 100 m increase in the Red Sea area the decrease is −0.08%. Jones et al. (2000) noticed that there are lower $\delta^{18}O$ values than the small rainfall events.

These include natural variations in their isotopic proportion depending on the fractionation of isotopes during phase transitions in addition to chemical and physical reactions. Due to chemical and physical changes in different isotopes of an element, the isotopic fractionations take place

and they are directly proportional to relative differences in their masses. The main environmental isotopes that are used in water studies are stable hydrogen (H), oxygen (O), carbon (C), and sulfur (S). Their average abundances in percentages and half-lives are presented in Table 5.23.

In this table radio-isotopes (^{3}H and ^{14}C) have very low abundance. The fractionation process is initiated by the large mass difference between heavy and light isotopes. The typical example of the fractionation process is the evaporation process from the surface of oceans, which enriches the heavy isotopes and the diffusion out of isotopically light molecules from seawater. According to Fritz and Fontes (1989), there is a striking isotopic content uniformity ($\delta^{18}O = 0 \pm 1$ and $\delta^2D = 0 \pm 5$ per mile) within the ocean waters relative to Standard Mean Ocean Water (SMOW)/Meteoric water is always depleted in $\delta^{18}O$ and δ^2D relative to seawater.

On the other hand, according to Dansgaard (1964), the degree of depletion depends on different factors such as the altitude, latitude

TABLE 5.23 Environmental Isotope Features

Element	Isotopes	Atomic Mass	Average Natural Abundance (%)	Half-Lives (year)
H	^1H = protium	1.00782	99.985	Stable
	^2H = deuterium	2.0141	0.015	Stable
	^3H = tritium	3.01603	0.00013	12.3
O	^{16}O	15.9949	99.759	Stable
	^{17}O	16.9991	0.037	Stable
	^{18}O	17.9991	0.204	Stable
C	^{12}C	12	98.892	Stable
	^{13}C	13.0033	1.108	Stable
	^{14}C	14.0032	$\sim 10^{-10}$	5730 ± 40
S	^{32}S	31.9721	95.02	Stable
	^{33}S	32.9715	0.75	Stable
	^{34}S	33.9679	4.21	Stable
	^{35}S	35.9671	0.02	Stable

effect, temperature, rainfall amount, and distance from the sea effects. He found that,

$$\delta^{18}O = 0.695T(^{\circ}C) - 13.6 \qquad (5.25)$$

On the other hand, Craig (1961) suggested a good relationship between the isotopic composition of $\delta^{18}O$ and $\delta^2 D$ as,

$$\delta^2 H = 8\delta^{18}O + 0.001 \qquad (5.26)$$

This expression is known as the average Global Meteoric Water Line (GMWL) or as (MWL). Later, Yurtsever (1975) used worldwide data for the possible relationships of these effects on $\delta^{18}O$. He found that the main effect was due to temperature variations among significant effects of other variations. He gave the relationship between $\delta^{18}O$ and mean surface temperature as,

$$\delta^{18}O = (0.521 \pm 0.014)T(^{\circ}C) - (14.96 \pm 0.21) \qquad (5.27)$$

On the other hand, Yurtsever and Gat (1981) found a good linear relationship between the weighted $\delta^{18}O$ and $\delta^2 D$ values in the precipitation samples of the International Atomic Energy Association (IAEA) network (Website: hhh. iaea. org/programs/ri/gnip/gnipmain.htm) as

$$\delta^2 H = (8.17 \pm 0.28)\delta^{18}O + (10.55 \pm 0.64)\text{‰} \qquad (5.28)$$

with a correlation coefficient 0.997. Rozanski et al. (1993) have updated the MWL by additional average precipitation data from World Meteorological Organization (WMO) as,

$$\delta^2 H = 8.13\delta^{18}O + 10.8\text{‰} \qquad (5.29)$$

The isotopic variations in natural waters help to determine the ages, sources, mixtures, and water–rock and water–gas interactions.

TABLE 5.24   Isotope Study Results in Saudi Arabia

Location	Sample Type	Regression Equation	Correlation Coefficient
Wadi Aqiq	Rainfall	$\delta^2H = 5.64\delta^{18}O + 5.87$	0.990
	Groundwater	$\delta^2H = 5.20\delta^{18}O + 3.32$	0.92
Wadi Khulays	Rainfall	$\delta^2H = 1.68\delta^{18}O + 0.37$	
	Groundwater	$\delta^2H = 8.13\delta^{18}O + 10.8$	
Wadi Wajj	Rainfall	$\delta^2H = 6.68\delta^{18}O + 7.80$	0.980
	Groundwater	$\delta^2H = 6.81\delta^{18}O + 8.45$	0.920
Wadi Turabah	Rainfall	$\delta^2H = 10.0\delta^{18}O + 15.67$	0.920
	Groundwater	$\delta^2H = 3.96\delta^{18}O + 0.83$	0.90
Wadi Abha	Rainfall	$\delta^2H = 5.85\delta^{18}O + 7.14$	0.55
	Groundwater	$\delta^2H = 8.13\delta^{18}O + 10.8$	—
Wadi Jizan	Rainfall	$\delta^2H = 7,83\delta^{18}O + 7.48$	0.99
	Groundwater	$\delta^2H = 6.20\delta^{18}O + 5.11$	0.99

Analysis of δ^2H and $\delta^{18}O$ data also helps to know focused or diffused recharge where in the former case there will be a higher isotopic gradient away from the recharge area. Hence samples from beneath the recharge are isotopically heavier than those collected away from it. It is assumed that the difference is a result of recharge temperature difference as a function of both latitude and altitude, which decrease in the flow direction. In the case of diffuse recharge, the isotopes of the recharged water would mix with those of the regional flow generating a low gradient in the isotopic ratios over the entire area (Wood and Sanford, 1995). δ^2H and $\delta^{18}O$ can be used to date waters between 1000 and 40,000 years old because long-term climatic changes are reflected in the values of these isotopes. Isotopic study results both for rain water and groundwater are shown in Table 5.24.

5.12 GROUNDWATER RISE AND QUALITY VARIATIONS

Adequate and reliable water supply to cities and removal of wastes has been among continuous problems in many civilizations. The chemical composition of groundwater can be identified based on its chemical concentrations or ratios of different concentrations. Such simple techniques are used in hydrogeology literature in order to identify the classification of groundwater quality (Lloyd and Heathcote, 1985). Similar ideas can also be used for the identification and classification of urban area groundwater. According to Barrett et al. (1999), an ideal recharge marker is an easily analyzed solute that is unique to one source and to one pathway at a constant concentration in the source and nonreactive in all conditions. It is not possible to find such solutes easily. They have categorized as potential marker solutes as follows.

1. Inorganic concentrations that are further grouped into major cations and anions nitrogen species (NO_3 and NH_4), metals (Fe, Mn, and trace metals), and other minor ions (B, PO_4, Sr, F, Br, and CN).
2. Organic concentrations starting from the most relevant ones are chlorofluorocarbons (CFCs), trihalomethanes (THMs), faucal compounds, such as coprostanol and I-aminopropanone, detergent-related compounds, such as optical brightness and EDTH, and industrial chemicals, including chlorinated solvents and many hydrocarbons.
3. Particulate matters including faecal microbiological species, and various colloidal particles.
4. Groundwater level rise and hydrochemical analysis results provide a basis for the characterization of groundwater variation within each hydrogeological unit. It is possible to interpret different processes that may occur within the aquifers; the effect of groundwater upon dewatering structures; the effect of groundwater abstraction on the aquifer host rock; and the suitability of groundwater for some uses.

The solutions of calcium sulfate and sodium chloride are important mechanisms in the groundwater environment. Ion exchange, removing calcium ions and releasing sodium ions to the groundwater, is also indicated as a potential mechanism for the modification to the groundwater chemistry at higher total dissolved solids concentrations. The factors that limit the rock solubility include the following points.

1. Solution of calcium and possibly magnesium carbonate by the infiltration groundwater.
2. Restrictions of the solution of calcium sulfate by availability within the aquifer rather than

other factors such as ion exchange or availability of sodium chloride in the groundwater.
3. Change of calcium for magnesium in dolomites within the aquifer releasing magnesium to the groundwater.
4. Solution of sodium chloride within the aquifer concentration being restricted by availability.
5. Ion exchange between calcium in the groundwater and sodium in the aquifer resulting in the release of sodium ions to the aquifer.
6. Solution of nitrates from the aquifer in higher total dissolved solids concentration waters.

It must be stated herein, that the above mentioned statements are tentatively valid and may be an oversimplification of a much more complex system.

5.13 STANDARD ION INDEX FOR GROUNDWATER QUALITY EVOLUTION

The main purpose of this section is to suggest a robust, dimensionless and standard groundwater quality index, which helps to trace the temporal and spatial groundwater variations by successive sampling. There is no restriction in the number of ions as it is in the classical trilinear (Piper) diagrams. The suggested method is referred to as the standard ion index (SII), which has zero arithmetic average and unit standard deviation, and hence, the first two statistical moments are out of order to represent the data and whatever the data units, they all become dimensionless, and hence have equal dimensional footing.

None of the classical techniques is useful in establishing robust groundwater characteristic blueprint identification. The association matrices

FIGURE 5.19 **Skewness–kurtosis relationships.**

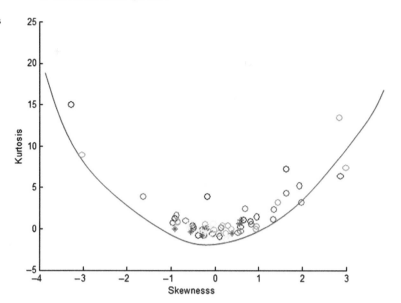

in Section 5.8.5 have indicated some characteristics but they cannot reflect the actual data, because their constructions are counted on the basis of relative frequency.

In order to achieve a reliable, robust, simple, objective, effective and all-ion dependent measure, the standard ion index (SII) conception provides a good tool. Standardization is a statistical procedure whereby the arithmetic average of a given sequence, say, $X_1, X_2, ..., X_n$ is rendered to zero mean and unit standard deviation through the following simple procedure.

$$x_i = \frac{X_i - \overline{X}}{S_x} \qquad (5.30)$$

where x_i is the standardized value, $\overline{X}$ is the arithmetic average, and S_x is the standard deviation of the original series. This simple transformation provides the following properties.

1. Standard series has zero mean.
2. Standard series has unit variance, and hence, unit standard deviation.
3. Standard series is dimensionless.

4. Standard series provide equal footing for comparison of different samples.

After standardization remaining statistical parameters are skewness and kurtosis for further distinctive representation of samples. Application of Eq. (5.30) to temporal or spatial groundwater samples leads to a distinctive pattern in the form of a parabola as a relationship between the skewness and kurtosis as in Figure 5.19. Such a relationship appears from a set of groundwater samples collected in the central western wadis of the Kingdom of Saudi Arabia. It implies systematic relationship between the skewness and kurtosis parameters.

The application is considered as two approaches depending on the purpose.

1. Individual SII approach, which provides the change of each ion separately on the same graph.
2. Successive SII approach, which provides temporal evolution of groundwater quality change by looking at the successive sample concentrations.

5.13.1 Individual SII Method

The question is to search for a possible variation pattern along the ions (cations and anions) by considering the SII value of each ion in each sample. This graph helps to make visual inspections whether the current sample fits with the general pattern of the previous samples. It is similar to the Schoeller diagram in Section 5.8.3 with the only difference where the vertical axis represents SII values instead of logarithms of ion values. Figure 5.20 indicates collectively the SII fluctuation of all the collected samples. Individual SII graph consists of 8-corner broken lines for each sample including major cations and anions. If needed, it is possible to increase the number of ions.

Before further explanations, one should keep in mind that in preparation of individual SII graph the following fundamental points are important for consideration.

1. The laboratory analyses of ions in ppm are converted into meq/l values, so as to see the hydrochemical equilibrium condition. All groundwater data are checked through this basic balance procedure.
2. It is known that in meq/l the summation of ions and cations must be equal to each other. This implies the requirement that the cations and anions groups contribute 50% to the overall equilibrium condition.
3. The meq/l values are then standardized so as to have zero mean and unit variance according to Eq. (5.30). In this way, the standard ion values will be dimensionless.
4. The individual SII graph is the plot of these standard values against each ion. There is no significance in the ion sequence order but the same sequence must be kept for comparison purposes. The most convenient way is to group cations on the left hand side and anions on the right hand side of the

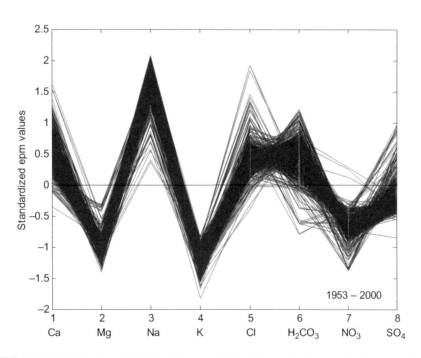

FIGURE 5.20 **Individual standard ion index graph for all groundwater sample ion concentrations.**

horizontal axis as shown in Figure 5.20. Hence, the arrangement of ions on the horizontal axis is rather arbitrary.

5. As a result of the defined standardization procedure some of the ion index values appear as positive and others as negative because their summation is equal to zero.

6. Positive (negative) SII dimensionless means that in the sample, this ion concentration exists more than (less than) the average value.

7. The zero line (perfect uniformity of each ion) is shown as a horizontal line that passes through the zero SII level. Hence, the closer the SII value to this line, the less active the ion in the overall quality balance.

8. Some of the ions may assume positive (negative) SII values only whatever the sample number is, these ions are considered as stable ions. However, if there are crossovers around the horizontal line then the concerned ion is rather unstable in the chemical composition evolution of the groundwater quality.

9. It is possible to see the variation range for each ion on the vertical axis. The smaller is the range the less variable is the concentration during the temporal evolution.

10. It is not possible for any water sample that all the cations are positive and the anions are negative. This is against the chemical (meq/l) balance and the SII concepts.

By consideration definition of these points, if one looks at Figure 5.20 for the overall groundwater quality variation one gets the impression that there appears a general pattern. The general features can be depicted as follows:

1. Mg^{2+} and K^+ assume negative values only whereas Na^{2+} has positive values. Although Ca^{2+} has positive values in majority of samples, but there are several cases with negative values.

2. On the anion part, Cl^- SII values are all positive, but the NO_3^- takes negative values.

However, both HCO_3^{2-} and SO_4^{2-} assume most of the time positive values with several negative values.

3. In general, the individual SII values assume alternative positive and negative SII values.

EXAMPLE 5.14 STANDARD ION INDEX DIAGRAM

In Table 5.25 hydrochemical analysis of eight major ions are given from eight well locations in the same area. Plot individual SII graphs and interpret with comparison to the Schoeller graph, which has already been explained in Section 5.8.3.

Solution 5.14

The Schoeller and SII graphs are presented in Figure 5.21. The former helps to identify the close water quality samples whereas the latter one provides additionally what are the dominant cations and anions among each sample.

The SII diagram provides comparison without any dimensionality in each ion concentration and they provide less variation domain for each ion among all the samples.

5.13.2 Successive SII Method

This is a procedure, which helps to control the association between two samples and possible deviations that may occur. If the two samples have exactly the same ion concentrations then their plot on an ordinary paper will result on a 45° straight line. Such a plot is shown in Figure 5.22.

The basic straight line is referred to as the "ideal similarity line," the closer are the scatter points to this line, the more the similarity between the samples. The template in Figure 5.22 is referred to as the successive standard ion index (SII) graph. It has four major parts with different interpretations. The following points

TABLE 5.25 Major Ion Concentrations

Sample Number (1)	Ca²⁺ (epm) (2)	Mg²⁺ (epm) (3)	Na⁺ (epm) (4)	K⁺ (epm) (5)	Cl⁻ (epm) (6)	HCO₃⁻ (epm) (7)	SO₄²⁻ (epm) (8)	CO₃⁻ (9)
1	81.63	46.46	47	2	7.57	166	1.82	1.51
2	75.6	43.42	41	2	10.57	146	1.87	2.45
3	64.58	37.61	52	2	13.77	138	1.69	3.12
4	85.08	51.39	14	6	5.79	138	2.2	1.12
5	101.56	44.16	51	3	8.79	177	2.31	1.92
6	95.52	18.02	41	8	14.67	143	2.72	5.13
7	70.83	54.72	102	2	55.86	171	1.62	3.21
8	39.1	36.01	76	46	42.44	146	4.42	4.2

can be stated about the features of the SII template.

1. The variables for plot do not have any dimensionality, and therefore, this graph provides a universal approach for any study.
2. If the purpose is to look for temporal groundwater quality evolution, it provides a dimensionless basis for the successive sample SII values in such a manner that there is no human interference as to the sequence of ions.
3. The scatter of points are visually inspected and preliminary interpretations can be deduced from the closeness of the scatter points to each other and to the ideal similarity line.
4. The first quadrant (I) includes low SII values, which implies similarity or closeness of the sample values. Similarly, the third quadrant, III, is also for close similarity between high ion concentrations.
5. It is possible to interpret quadrants I and III such that low (high) ion concentrations follow low (high) ion concentrations. This implies that for the two samples to become similar to each other, first condition is the low (high) concentrations in the previous sample must be followed by the low (high) concentrations.

Otherwise they cannot be similar, in other words they cannot be originating from the same characteristics.
6. The quadrant II (IV) includes the high (low) ion concentration values of the previous sample with low (high) concentrations of the current sample.
7. It can be concluded from the aforementioned discussions that quadrants I and III are generally for the similarity of the overall water quality, whereas quadrants II and IV are for dissimilarity.

Another elegant feature of the successive SII graph is that many ions can be plotted on the horizontal and vertical axes, so it is not restricted with the number of ions as in the Piper or Durov diagrams where at the maximum three ions can be represented.

1. In successive SII templates two perpendicular deviations or scatters can be considered for meaningful interpretations. These are along the 45° ideal straight line and perpendicular to it. They will be referred in this paper as the longitudinal and lateral dispersions, deviations or better scatters, respectively. This concept is very similar to the principle

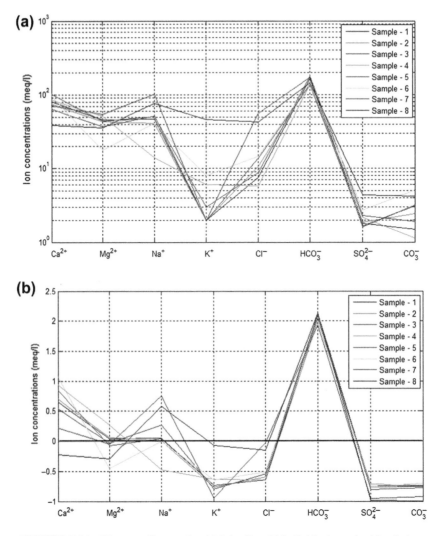

FIGURE 5.21 **Water quality graphs.** (a) Schoeller, (b) individual standard ion index.

component analysis (Davis, 1986), but without assumptions such as the normal (Gaussian) distribution of the data.

2. The longitudinal dispersion (along the ideal line) is significant for identification any different ion groups whereas the lateral scatter is meaningful for the deviations from the similarity within each group.

3. In an ideal situation, it is expected that along the similarity line, there are distinct ion groups with small lateral deviations. Hence, rather than the longitudinal dispersion, the lateral one is more significant in deciding whether the current sample transgresses allowable limits.

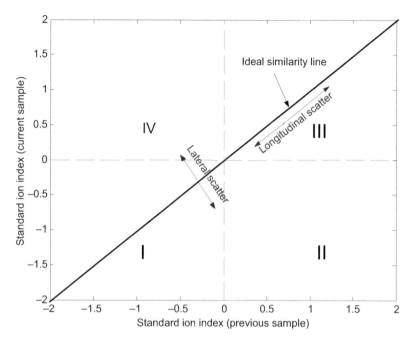

FIGURE 5.22   **Successive standard ion index graph.**

In the following sequence, some representative groundwater quality SII values are plotted on successive SII graph for different sampling periods and the relevant comparative interpretations are deduced. For instance, in Figure 5.23 the plots of eight major ions on the successive SII graph appear as scatter points with dominant points in quadrants I and III, which implies that on the overall the samples are similar to each other.

Although there are not distinctive groupings, but blurred i.e., fuzzy groupings are observable (Section 5.17). For instance, the K^+ and Na^+ are at low and high concentration locations at the lower and higher ends of the ideal similarity line. It is obvious that the Na^+ and Cl^- longitudinal dispersions are more than any ion. The number of scatter points in dissimilarity regions (namely II and IV) is negligibly small which indicates that there are significant associations or correspondences or similarities between successive sampling periods.

5.14 GROUNDWATER QUALITY VARIATION ASSESSMENT INDICES

Each groundwater well has quality signature that distinguishes it from other if not in macro scales but in micro scales. This implies that there should be certain way(s) of expressing the signature of each groundwater quality and to compare them on this basis. New and effective groundwater assessment methodology is presented with different indices, which do not involve complicated mathematical or statistical formulations but simple procedures provide possibility of visual and linguistic interpretations. Statistical methods such as the multivariate analysis, principle component analysis, cluster analysis, and alike cannot be useful in identifying the characteristic of any water quality due to the following reasons.

1. These are global techniques and their application requires a set of assumptions even

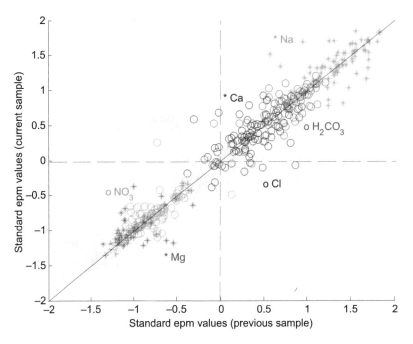

FIGURE 5.23   **Successive standard ion index graph for 2002.**

if the data is available. For instance, the data do not abide with normal (Gaussian) distribution the applications cannot be considered as reliable.

2. Statistically, the reliability of such techniques requires that there must be at least around 30 data values for each ion.
3. These techniques depend on dimensional data, and therefore, the results may not be comparable.
4. The statistical methods like the multiple regression analysis are not robust approaches and depend on the statistical parameters, which may not imply any physical significance in the interpretations.

Due to these difficulties in practical applications one should not depend on the classical techniques only. New and innovative approaches must be developed for robust and reliable results with simple foundations.

5.14.1 Water Quality Index (WQI)

Since it is expected to have only the chemical constituents that originate from the geological formations, the water quality index (WQI) of groundwater can be based on eight major ions. Supplementary, composite ionic quantities (EC or TDS) can also be used in the assessments. It is rather straightforward to suggest a few WQIs for groundwater, which in general take the following weighted average formulation form.

$$W_I = \frac{\sum_{i=1} w_i C_i}{\sum_{i=1}^{n} w_i} \tag{5.31}$$

where W_I is water quality index, and C_i is concentration of i-th ion with w_i as weight. The only question in this formulation is the determination of weights, which may be based on different causes, such as environmental effects and international quality guidance limitations as maximum admissible limit, optimum allowable limit or internal source of ion variability.

For instance, according to SAS (1984) the major ion concentrations should have the permissible limits as $Ca < 200$ ppm; $Mg < 150$ ppm; $Na^+ < 200$ ppm; $K^+ < 12$ ppm, $Cl^- < 600$ ppm; $H_2CO_3^- < 300$ ppm; $NO_3^- < 40$ ppm; and $SO_4^- < 400$ ppm. By regarding these maximum permissible levels as the weights for each ion, Eq. (5.31) can be written for groundwater quality control as,

It is obvious from this figure that there is hardly any consistency between two following samples. The sample sequence may be on time or distance axis. This is enough to conclude that WQI as defined above cannot be a good indicator for the groundwater signature determination. Especially, the global representation of all ions through WQI cannot be helpful in the signature identification. It is better to think of some figures, which may include all the ions explicitly rather than a single value for each sample. In such a case, even the TDS concentrations

$$W_I = \frac{200C_{Ca} + 150C_{Mg} + 200C_{Na} + 12C_K + 600C_{Cl} + 300C_{H_2CO_3} + 40C_{NO_3} + 400C_{SO_4}}{200 + 150 + 200 + 12 + 600 + 300 + 40 + 400} \qquad (5.32)$$

Now it is possible to calculate the WQI for groundwater for any sample. The following points are necessary in the application of this expression.

1. Division of each weight in front of each ion by the denominator gives the percentage contribution of the ion to WQI.
2. The unit of WQI is the same as the unit of ions, which may be in ppm or meq/l.
3. The summation of all the percentages should be equal to 1.

Due to these restrictive conditions, it is noticed that the adaptation of weights as standard limits, averages, allowable maxima, or the standard deviation of each ion does not bring any additional benefit.

In order to show the fluctuations in detail, the WQI variations may be presented as in Figure 5.24, where there is fluctuations even more than one standard deviation around the average WQI. On the basis of WOI standard deviation one can identify different regions. If there is a set of groundwater sample ionic analysis results at a particular study area, one can then apply Eq. (5.32), and hence, obtain a single WQI value for each sample.

could be used for signature establishment. However, it is used rather for coarse classification of the ground waters such as the criterion presented in Table 5.7. Statistically, different WQI definitions may be considered based on the arithmetic averages, maxima, and the standard deviation of each ion but they yield almost the same result. Figure 5.25 presents the WQI pattern for the same set of groundwater samples. Average-based WQI can be obtained from Eq. (5.31) by considering all weights equal to 1, hence, without any distinction between ion concentrations.

In order to confirm that almost all the WQIs yield similar plots, standards based and maximum based WQI values are plotted against each other for the groundwater samples in Figure 5.26. It is obvious that especially at low and moderate WQI values they appear align the 45° straight line, which implies their similarity.

5.14.2 Similarity Models

In search for groundwater quality variation and comparison successive time or distance sequenced chemical data similarity measurements can be used. Such an approach can be

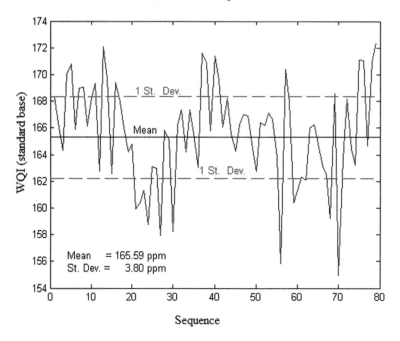

FIGURE 5.24   Water quality index (WQI) fluctuations based on standards.

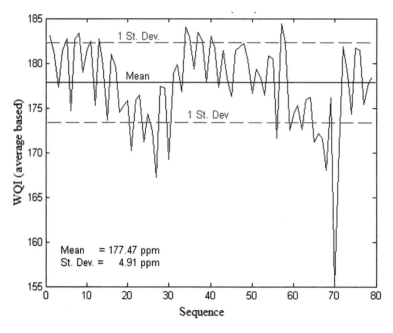

FIGURE 5.25   Water quality index (WQI) fluctuations based on averages.

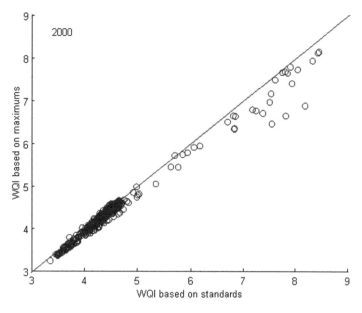

FIGURE 5.26 **Standard and maximum water quality index (WQI) comparison.**

achieved with adaptation of a similarity criterion. If eight major ions are considered then each sample is assumed to represent a point in an eight dimensional domain. Four of the dimensions are cations and others are anions. Hence, the Euclidian distance similarity, S_E, between two successive samples i and j can be expressed as,

$$S_E = \sqrt{\left(Ca_i - Ca_j\right)^2 + \left(Mg_i - Mg_j\right)^2 + \left(Na_i - Na_j\right)^2 + \left(K_i - K_j\right)^2 + \left(Cl_i - Cl_j\right)^2 + \left(HCO_i - HCO_j\right)^2 + \left(NO_i - NO_j\right)^2 + \left(SO_i - SO_j\right)^2}$$

(5.33)

The unit of the similarity coefficient, S_E, is in meq/l. A representative sample is given in Figure 5.27, which has rather chaotic behavior.

The similarity values are so erratic that it is not possible to deduce a general conclusion from its appearance. This is also a global value of all the ions included together, and therefore, cannot be regarded as a robust indicator for the sought signature. There are small values more frequently than the medium and high similarity values. The frequency diagram of the same similarity values, i.e., the histogram is shown on Figure 5.28.

Since the mean (0.53 meq/l) and the standard deviation (0.56 meq/l) are practically equal to each other the similarity law abide with the theoretical exponential probability distribution function, and therefore, randomness is imbedded in the similarity value changes (Şen, 2009a).

5.14.3 Water Quality Association Matrix

None of the aforementioned methodologies provide sample by sample successive calculation based on each ion. In search for a quality

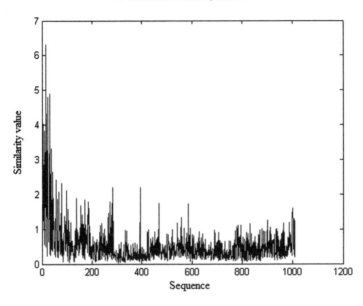

FIGURE 5.27   Similarity coefficient time variations.

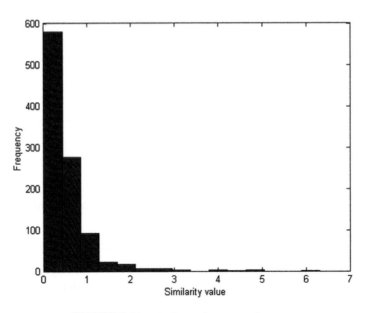

FIGURE 5.28   Similarity frequency diagram.

TABLE 5.26 Associative Matrix Template

	Cl^-	$H_2CO_3^-$	NO_3^-	SO_4^{--}
Ca^{++}	V_{11}	V_{12}	V_{13}	V_{14}
Mg^{++}	V_{21}	V_{22}	V_{23}	V_{24}
Na^+	V_{31}	V_{32}	V_{33}	V_{34}
K^+	V_{41}	V_{42}	V_{43}	V_{44}

signature, it is necessary to be able to visualize all mutual associations between each pair of the ions. An association is sought only between individual cations and anions as in Section 5.8.5.

Among the few association matrices (AM) are maximum AM (MAAM), minimum AM (MIAM), correlation AM (COAM), and similarity AM (SIAM). In general, an AM representation associates ions with cations and its matrix form is shown in Table 5.26. It is possible to interpret this figure by associating each one of the cations in the first column with each of the anions in the first row and then the corresponding joint value can be found accordingly. It is similar to the work done about the groundwater chemistry by Şen and Al-Dakheel (1986). For instance, say $NaHCO_3$, it will have an associative value of V_{32}. In this manner there are 16 different types of ion association values.

The filling of the values in this matrix serves the desired goal according to type of work. For instance, MAAM includes the numbers of maximum association between a cation with an anion given a number of data. Hence, if there are n water samples, then the summation of all values in the AM should be equal to n.

5.15 FUZZY GROUNDWATER CLASSIFICATION RULE DERIVATION FROM QUALITY MAPS

Hydrochemistry imparts its usefulness also in determining the geologic characteristics of the host rocks and direction of water movement. It gives rather rough and vague (fuzzy) idea about the time period for which water has remained in contact with the surrounding rocks matrix. It also helps in evaluating hydrogeochemical processes responsible for changing the chemistry of groundwater. Information about variations in groundwater chemistry and hydrogeochemical processes is available in literature (Miller, 1991; Kimblin, 1995; Mayo and Loucks, 1995; Hudson and Golding, 1997). Groundwater chemistry is affected from many factors such as movement through the rocks, recycling by irrigation practices, natural or artificial recharge and discharge.

Linguistic variables are the most fundamental elements in human knowledge exposition and dissemination. The introduction of linguistic variables gives the opportunity to formulate vague descriptions in natural language in a precise mathematical manner. In almost all the hydrochemical classification tables a range of numerical values are matched with a linguistic word, which is vague, uncertain, or fuzzy (Table 5.27).

The fuzzy concepts have linguistic basis, and therefore, the foundation elements are words and their derivatives with adjectives. In some publications the fuzzy operations are called as "word computation" (Zadeh, 1965, 1999). It might seem rather strange to many readers but some words in daily life imply models, numerical values, ranges, and percentages. The transition from the words to numerical values is presented through the membership degrees (MDs) and membership functions (MFs). Classes with noncrisp boundaries are the basis of fuzzy sets. The transition between the classical crisp boundary sets and the fuzzy sets are interchangeable. Fuzzy logic (FL) is a universal principle and concept.

The degree of fuzziness of a system analysis rule can vary between being very precise, in which case one would not call it fuzzy, to being based on an opinion held by a human, which would be fuzzy. Being fuzzy or not fuzzy,

TABLE 5.27 Joint Water Quality Indicator

Parameters	Range	Classification
RSC	<1.25	Good
	1.25–2.50	Doubtful
EC	<250	Excellent
	250–750	Good
	750–2000	Permissible
	2000–3000	Doubtful
	>3000	Unsuitable
TDS	<200	Fresh water
	200–500	Brackish water
	500–1500	Saline water
	>1500	Brine water
TH	0–75	Soft
	75–150	Moderately hard
	150–300	Hard
	>300	Very hard
pH	3–3.5	High acid water
	3.5–5.5	Acid water
	5.5–6.8	Low acid water
	6.8–7.2	Neutral water
	7.2–8.5	Low water
	>8.5	Alkaline water
SAR	SAR (epm)	Classification
	<10	Excellent
	10–18	Good
	18–26	Permissible
	>26	Unsuitable
SC	<20	Excellent
	20–40	Good
	40–60	Permissible
	60–80	Doubtful
	>80	Unsuitable

TABLE 5.27 Joint Water Quality Indicator (cont'd)

Parameters	Range	Classification
RSC (meq/l)	<1.25	Generally safe for drinking
	1.25–2.50	Marginal as an irrigation source
	>2.5	Usually unsuitable for irrigation without amendment
CC	<0.14	Extremely fresh
	0.14–0.84	Very fresh
	0.84–4.23	Fresh
	4.23–8.46	Fresh brackish
	8.46–28.21	Brackish
	28.21–282.1	Brackish-salt
	282.1–564.1	Salt
	>564.1	Hyperheline
Na (%)	<20	Excellent
	20–40	Good
	40–60	Permissible
	60–80	Doubtful
	>80	Unsuitable

RSC, residual sodium carbonate; EC, electrical conductivity; TDS, total dissolved solid; TH, total hardness; SAR, Sodium adsorption ratio; SC, Sodium content; CC, chloride classification.

therefore, has to do with the degree of precision of a system analysis rule. A hydrologic system analysis rule can be based on human fuzzy perceptions under the light of incomplete and vague information and personal experience even though it may be subjective (Şen, 2009b). For example, one could state as a rule.

"IF the chloride content rises to a dangerous point THEN stop water pumping from the plant."

This rule is not fuzzy because "stop water pumping" is a crisp expression. However, its fuzzy counterpart can be expressed as.

"IF the chloride content rises to a dangerous point THEN start to stop the pumps slightly."

Addition of word "slightly" renders the crisp sentence into a fuzzy statement. Similar rules map the input variables onto the output through a collection of plausible IF-THEN statements for the problem at hand. Rules are very puzzling since they look like they could be used without bothering with FL, but remember the decision is based on a set of rule as,

1. All the rules that apply are invoked using the MF and value obtained from the inputs to determine the result of the rule.
2. This result in turn will be mapped into an MF and truth-value controlling the output variable.

These results are combined to give a specific (crisp) answer, through a procedure known as defuzzification. In some cases, the MFs can be modified by "hedges" that are equivalent to adjectives. Common hedges include "about," "near," "close to," "approximately," "very," "slightly," "too," "extremely," and "somewhat." These operations may have precise definitions, though the definitions can vary considerably between different implementations. For instance, "very," squares the MFs; since the MDs are always less than 1, this narrows the MF. "Extremely" cubes the values to give greater narrowing, while "somewhat" broadens the function by taking the square root of the MF.

An algorithm is a procedure, such as the steps in a computer program. A fuzzy algorithm, then, is a procedure, usually a computer program, made up of rules relating linguistic variables. For instance, IF "water quality" is rather poor THEN make some "treatment."

If the rate of change of temperature of the air is much too high then turn the heater down a lot.

The rules use the input MDs as weighting factors to determine their influence on the fuzzy output sets of the final output conclusion. Once the functions are inferred, scaled and combined, they are defuzzified into a crisp output, which drives the system. There is different MF associated with each input and output response. Some features to note are as follows.

1. Shape: Although triangular MF are very common especially at the early stages of fuzzy works, there are also Gaussian (bell shaped), trapezoidal and exponential MFs.
2. Height or magnitude (usually normalized to 1).
3. Width (of the base of function) which is also called as the support of the fuzzy set.
4. Center points (center of the MF shape).
5. If there are at least two fuzzy sets they should overlap each other for smooth transition from one fuzzy state to the neighboring one.

There is a unique MF associated with each input parameter. The MFs associate a weighting factor with values of each input and the effective rules. These weighting factors determine the degree of influence or MD for each active rule. By computing the logical product of the membership weights for each active rule, a set of fuzzy output response magnitudes are produced. It is necessary to combine and defuzzify these output responses (Ross, 1995).

This simple idea of different MDs in a fuzzy set is extremely helpful in constructing expert systems. It is beneficial to deal with fuzzy sets of descriptive words, where the MD represents confidence that the descriptor is true of whatever the expert considers. Usually, he/she will use the term confidence rather than MD. This permits him/her to use ordinary language in describing things in a precise way.

EXAMPLE 5.15 FUZZY WATER QUALITY CLASSIFICATION

Propose quality MFs for the EC classifications in Table 5.2 and provide crisp quality indices for values 480 ppm and 1400 ppm.

Solution 5.15

Table 5.6 has five crisp classes, where at the boundaries there is no elasticity or fuzziness. Each class is in the form of rectangular shape as in

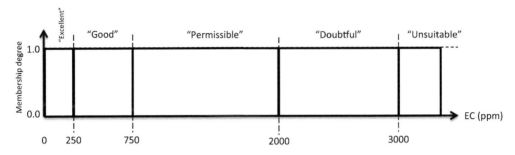

FIGURE 5.29 **Crisp classification.** EC, electrical conductivity.

Figure 5.29, where the belongingness values are equal to 1.0.

In this figure the problematic locations are at the sharp boundaries, which imply that a very small increase in the EC gives the quality in an entirely different class. For instance, if the EC values are 743 ppm and 755 ppm then they will fall into entirely different classes, namely, "good" and "permissible" with the same belonging (membership) degrees equal to 1. This seems rather illogical, because the qualities of these two groundwater samples are very close to each other.

Linguistic (fuzzy logic) expression of these quality regions should have elastic (mutually inclusive) boundaries. Consideration of the quality classes from Table 5.6 leads to the set of triangular membership functions as in Figure 5.30, which

implies that there are not sharp transitional boundaries.

The groundwater quality transition between adjacent classes from one to the other is rather gradual and therefore, the qualities of the boundary values are given in Figure 5.31.

The entrance of the EC values along the horizontal axis in this figure yields two types of quality classifications each with membership degree as follows.

1. For 480 ppm the quality classifications are "excellent" and "good" with respective membership degrees of 0.14 and 0.80.
2. For 1400 ppm the quality classifications are "permissible" and "doubtful" with respective membership degrees of 0.10 and 0.85.

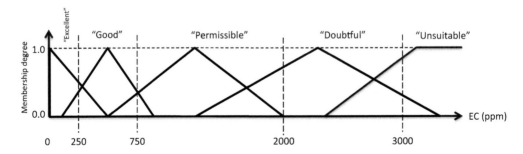

FIGURE 5.30 **Fuzzy classification.** EC, electrical conductivity.

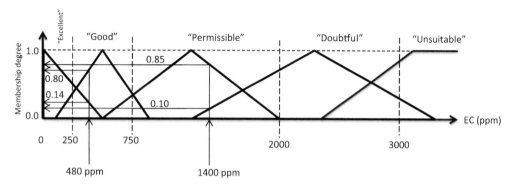

FIGURE 5.31 **Fuzzy classification applications.** EC, electrical conductivity.

EXAMPLE 5.16 FUZZY GROUNDWATER CLASSIFICATION RULE DERIVATION FROM QUALITY MAPS

The groundwater hydrochemistry records are used for the implementation of the Kriging methodology so as to obtain triple diagrams that give the common behavior of three variables. Herein, three distinctive but complementary investigations are considered. These are (Şen, 2009a),

1. Triple diagrams are constructed directly from major anions and cations, so as to consider three major anions and/or cations common behaviors within the study area. For example, the triple diagram of equal TDS concentrations based on Cl and HCO_3 values is presented in Figure 5.32.
   It is also helpful to look at the three-dimensional (3D) surface relationship between these chemical constituents as in Figure 5.33. The interpretation of the triple diagram map and 3D surface leads to the following logical inferences concerning high TDS concentration rates based on Cl and HCO_3.

   R1: IF Cl is "medium" AND HCO_3 is "low" THEN TDS is "high"

OR

R2: IF Cl is "medium" AND HCO_3 is "medium" THEN TDS is "very high"

OR

R3: IF Cl is "high" AND HCO_3 is "low" THEN TDS is "high"

These logical statements lead hydrogeologists to think about the possibilities of each IF ... THEN rule on the basis of geological subsurface composition of the study area, in addition to the hydrological and hydrogeological features' interactions. In this manner, it is possible to obtain clues for reasons of groundwater quality variations. On the other hand, these logical statements provide a common basis for the general variability description of ions within the study area. Such rule bases are prerequisites for fuzzy logic modeling as suggested by Zadeh (1965).

2. Similar triple diagrams can also be obtained among the mille-equivalent percentages of the anions and cations as they are used in the construction of the classical trilinear diagrams. This approach brings a restriction as the summation of the percentages is equal to 100%, which is the basis of the classical trilinear diagrams (Piper, 1944). If, for instance, the percentages of $(SO_4 + HCO_3)$, Cl and NO_3 [or Ca, Mg and (Na + K)] are α, β,

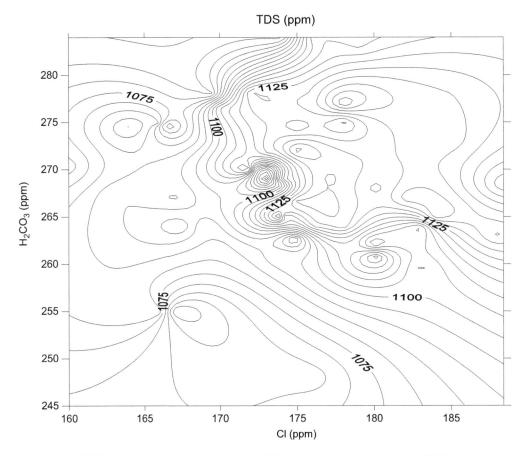

FIGURE 5.32   **Equal total dissolved solid (TDS) lines based on Cl and HCO₃.**

and γ, respectively, then by definition $\alpha + \beta + \gamma = 1.0$, which implies that $0 < \alpha, \beta, \gamma < 1$. It is obvious that this expression gives points on the equilateral inclined triangular surface shown in Figure 5.34. This triangle is identical with the Piper diagram basic ionic triangles.

3. However, the conventional equilateral triangle representation of ions is considered similar to diagram in the first step, but with percentages. Figures 5.35 and 5.36 show the percentage change of TDS with of Cl and HCO₃ percentages in two and 3D maps, respectively.

4. It is also helpful to construct triple diagrams and 3D surfaces in terms of any ion representing dependent variable with two independent composite variables electrical conductivity (EC), total dissolved solid (TDS) or pH.

The application of these four steps yields to a bundle of triple diagrams that can be interpreted leading to common logical and scientific statements about the NO₃ changes with respect to two other ions. Figures 5.37 and 5.38 indicate TDS concentration changes with H₂CO₃ and SO₄ as triple diagrams and 3D map, respectively.

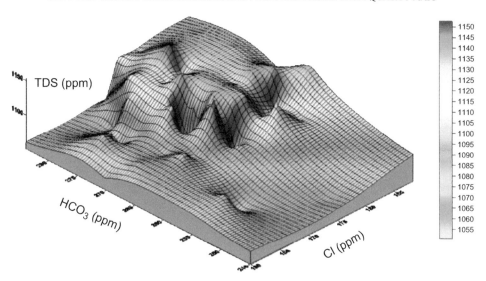

FIGURE 5.33   **Three-dimensional total dissolved solid (TDS) change with Cl and HCO₃.**

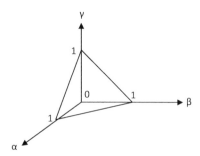

FIGURE 5.34   **Equilateral inclined triangular surface.**

Consideration of these figures for the "high" NO₃ concentration occurrences leads to the following verbal rules.

R1: IF Cl is "medium" and TDS is "high" THEN TDS is "very high"
    OR
R2: IF Cl is "low" and TDS is "high" THEN TDS is "high"
    OR
R3: IF Cl is "high" and TDS is "high" THEN TDS is "high"
By considering the aforementioned and

similar other rules, it is possible to develop a linguistic model based on the fuzzy logic and systems for groundwater quality control coupled with the continuous monitoring network. Such a model helps to reduce time and effort in routine measurements and leads toward an optimum monitoring network management with restrictive budget conditions.

1. The triple diagrams help to identify the high TDS, and especially, NO₃ concentrations with respect to two other anions or cations. Additionally, three-dimensional surfaces help to identify such maximum occurrences. It is observed that especially Cl, Na, and K give rise to such high values according to the linguistic specification of basic anions and cations.

2. The linguistic rules for maximum NO₃ concentration occurrences are advised to be taken into consideration in the design of

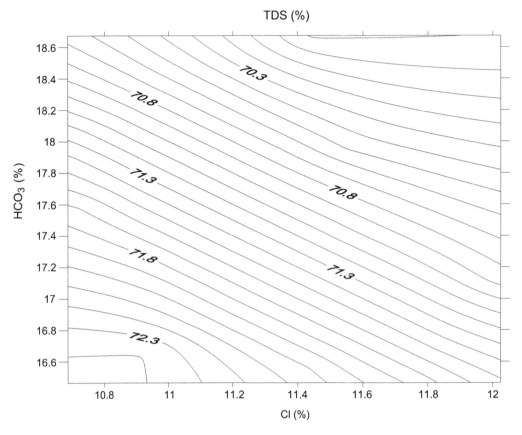

FIGURE 5.35 **Equal percentage total dissolved solid (TDS) lines based on percentages of Cl and HCO₃.**

optimum monitoring system for the groundwater quality control.

3. For reduction of NO_3 concentration, three different artificial mixture procedures are suggested. These are either the mixture with desalination water or with purified water or groundwater with low NO_3 concentrations from the nearby well fields.

5.16 CLIMATE CHANGE AND GROUNDWATER QUALITY

Indirect effects of climate change on groundwater quantity can result from climate-induced changes of groundwater withdrawals or land use. The former may increase because

1. Irrigation water requirements increase.
2. River discharge decreases or its temporal variability increases, so that the reliance on surface water goes down.

If irrigated areas decrease due to less available surface waters, groundwater recharge via leached irrigation water decreases. Climate change may lead to vegetation changes, affecting groundwater recharge.

With respect to groundwater quality, climate change is likely to have a strong impact on coastal salt water intrusion as well as on

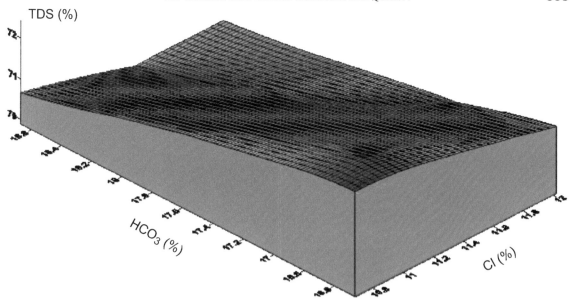

FIGURE 5.36 Three-dimensional percentage total dissolved solid (TDS) change with Cl and HCO₃.

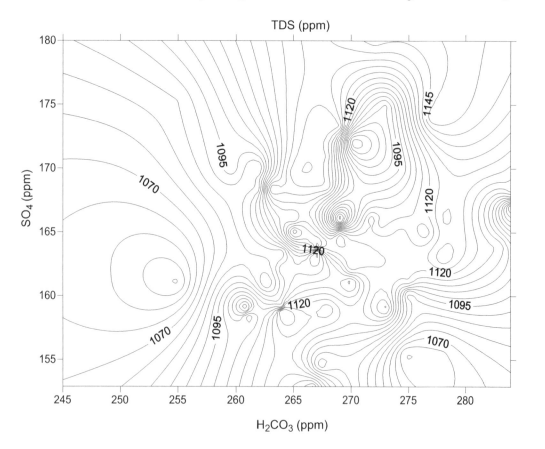

FIGURE 5.37 Equal total dissolved solid (TDS) lines based on Cl and TDS.

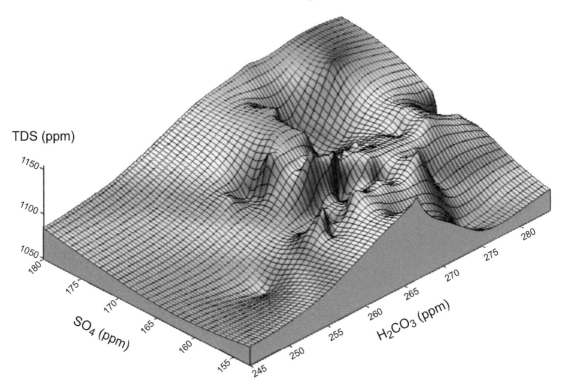

FIGURE 5.38 Three-dimensional total dissolved solid (TDS) change with H_2CO_3 and SO_4.

salinization of groundwater. For two small and flat coral islands off the coast of India, Bobba et al. (2000) computed the impact of sea level rise on the thickness of the freshwater lenses. With a sea level rise of only 0.1 m, the thickness of the freshwater lens decreased from 25 to 10 m for the first island and from 36 to 28 m for the second island. In addition to the sea level rise, any change in groundwater recharge affects the location of the freshwater/saltwater interface, and saltwater intrusion is expected to increase if less groundwater recharge occurs. This can also happen inland where saline water is located next or below freshwater (Chen et al., 2004). For many semiarid areas, a decrease in precipitation is projected and enhanced evapotranspiration in the warmer world might cause a salinization of groundwater.

It is therefore necessary to develop water resources managements by considering regional variations in water quality. Şen et al. (2004) suggested such an artificial mixture programs through simple management rules for groundwater resources in the central western parts of Saudi Arabia along the Red Sea coast where there is also salt water intrusion into the unconfined aquifers.

Drought may reduce the chances of grass encroachment but may increase the chances of disturbances to heather vegetation. Similarly, warming increases and drought decreases the chances of nitrate pollution to the groundwater, which is often used as a drinking water source. Warming of the upland heathland in the United Kingdom increases its productivity, which might enable higher grazing densities leading to improved agricultural production. However, complex interactions between heather and invading species may be affected. Furthermore, nitrate production is increased, which may lead to groundwater pollution. Under drought conditions, productivity decreases and agricultural production capacity drops. In the Mediterranean shrub land in Spain, both warming and drought led to a shift in the species composition of seedlings and recruitment, which might lead to a change in the plant community and a reduction in biodiversity. In the drought treatment, decreasing soil carbon content may lead to a loss of biodiversity, recreational possibilities and an increased threat of wildfires and erosion.

References

Ahmad, N., 1998. Evaluation of groundwater resources in the upper middle part of Chaj Doab Area, Pakistan (Unpublished Ph.D. thesis). Department of Meteorological Engineering, Technical University of Istanbul, Turkey.

Ahmad, N., Şen, Z., Ahmad, M., 2003. Ground water quality assessment using multi-rectangular diagrams. Ground Water 41 (6), 828–832.

Al-Bassam, A.M., Awad, H.S., Al-Alawi, J.A., 1997. Durov Plot: A computer Program for Processing and Plotting Hydrochemical Data. Ground Water 35 (2), 362–367.

Al-Yamani, M.S., 2001. Isotopic composition of rainfall and ground water recharge in the western province of Saudi Arabia. J. Arid Environ. 49, 751–760.

Appelo, C.A.J., Postma, D., 1999. Geochemistry, Groundwater and Pollution. Balkema, Rotterdam.

Ayers, R.S., Wescot, D.W., 1976. Water Quality for Agriculture. Food and Agriculture Organization of the United Nations. Irrigation and Drainage Paper 29, Rome, Italy.

Back, W., 1960. Origin of hydrochemical facies and ground water in the Atlantic Coastal Plain. Rep. XXI Int. Geol. Congress, Nordend (1), 87.

Barrett, M., Nalubega, M., et al., 1999. On-site sanitation and urban aquifer systems in Uganda. Waterlines 17 (4), 10–13.

Bobba, A., Singh, V., Berndtsson, R., Bengtsson, L., 2000. Numerical simulation of saltwater intrusion into Laccadive Island aquifers due to climate change. J. Geol. Soc. India 55 (2000), 589–612.

Boyd, C.E., 2000. Water Quality. An Introduction. Kluwer Academic Publishers, Boston, Dordrecht, London, 325 pp.

Chen, Z., Grasby, S., Osadetz, K., 2004. Relation between climate variability and groundwater levels in the upper carbonate aquifer, southern Manitoba. Can. J. Hydrol. 290, 43–62.

Chow, V.T., 1964. Handbook of Applied Hydrology. Sec. 11: Evapotranspiration. Mc-Graw Hill Co., New York, pp. 11–38.

Clark, I., Fritz, P., 1997. Environmental Isotopes in Hydrogeology. CRC Press/Lewis Publishers, 328 pp.

Craig, H., 1961. Isotopic variations in meteoric waters. Science 133, 1702–1703.

Dansgaard, W., 1964. Stable isotopes in precipitation. Tellus 16, 436–468.

Davis, J.C., 1986. Statistics and Data Analysis in Geology, second ed. John Wiley, New York. 646 pp.

Doneen, L.D., 1964. Notes on Water Quality in Agriculture. Published as a Water Science and Engineering Paper 4001. Department of Water Science and Engineering, University of California.

Drabbe, J., Baydon-Ghyben, W., 1889. Nota in verband met de voorgenomen putboring nabij Amsterdam. In: Tijdschrift van het Koninklijk Institut van Ingenieurs. The Hague, Netherlands.

Durov, S.A., 1948. Natural waters sand graphic representation of their composition. Dok. Akad Nauk SSSR 59, 87–90.

Fair, G.M., Geyer, C.J., Okun, D.A., 1971. Elements of Water Supply and Waste Water Disposal, second ed. John Wiley and Sons, Inc., New York, New York. 752 pp.

FAO, 1985. Water Quality for Agriculture. Irrigation and Drainage Paper No. 29, Rev. 1. Food and Agriculture Organization of the United Nations, Rome.

Freeze, R.A., Cherry, J.A., 1979. Groundwater. Prentice-Hall International, Inc., London, 605 pp.

Fritz, P., Fontes, J., 1989. Handbook of Environmental Isotope Geochemistry: The Marine Environment B, vol. 3. Elsevier, pp. 425.

Hem, J.D., 1970. Study and Interpretation of the Chemical Characteristics of Natural Water. U. S. Geol. Survey. Water Supply Paper 1473. pp. 363.

Herzberg, A., 1901. Die Wasserversorgung einiger Nordseebäder. J. Gasbeleucht. Wasserversorg. 44, 815–819, 842–844.

Hill, R.A., 1940. Geochemical patterns in the Coachella Valley, California. Trans. Am. Geophys. Union 21, 9–46.

Hubbert, M.K., 1940. The theory of groundwater motion. J. Geol. 48 (8), 785–944.

Hudson, R.O., Golding, D.L., 1997. Controls on groundwater chemistry in subalpine catchments in the southern interior of British Columbia. J. Hydrol. 201, 1–20.

Jones, I.C., Banner, J.L., Humphrey, J.D., 2000. Estimating recharge in a tropical karst aquifer. Water Resour. Res. 36 (5), 1289–1299.

Karanth, K.R., 1989. Groundwater Assessment, Development and Management. Tata McGraw-Hill, New Delhi.

Kimblin, R.T., 1995. The chemistry and origin of groundwater in Triassic sandstone and Quaternary deposits, northwest England and some UK comparisons. J. Hydrol. 172, 293–308.

Kirk, S.T., Campana, M.E., 1990. A deuterium-calibrated groundwater flow model of a regional carbonate-alluvial system. J. Hydrol. 119, 357–388.

Langelier, W.F., 1936. The analytical control of anti-corrosion water treatment. J. Am. Water Works Assoc. 28, 1,500 pp.

Langmuir, D., 1997. Aqueous Environmental Geochemistry. Prentice Hall, 600 pp.

Lloyd, J.W., Heathcote, J.A., 1985. Natural Inorganic Hydrochemistry in Relation to Groundwater. Clarendon Press, Oxford, England.

Lyles, B.F., Hess, J.W., 1988. Isotope and Ion Geochemistry in the Vicinity of the Las Vegas Valley Shear Zone. University of Nevada, Desert Research Institute Publication 41111, 78 pp.

Mandel, S., Shiftan, Z.L., 1981. Ground-Water Resources—Investigation and Development. Academic Press, New York, 269 pp.

Mayo, A.L., Loucks, M.D., 1995. Solute and isotopic geochemistry and groundwater flow in the central Wasatch range, Utah. J. Hydrol. 172, 31–59.

Miller, J.A., 1991. Summary of the Hydrology of the Southeastern Coastal Plain Aquifer System in Mississippi, Alabama, Georgia and South Carolina. U. S. Geological Survey Professional Paper 1410-A, pp. 33–36.

Musgrove, M., Banner, J.L., 1993. Regional groundwater mixing and the origin of sline fluids. Midcontinent, United States. Science 259, 1877–1882.

Noack, R.E., 1988. Sources of Ground Water Recharging the Principal Alluvial Aquifers in Las Vegas Valley, Nevada (unpublished M.S. thesis). University of Nevada, Las Vegas, 167 pp.

Piper, A.M., 1944. A graphic procedure in the geochemical interpretation of water analysis. Trans. Am. Geophys. Union 25, 914–923.

Richards, L.A., 1954. Diagnosis and Improvement of Saline and Alkali Soils, U. S. Department of Agriculture Handbook, vol. 60. Washington D.C., USA. p. 160.

Ross, J.T., 1995. Fuzzy Logic with Engineering Applications. McGraw-Hill, New York.

Rozanski, K., Araguas-Araguas, L., Giofiantini, R., 1993. Isotopic patterns in modern global precipitation. Geophys. Monogr. 78, 1–36.

Salomons, W., Forstner, U., 1984. Metals in the Hydro Cycle. Springer-Verlag, Berlin.

SAS, 1984.

Schoeller, H., 1967. Geochemistry of ground water. An international guide for research and practice. UNESCO 15, 1–18.

Scholl, M.A., Ingebritsen, S.E., Janik, C.J., Kauahikaua, J.P., 1996. Use of precipitation and groundwater isotopes to interpret regional hydrology on a tropical volcanic island: Kilauea volcano area, Hawaii. Water Resour. Res. 32 (12), 3525–3537.

Şen, Z., 2009a. Spatial Modeling Principles in Earth Sciences. Springer, New York, 351 pp.

Şen, Z., 2009b. Fuzzy groundwater classification rule derivation from quality maps. Water Qual. Expo. Health 1, 115–122.

Şen, Z., Al-Dakheel, A., 1986. Hydrochemical facies evaluation in Umm Er Radhuma Limestone, Eastern Saudi Arabia. Ground Water 24 (5), 626–635.

Şen, Z., Saud, A.A., Altunkaynak, A., Özger, M., 2004. Increasing water supply by mixing of fresh and saline ground waters. JAWRA J. Am. Water Resour. Assoc. 39 (5), 1209–1215.

Stallard, R.F., Edmond, J.M., 1981. Geochemistry of the Amazon: 1. Precipitation chemistry and the marine contribution to the dissolved load at the time of peak discharge. J. Geophys. Res. 86, 9844–9858.

Stiff Jr, H.A., 1951. The interpretation of chemical water analysis by means of patterns. J. Petr. Technol. 3 (10), 15–17.

Stuyfzand, P.J., 1989. Hydrology and water quality aspects of Rhine bank groundwater in the Netherlands. J. Hydrol. 106 (3/4), 341–363.

Thomas, J.M., Mason, J.L., Crabtree, J.D., 1986. Groundwater Levels in the Great Basin Region of Nevada, Utah, and Adjacent States: U.S. Geological Survey Hydrologic Investigations Atlas HA-694-b, 2 Sheets, Scale 1: 1,000,000.

Todd, D.K., 1980. Ground Water Hydrogeology. John Wiley and Sons.

USSL, 1954. Diagnosis and Improvement of Saline and Alkali Soils. USDA, Handbook, vol. 60, p. 147.

Vengosh, A., Gill, J., Davisson, M.L., Hudson, G.B., 2002. A multi-isotope (B, Sr, O, H, and C) and age dating (^{3}H-^{3}He and ^{14}C) study of groundwater from Salinas Valley, California: hydrochemistry, dynamics, and contamination processes. Water Resour. Res. 38, 9-1–9.17.

World Health Organization, WHO, 2004. Global Strategy on Diet, Physical Activity and Health. Resolution WHA55.23.

WMO, 1977. The Use of Satellite Imagery in Tropical Cyclone Analysis, TD-No. 473, Technical Note No. 153.

Wilcox, L.V., 1955. Classification and Use of Irrigation Waters. US Department of Agriculture. Cire. 969, Washington D.C., USA, p. 19.

Wood, W.W., Sanford, W.E., 1995. Chemical and isotopic methods for quantifying groundwater recharge in a regional, semiarid environment. Ground Water 33 (3), 458–468.

Yurtsever, Y., 1975. Worldwide survey of stable isotopes in precipitation. Rep. Sect. Isotope Hydrol. IAEA, 40 pp.

Yurtsever, Y., Gat, J.R., 1981. Atmospheric Waters. In: Stable Isotope Hydrology, Deuterium and Oxygen-18 in the Water Cycle. IAEA Technical report series 210, 339 pp.

Zadeh, L.A., 1965. Fuzzy Sets, Information and Control, vol. 8, pp. 338–353.

Zadeh, L.A., 1999. From computing with numbers to computing with words—from manipulation of measurements to manipulation of perceptions. IEEE Trans. Circ. Syst. I Fund. Theor. App. 4 (1), 105–119.

6

Groundwater Management

Practical and Applied Hydrogeology
http://dx.doi.org/10.1016/B978-0-12-800075-5.00006-6

Copyright © 2015 Elsevier Inc. All rights reserved.

6.1 GENERAL

Pressure on global water resources is increasing in an unprecedented manner, and therefore, sustainable groundwater management practices gain importance more than ever before. Although there are many procedures, formulations, and algorithms for sustainable aquifer management (AM), still there is a great need for effective, practical, and applied methodologies. A large number of different approaches including numerical models and their software are available for ready uses, but sustainable water management needs special attention for successful local applications. It is practically impossible to provide a general definition and solution to cover different management strategy, because each AM requires solutions under special, local, and environmental conditions, which may not be exactly valid for other regions. Among the nonsustainable practices are quantitative (overpumping, destruction of wetlands, salinization of soils, and water resources) and qualitative (contamination, salt, chlorinated hydrocarbons) issues. Although the role of numerical modeling as a tool is significant for sustainable management practices, its pros and cons must be identified for each case prior to mechanical applications.

Water management requires a hectic effort for achieving the preset targets taking into consideration a multitude of aspects, including hydrometeorology, suitable geographical location, climate, hydrology, physiography, geology, hydrogeology, hydrochemistry, depositional environments, and storage and transmission properties of aquifers in addition to its availability. Groundwater exploitation without management program leads to depletion of present-day storages, which are necessary for reliefs in emergency situations (Şen, 2008).

Water resources management problems are unique in many regions and not only the physical aspects of intermittent or no-surface runoff, depletion of existing groundwater storages, overexploitation and consumption, salt water intrusion, and pollution of unconfined aquifers are among the major problems, but also there are managerial aspects of lack of trained personnel, deficient institutional arrangements, and poor or nonexistent resource management rules, regulations, programs, and software. In many regions, groundwater resources are exploited almost free of charge.

Especially in drainage basin, the scale studies in water resources assessment concerning meteorological, geological, morphological, and hydrological studies have interconnected importance. Based on sufficient meteorological and hydrological data, there are different methods developed for the conversion of rainfall to run off, i.e., rainfall-runoff models, rainfall as input and runoff as output. It is necessary to depend on simple and practical formulations and approaches based on rational thinking and logic in addition to the most sophisticated scientific and academic interest models that provide applicable procedures.

6.2 MANAGEMENT PLANNING

In different parts of the world, groundwater resources are not managed by considering optimum common benefit for the present and future generations. Simple and effective management rules and regulations must be put forward prior to the preparation of a dynamic, optimum, and adaptive management program planning, operation, and maintenance tasks. Such rules and regulations should be based on local knowledge and experiences supported by scientific information sources. The main concern of management must be planned in such a way that,

1. Groundwater abstractions must be adjusted according to sustainable supply and demand stipulations in the future.

2. The renewable (recharge) character of potential aquifers must be considered with optimum aquifer pump rates (discharges), well-field locations, and especially, priority of subaquifer units within the integrated aquifer-system management.
3. The economy, social impact, local administration strategies, and water quality variations should also be considered.

Due to nonscientific applications and nonexistence of proper management programs, some of the significant aquifers are almost mined in the world. A good example for this is the Ogallala aquifer in the United States, which is overexploited. Groundwater resources may set the limit on the use of other resources and thus ultimately on the density of population and the standards of living that can be sustained in areas of interest (Walton, 1970).

Water resources are subject to various external effects that may cause quantitative decrease and/or qualitative deterioration by time. Among the undesirable events, the following points are worth to consider:

1. Population increase not only by birth but equally important by migration due to economic, political, military, agriculture, social, and hazardous situations.
2. Industrial investments, development, and processing in potential catchment areas, and especially, groundwater recharge areas lead to water resources pollution, and hence, water potentiality for domestic use decreases.
3. Recent climate change, global warming, and greenhouse events give rise to groundwater resources' quantity and quality deteriorations.
4. Mismanagement of water resources, in general, and groundwater reservoirs in particular, may lead to overpumping toward aquifer mining causing consequent groundwater quality changes, sinkholes, and subsidence.

Groundwater resources are the most conservative reservoirs and have the most strategic significance due to the following facts:

1. They are replenished by rainfall after infiltration and penetration processes through subsurface geological layers, and therefore, their quality is better than that of surface waters. Especially, geological formations including quartz, granite, sandstones, and alluvial deposits provide the best-quality groundwater reservoirs.
2. They are protected from any atmospheric pollution dangers and can preserve the quality and temperature throughout the year.
3. They are available anywhere in the world, and therefore, they are the most dependable water resources especially in the cases of emergency such as natural or anthropogenic disasters.
4. Aquifers are natural and dependable reservoir spaces as future exploitable water resources without much cost. They reduce evaporation losses almost to zero even in very hot climate regions, and therefore, they are most dependable and preferable in arid regions.

It is a general rule that groundwater modeling software programs are commercially available and one can buy and apply them to solve his/her problem. In cases of management problems, each region has its own circumstances that must be taken into account, and therefore, ready management programs may not be sufficient. They need basic philosophy and logical statements for management under the light of particular and local sets of social, economic, and administrational rules. In many parts of the world, groundwater storages are exploited without rules and regulations for commercial benefits (Şen, 2008).

The exploitation and consumption of groundwater must be so adjusted that the present storages sustain critical periods until the next rainfall event occurrences with sufficient groundwater recharges. It has been noticed that the indigenous plants and animals have adapted to this paradigm by various methods of drought evasion. Human adaptation to this situation cannot be thought without an effective management program. The purpose of integrated AM is to understand the human, ecological, economical, hydrological, hydrogeological, and hydrochemical interactions and to apply these concepts for groundwater storage improvements in addition to other essential resources such as soil and land use for long-term sustainable development and productivity. For an effective and strategic integrated groundwater resources management in two or more aquifers, the first step is to get a common agreement including each authority concerned. The implementation of any effective scientific management program can be achieved with the support of administrational and/or political authorities.

6.3 MANAGEMENT ENVIRONMENTS

In general, there are seven different but mutually inclusive management planning components depending on the final purpose. The following points should be considered individually or collectively in an integrated groundwater management study:

1. Spatial environment: The groundwater resources may cover more than 4 km depth within the Earth's crust, but exploitation cannot be done practically at more than 2 km depth. Most often the groundwater wells and boreholes have depths varying from a couple of meters (shallow) to 100–1500 m (deep). The depth is one of the restrictive factors in strategic groundwater resources planning. The aquifer extent is important as another spatial factor. For instance, it is essential to know the surface area of recharge exposure in strategic groundwater resources planning.

2. Temporal environment: The connection of aquifers with the present-day hydrological cycle must be considered, if any, and according to replenishment rates, groundwater withdrawal amounts should be planned and managed. Aquifers may be classified as replenishable and nonreplenishable according to their connections with the hydrological cycle. Quaternary deposit formation unconfined aquifers are in contact with the present-day hydrological cycle. Occasional floods, flash floods, or large amounts of runoff volumes provide additional rich natural and artificial recharge possibilities. Rainfall events provide temporal recharge occasions, and therefore, it is necessary to take into account the haphazard temporal variability of rainfall in any groundwater recharge study (Chapter 4).

3. Hydrogeological environment: The geological formations play a dominant role in any groundwater resources evaluation, management, assessment, and exploitation. Such a role is dependent on the geological formation composition with voids, fissures, fractures, crevices, solution cavities, faults, and other structural geological features. The amount of porous, fractured, and possible karstic media should be considered in an effective and strategic management planning.

4. Hydrochemical environment: The groundwater quality, ionic exchange, salt water intrusion, upconing, and closeness to sea shore are additional factors that must be taken into consideration in strategic planning and management of an aquifer. Comparatively worse-quality water zones within the study area cannot be overlooked, since an artificial mixture of different groundwater qualities may give rise to acceptable quality levels (Chapters 4 and 5).

5. Alternative strategies: The planner or decision maker may have different alternatives for the same problem, so that s/he can decide on the most suitable solution under local

circumstances. If there are two or more aquifers, then the question is which groundwater reservoir should have exploitation priority? For how long it should be exploited before the second alternative enters the circuit for joint strategic planning?

6. Emergency situations: The planner must keep in mind that in emergency cases such as diseases, terrorist attacks, wars, and natural destructions including earthquakes, landslides, etc. the infrastructure of alternative water resources supply system may malfunction, but the groundwater supplements remain at their highest precious levels.

6.4 LOCAL CONDITIONS

The groundwater resources are location dependent in terms of local geological, hydrogeological, geochemical, morphological, tectonic, climatological, social, and economic factors. In humid regions, the groundwater resources are not exploited extensively due to surface water resources alternatives. For strategic planning, the following specific points must be considered:

1. Rainfall events, aquifer recharge areas, and following infiltration rates become important. It is, therefore, necessary to know the surface geological and morphological features so as to calculate the runoff and groundwater recharge (Chapter 4).

2. The plain areas at lower lands are the main courses of surface flow that originates from relatively higher surroundings. It is important to know the areal extent and infiltration properties of these plains. The rainfall and runoff relationships must be taken into consideration by simple techniques (Chapter 1) including flood and flash flood occurrences. Such simple considerations may lead to further extensive evaluations that may furnish a

simple strategy rather than the use of complicated software models, which are developed mostly for humid regions.

3. The geometric dimensions of subsurface cross-section must be determined through geophysical prospecting. Earth surface and especially piezometric surface (water table) slopes are also important for surface and subsurface groundwater flow movement evaluations.

4. Groundwater management practices require determination of different hydrogeological parameters such as porosity, specific yield (storage coefficient or effective porosity), and hydraulic conductivity for actual water volume calculations. It is preferable to obtain parameter estimations by field techniques, and especially, through the aquifer tests and their proper interpretations (Chapters 2, 3, and 4).

5. In replenishable groundwater reservoir strategic planning, the groundwater quality is significant not only for management, but also for controlling the excessive exploitation possibilities. Naturally, groundwater does not have uniform but different water types existing in the same storage. In general, the bottom layers in aquifers are saline, and therefore, its upconing to pumping wells must be avoided.

6. Safe yield exploitation must be considered for natural replenishment and groundwater abstraction balance. A certain amount of water must be left within the aquifer for uses in cases of possible future unexpected emergency cases.

7. The population and its future growth are strategical factors for groundwater allocation. The minimum requirements, say, 50 l/day/capita or 70 l/day/capita can be taken into account in emergency situations (Şen et al., 2013).

8. Private well owners must be convinced to allow for groundwater strategic planning by avoiding haphazard and unnecessary exploitations.

6.5 PRELIMINARY MANAGEMENT REQUIREMENTS

An effective and well-organized development plan for integrated groundwater management for different aquifers is very essential because such a plan relates all necessary tasks, resources, and time. Groundwater management studies should include the following three fundamental requirements. Their completion may give way to answer many simple and essential questions about the availability of groundwater resources in an area.

1. Database: Data must be assembled from different sources such as previous reports and studies, reconnaissance field surveys, and through discussions with local experts on the subject as well as with experienced settlers. It is fruitful to collect linguistic (verbal) information from the settlers about the current and past practices of groundwater exploitation, rainfall regime, and recharge possibilities. Such a linguistic database is missing or not cared for in many groundwater resources development and management programs. Recently, fuzzy logic and system approach have been helping to treat the verbal data for the purpose of management in an effective manner (Şen, 2010). Most often, numerical data are derived from various maps such as topographic, geographic, and geologic maps in addition to photogrammetry, digital elevation model, and satellite images. Meteorological, climatologic, and hydrological records are the backbone of the database prior to groundwater management studies.

   In order to gain a preliminary overall view, it is necessary to have a reconnaissance phase, where all available maps, data, and numerical or verbal information are reviewed for appreciation of the ultimate management work with a set of relevant objectives and restrictions. This phase also provides basic data and information on hydrologic, geologic, hydrogeological, groundwater quality, and other related aspects. Qualitative reconnaissance methods are only an expedient and they cannot substitute adequately and carefully designed qualitative groundwater management studies.

2. Restrictions: It is necessary to define the framework of groundwater assessment from different problematic views including possible restriction impositions such as,
   a. Water demand quantity and types (mainly domestic, agriculture, industrial),
   b. Safe yield,
   c. Possibility of well numbers and locations,
   d. Water supply quality,
   e. Recharge possibilities and restrictions,
   f. Pump levels and capacities,
   g. Unsaturated volume for future water storage,
   h. Saturated water volume for availability,
   i. Rainfall water volume for recharge,
   j. Groundwater potentiality,
   k. Any limitation on well-completion methods,
   l. Groundwater contamination and pollution possibilities,
   m. Economic situation under available budget restrictions,
   n. Other jointly manageable nearby groundwater storages.

   Decomposition of AM program into subprograms is helpful so as to assess the economic, hydrologic, hydrogeological, and hydrochemical supply and demand aspects of each unit toward an integrated decision making.

3. Effective field survey: It is necessary to undertake a field survey with the aforementioned information and queries at hand in order to appreciate the effect of each groundwater occurrence, movement, and quality factors so as to achieve a complete,

sustainable, and successful management program. A set of management options can be generated according to different scenarios, which are useful for identifying the optimum solution in the uncertainty domain of the problem especially in a multiple-aquifer system management.

After these three fundamental steps, one can imagine a rather clear picture about the groundwater resources situation in the study area. The following questions can be asked, which provide clues about the solutions:

a. Which locations have the priority for groundwater development in the area?

b. What other groundwater resources in the nearby locations can be managed in an integrated manner so as to reach a better optimum solution?

c. What are the possibilities of additional well drilling, location, and pumping in the study area?

d. Are there adequate groundwater resources quantitatively to meet the demand in the area?

e. Is the groundwater quality suitable for developmental activities in and nearby areas?

f. What are the discharge and recharge possibilities and rates temporally and spatially in the area?

g. Are there groundwater mixture possibilities so as to enhance the groundwater quality and quantity? (Chapters 4 and 5)

h. What are the potentialities in the area? Are there different aquifers that may be exploited in an integrated manner?

i. Are there possibilities of quick drawdown drops, thus an increase in the pumping lift and cost?

j. Are there any potential hazards either to the groundwater reservoir from pollution or to any infrastructure due to excessive pumping and subsidence?

k. What is the number of population that can be supported by the available groundwater resources?

l. If already there are groundwater developments in the study area, is it possible to integrate them into the management program?

It is possible to provide simple preliminary, rational, and logical answers to these questions on the basis of collected numerical and verbal databases. Prior to any detailed and refined management study, the planner should try to provide simple answers for the study area.

It is not possible to avoid uncertainties in groundwater management programs, but through an effective planning one can reduce them to a minimum by developing an adaptive management program, which renews itself each time with the coming of new information.

6.6 GROUNDWATER MANAGEMENT OBJECTIVES

At the regional scale, it is extremely important to undertake identification of priorities according to the following steps:

1. Groundwater system susceptibility assessment to degradation through inadequately controlled exploitation,

2. Identification of critical areas and their declaration as resource conservation zones, a concept which can play a critical role in the development of effective groundwater management.

Any groundwater resources management involves planning, implementation, and operation that are necessary to provide safe and reliable groundwater supplies. The groundwater management objectives are focused on aquifer yield, recharges, and water quality in addition to legal, socioeconomic, and political factors. Although formal groundwater management is important in large-scale developments, it may also be applied to smaller-scale or even individual well projects. In any groundwater management, safe and sustainable yields are among the main objectives.

6.6.1 Safe Yield

In groundwater management, the safe yield is the rate at which groundwater can be withdrawn from an aquifer without causing an undesirable adverse effect (Dottridge and Jaber, 1999; Heath and Spruill, 2003). The traditional definition of the safe yield assumes the pumping rate equal to the total recharge. Nian-Feng et al. (2001) assume that the safe yield is 50% of total natural recharge of groundwater.

Any groundwater management should include objectives of renewability (replenishment, recharge) of the resources, practical exploitations, and consumption rationality. The concept of "safe yield" comes into view as a basic objective, which is associated with the amount of supply that a water user can sustainably depend upon. Meinzer (1934) defined the safe yield of an aquifer as "the practicable rate of perennially withdrawing water from it for human use." In this definition, "perennial withdrawing" does not suit the aquifers in arid regions, and therefore, the safe yield can be modified as that "amount of groundwater storage, which can be abstracted to cover the supply needs during a stipulated period of time without damaging the aquifer, which is replenished only ephemerally each year after occasional, sporadic, and sufficiently intense rainfalls and their consequent floods." In general, the safe yield must be adjusted in such a way that neither the quantity nor the quality of groundwater is allowed to reach unacceptable limits. In practical studies, the safe yield should be less than the annual average recharge in order to compensate minor groundwater losses. If the safe yield is overtopped for some time then the aquifer is bound to be mined, which is an undesirable situation in groundwater management strategies. The main

reason for such mined aquifers in some parts of the world is the nonexistence of an effective groundwater management system.

EXAMPLE 6.1 SAFE YIELD CALCULATION

A confined aquifer has 25 km^2 areal extent and its thickness is 12.6 m. The aquifer hydraulic conductivity and the storage coefficient are 3.2×10^{-1} and 2.8×10^{-3} m/min, respectively. An observation well monitoring for several years indicated piezometric level changes between 27 and 24 m. Lateral flow rate due to infiltration from far distances is 1.5 m^3/s. Under these circumstances,

1. What is the aquifer safe yield?
2. If each well pump is 5 l/s, how many wells are needed for safe exploitation?

Solution 6.1

In this case, the lateral flow does not enter calculations because the piezometric level fluctuation already assumes its effect.

1. Since long-term piezometric level fluctuation difference is $27 - 24 = 3$ m, the total amount of water that can be withdrawn from this aquifer can be calculated according to Eq. (2.22), which yields annually safe yield water volume as,

$$V_W = 2.8 \times 10^{-3} \times 25 \times 10^6 \times 3$$
$$= 0.21 \times 10^6 \text{ m}^3/\text{year}$$

2. The pump discharge from a single well is $Q_w = 5 \text{ l/s} = 5 \times 10^{-3} \times 365 \times 24 \times 60 \times 60 \approx 0.16 \times 10^6 \text{ m}^3/\text{year}$. Hence, the number, n, of wells is,

$$n = \frac{V_W}{Q_W} = \frac{0.21 \times 10^6}{0.16 \times 10^6} \approx 2 \text{ wells}$$

In practical applications, an additional standby well must be taken for operation in case of any one of these two wells' failure. Hence, three wells are enough for sustainable management.

6.6.2 Sustainable Yield

Sustainable development is defined as the enhancement of the economic, social, and ecological well-being of current and future generations. In practice, the sustainable yield corresponds to water abstraction from an aquifer during one year such that undesirable effects are not allowed in the area. This yield can also be limited by an amount less than recharge but should not exceed the long-term mean annual discharge. It can also be limited by the physical size of the aquifer including saturated and unsaturated volumes or by the rate at which water moves to the withdrawal area. For instance, groundwater entrance velocity into the well can be used as a guide for sustainable yield calculations (Section 6.14). In the long term, the sequences of wet and dry spells tend to balance out average recharge rate, but management ignorance may lead to overdraft.

It is not possible to maintain a constant sustainable yield in a management program, due to various reasons. The sustainable yield may change seasonally and interannually.

The rearrangement of the pumping pattern, urbanization, and changes in vegetation and crop patterns reduce natural recharge. The change in the groundwater exploitation pattern such as from domestic to agriculture or vice versa or to industrial affects sustainable yield pattern.

The groundwater storage volume, water quantity, and quality may change due to some nonpumping effects as a result of agricultural and industrial developments. The irrigation return water may increase the groundwater storage volume, but in the meantime may cause deterioration in water quality. At some places, illegal wells may pump groundwater for commercial benefits without management considerations and this invites groundwater depletion and salinization. All such activities endanger the sustainability of groundwater resources.

EXAMPLE 6.2 SUSTAINABLE YIELD CALCULATION

Extensive study in an area has indicated that the groundwater recharge is about 1.20×10^6 m^3/year. In the meantime, there is a need for water supply for domestic and agricultural activities in nearby areas. The population of the town is 15,000 and per capita daily consumption is planned at 110 l/capita/day. Yearly agricultural water demand is calculated as 0.70×10^6 m^3/year.

1. Check whether the groundwater storage is enough for sustainable development.
2. If not, what could be suggested for sustainability?

Solution 6.2

Sustainability in this case is dependent on groundwater abstraction as equal to yearly replenishment of the aquifer.

1. The total annual demand (groundwater supply) has two components: domestic and agricultural demands. Since per capita consumption is restricted by 110 l/capita/day, the annual domestic demand, Q_{AD}, is

$$Q_{AD} = 15,000 \times \left(110 \times 10^{-3} \times 365\right)$$
$$= 0.60 \times 10^6 \text{ m}^3/\text{year}$$

The total annual demand, Q_{DT}, is the summation of this value with the agricultural need.

$$Q_{AT} = 0.60 \times 10^6 + 0.70 \times 10^6$$
$$= 1.30 \times 10^6 \text{ m}^3/\text{year}$$

Since the total annual demand is greater than groundwater recharge ($1.30 \times 10^6 > 1.20 \times 10^6$), it is not possible to sustain the groundwater storage without additional water level drop.

2. For the sustainability of the aquifer $1.30 \times 10^6 - 1.20 \times 10^6 = 0.10 \times 10^6$ m^3/year must be supported by some other water resources. Additional water support may be solved by several alternatives depending on the

circumstances among which one or a few of the following solutions might be adopted.

a. If the above calculations are valid for a dry year or for a short duration drought period, then additional groundwater may be taken from the same aquifer, in the hope that future wet periods may refill the aquifer to its normal levels. Hence, long duration sustainability may be satisfied by local groundwater management. Such a decision should be based on dry period rainfall analyses and its groundwater recharge impact studies. In some areas, the return period of dry seasons may be established through past records' examination and for instance it may have 2-year, 3-year, or 5-year periods. Depending on the aquifer potentiality and water storage volume, the final decision can be made by considering the convenient return period.

b. For local and temporary dry periods, the per capita water consumption may be reduced to such a level that the water deficit, $0.10 \times 10^6 \, \mathrm{m^3/year}$, can be covered. For instance, if the consumption rate is reduced by 20 l/capita/day, then the water that can be saved is

$$15,000 \times 20 \times 365 = 0.109 \times 10^6 \, \mathrm{m^3/year}$$

In this way, local administrators may avoid unsustainability for one year with the contribution from the local people with their endurance and patience. This is, in a way, a good strategy because rather than having expensive solutions, the temporary water shortage can be alleviated by water saving. The success of this alternative depends on public awareness.

c. It may also be possible to save agricultural water demand to a certain extent by changing crop types that require less water. Furthermore, the local experts and administrators may lay down plans for joint water savage from per capita and agriculture demands.

d. If the aforementioned solutions are not sufficient or cannot be applicable for some reasons, then additional water supply should be sought from outside the area. Among such alternatives are sources from nearby aquifers, lakes, and surface or subsurface dams. In case of none of such alternatives, as in arid regions, for long term sustainability desalination plants may be established.

6.6.3 Safe Yield and Sustainability

After Meinzer's (1934) definition of "safe yield" as the perennial water abstraction from an aquifer without producing any undesirable results, Todd (1980) suggested "perennial yield" as the flow of water that can be abstracted from a given aquifer without producing adverse results. Each aquifer has a different safe yield, depending on its recharge and hydrogeological parameters. The key point is the consensus on the "undesirable" or "adverse" results. In any groundwater management task, there are different expectations by consumers, administrators, scientists, government officers, farmers, society, and environmentalists. In some regions, the deeper the well the more is the groundwater potentiality, which may cause friction between rich and poor people. All these points are from the quantity point of view; however, the groundwater quality becomes more of concern in many cases. Quality variations are more difficult to control and manage because they are not well perceived and understood by the public, mass media, and even by decision makers, engineers, and hydrogeologists. Once the aquifer is depleted it may take years to recover.

Groundwater management deals with the complex interaction between human societal activities and the physical environment, which pose an extremely complex and difficult problem to solve for the benefit of all parties involved.

Those using aquifer are little motivated to preserve it; any preservation may simply be exploited in future by other individuals. Consequently, there are rivalries between the exploiters without care about management programs. Such a competition might lead to inflicting damages on the third party, which is the society itself. Among the management strategies are the following points:

1. Adjustment of the annual rates of pumping, generally, based on monthly rates,
2. Adjustment of the well configuration,
3. Augmentation of water supply from other sources or groundwater recharge enhancement,
4. Awareness of groundwater beneficiaries,
5. Limitation of exposition according to monitoring results.

Strategic planning does not mean aquifer overexploitation or storage of groundwater more than necessary. The following points must be taken into consideration for a successful management policy:

1. Physical dimensions of exploitable water quantity and possible groundwater recharge amounts. It is necessary that a hydrodynamic equilibrium is held between the natural and/or artificial recharge and groundwater abstraction rates.
2. Chemical dimensions of groundwater quality variations during recharge or exploitation processes. Degradation in groundwater quality must not be allowed.
3. Economic consequences over short- and long terms.
4. Social impacts are also important. For instance, detrimental effects on third-party users must be avoided.
5. Environmental sustainability consequences over short and long terms.
6. Social impacts are also important. Damage to the natural environment, especially sensitive aquatic ecosystems, must be avoided.

In any strategic planning of groundwater resources, answers to the following questions are the preliminary requirements for efficient planning:

1. How to exploit the groundwater resources? Is it from physical, hydraulic, social, or engineering points of view?
2. Is it necessary to maintain a minimum groundwater table level especially in unconfined aquifers?
3. Is it necessary to maximize the annual benefits and overturn from economic points of view?
4. Do the groundwater quality variations play a significant and perhaps sole role in the groundwater allocation in emergency cases?
5. Is it necessary to maintain a certain amount of groundwater volume in the aquifer so as to cover the emergency needs of population of a certain size?

After what have been explained in the previous sections, it is clear that each area needs its own management and development program by considering the site-specific conditions, restrictions, and data availability. The groundwater usage may be curtailed through some of the following effective mechanisms:

1. The groundwater reservoir monitoring requires extensive records and investigations. This may be one of the key reasons why overdraft may escape recognition until it is too late, because the effects of development are commonly obscure and complex, and becomes understandably slow by time.
2. The causes and effects of groundwater overdraft are not reversible immediately or fully. Consequently, cutting of withdrawals in the reverse order of their priorities does not assure that the overdraft trend will be reversed. There may be two reasonable doubts that the available facts would suffice to sustain against any appeal from an order for reduction, or that the statutory procedure

would recapture the status of the earlier appropriators.

3. Even though the perennial or sustainable yield and appropriations for use may be almost balanceable, some further development may also be feasible.

6.7 INTEGRATED GROUNDWATER MANAGEMENT

In a conjunctive use of different groundwater storages on the basis of a proper management framework, the planner or decision maker should try and maximize the use of all available storages in the best, convenient (to the society), and optimum ways. There are different alternatives such as inter- and intraaquifer groundwater transfers by simultaneous consideration of recharge, abstraction, and distribution characteristics within an integrated system, which provides minimization of water losses, optimization, and conservation of groundwater storage at convenient times and locations, and efficient distribution networks with a convenient number of wells. Assessment and management are the key factors in integrated development and management strategies for water resources. Without a correct and detailed assessment, it is almost impossible to plan, design, realize, and manage water resources development projects. The results of the assessment and subsequent management are the bases for any decision-making process, since they can lead to large investments and serious consequences for the environment. Proper management and utilization of aquifer flow depend essentially on the availability of data and rational, logical, and scientific analysis techniques. Water resources assessment consists of determining the quantity, quality, and variability for sustainable development and rational management. The following steps are necessary to have the final solution with confidence:

1. Individual considerations: These include detailed hydrologic and hydrogeological works within each drainage basin and aquifer independently from each other. For strategic planning, the major questions are:
   a. What are the climatological, meteorological, and hydrological recipients from the rainfall events, which have direct impacts on water resources in the region? (Chapter 1)
   b. What are the recharge possibilities and present abstraction rates from each aquifer?
   c. What are the volumes of saturation and unsaturated layers in each aquifer with respect to future groundwater storages? (Chapter 4)
   d. What are the natural safe yield level and volume in addition to permissible abstraction volume?
   e. What are the hydrochemical facies and groundwater quality zonation in relation to host geology and regional hydrological regime? (Chapter 5)
   f. What are the artificial groundwater recharge enhancement practices that can be applied efficiently? (Chapter 4)

2. Conjunctive considerations: Conjunctive groundwater resources exploitation must be sought in cases of two or more different aquifer availability. For this purpose, the following questions are valid:
   a. What are the two best aquifers that will result in the most optimal conjunctive usage?
   b. What is the best manner of groundwater withdrawal from the aquifers during different months of the year, and what is the ratio of supply from each aquifer for meeting the demand?
   c. What is the mixing ratio based on the water quality of each aquifer for the satisfaction of demand quality?
   d. What are the water withdrawal preference locations in each aquifer and volume (discharge) each month?
   e. What is the suggested plan for necessary infrastructure (wells, pipes, storage tanks, pumps, etc.)?

3. Desalination consideration: The strategic planning of groundwater resources especially in arid and semiarid regions must also consider the possible supports of desalination plants, if any. Herein, the following questions can be asked:

   a. Which are the prior aquifers for groundwater mixture with desalinization water?

   b. What percent of desalination water should be mixed with which aquifer?

   c. Is it possible to store desalination water in the unsaturated zone, and if possible, in which aquifers?

6.8 BASIC MANAGEMENT VARIABLES

Drainage basin hydrologic systems have significant spatial and temporal variability. Long dry periods may continue for several years causing severe droughts; short-duration heavy rainfalls may cause floods with damages on human life and property but also provide groundwater recharge opportunities in arid and semiarid regions. Recently, due to urbanization, agriculture, and industrial activities, water resources mobilization has increased unprecedentedly for water supply and demand purposes. Such increases are observable in many countries for satisfying the expected socioeconomic development and to ensure food security. Some countries have had opportunity to benefit from nonconventional water resources, such as desalination units despite the high cost.

6.8.1 Water Demand

Accurate and up-to-date population data are a prerequisite for many different applications, including risk and vulnerability management. There is, however, a shortage of data with a high spatial resolution, particularly in developing countries. Global population data sets are designed for global modeling studies, including climate change research, and their resolution is generally too low for local or community purposes. Population data with an appropriate format and resolution is required for a variety of applications such as spatial planning processes, disaster and emergency management, and risk and vulnerability assessment (Aubrecht et al., 2010; Hall et al., 2008; Schneiderbauer, 2007; Sweitzer and Langaas, 1995; Tatem and Linard, 2011).

Different shareholders request water for various purposes, including domestic, agricultural, and industrial and production activities. Domestic use depends on the per capita water consumption per time interval, say, a day. In order to calculate expected total water request for a population, first the projection of population growth is necessary and then per capita per day water demand multiplication by the projected population number yields total water demand.

Simple models for population growth assume that there are no migrations, natural disasters, epidemic sicknesses, and therefore, an initial population size, N_0, is bound to increase by time, $N(t)$. Provided that there are no water and food restrictions, rationally and logically the population growth will have a proportional but nonlinear increase by time, which can be appreciated as the curves in Figure 6.1.

It is well known that a curve with an intercept on one axis and asymptotic to another has an exponential function shape (Şen, 2013). Hence, it is possible to write the mathematical form of this curve as,

$$N(t) = N_0 e^{\gamma t} \qquad (6.1)$$

where γ is the population growth rate. Depending on the sign of γ, the population growth ($\gamma > 0$) or decay ($\gamma < 0$) takes place. In practical studies and future planning, population increase is always taken into consideration.

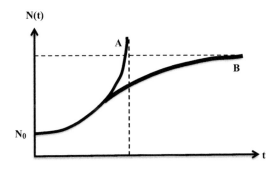

FIGURE 6.1 Natural population growth.

On the other hand, in water and food restrictive cases although the population increase will be initially similar to A in Figure 6.1, but later on it will have to approach a limitation as for curve B. Water demand problems in many parts of the world can be summarized along the following points, which are not exhaustive:

1. Demand for groundwater is in steady increase all over the world, hence each day there are well diggings or drillings.
2. Uncontrolled groundwater abstraction is in operation because of missing management programs.
3. Sea water intrusion takes place along the coastal areas due to overpumping, and hence, water quality deteriorates steadily.
4. Upconing of saline water due to overpumping leads to groundwater quality deteriorations in deep aquifers.
5. Extravagant groundwater use for agriculture through old-fashioned irrigation practices wastes groundwater. This leads also to top fertile soil degradation as a result of salinization.
6. Insufficient and inefficient city water distribution networks cause leakages from the pipes. These leakages may be thought of as augmentation in the city groundwater storages, but they cause water table rise and damage to infrastructure (foundations, cables, etc.).

7. There are impacts of man's activities on groundwater duality as a result of contamination.
8. Water migration between fresh and saline aquifers due to bad well construction.
9. Due to overpumping, groundwater levels drop, and hence, groundwater haulage expenses increase.

EXAMPLE 6.3 POPULATION PREDICTION AND WATER DEMAND

In a town of 10,600 persons, presently available sustainable groundwater resources are about 0.530×10^6 m^3 per year. The population growth rate is 0.04 per year. Usual per capita consumption is 120 l/day. However, in dry periods the least per capita is planned as 70 l/capita/day.

1. Is the present groundwater resources enough for water supply?
2. What is the amount of remaining yearly water? For how many years it is sufficient without any other water supply alternative entrance into the scene?
3. What are the water supply amounts after 5 and 10 years? Is the existing resource sufficient for their supply? If not, what to do?
4. How many people can be supplied by groundwater resources during a drought period?

Solution 6.3

It is assumed that in this town the water demand is only for domestic use, and agricultural products are imported from other areas and there is no industrial development.

1. In order to check whether the existing water supply is enough, the annual water consumption rate, Q_A, can be calculated as,

$$Q_A = 10,600 \times 120 \times 10^{-3} \times 365$$
$$= 0.464 \times 10^6 \text{ m}^3/\text{year}$$

Since, this is less than the existing groundwater supply, $(0.464 \times 10^6 < 0.530 \times 10^6)$, the annual groundwater supply is sufficient to satisfy the domestic water supply.

2. The amount of remaining annual water volume is $0.530 \times 10^6 - 0.464 \times 10^6 = 0.066 \times 10^6 \, \text{m}^3$. For future simple management prediction, with the same per capita water consumption rate, the question is for how many years sustainability can be continued? Since future available water amount is $0.186 \times 10^6 \, \text{m}^3/\text{year}$ and the number of years to consume this amount is not known, let us denote it by y. With the given population growth rate, the number of future years can be calculated simply as follows.

   a. Let us consider one year, then the population will increase to $10{,}600 \times (1 + 0.04) = 11{,}024$ capita. Their annual water consumption is,

   $$Q_A = 11{,}024 \times 120 \times 10^{-3} \times 365$$
   $$= 0.482 \times 10^6 \, \text{m}^3/\text{year}$$

   This is still less than the available groundwater resource, and hence, one can make similar calculations for the second year with the new population, which is $11{,}024 \times (1 + 0.04) = 11{,}465$ capita and their water consumption is,

   $$Q_A = 11{,}465 \times 120 \times 10^{-3} \times 365$$
   $$= 0.502 \times 10^6 \, \text{m}^3/\text{year}$$

   Still there is a room for available water cover. Hence, the third-year population is $11{,}465 \times (1 + 0.04) = 11{,}924$ capita and the water consumption is,

   $$Q_A = 11{,}924 \times 120 \times 10^{-3} \times 365$$
   $$= 0.522 \times 10^6 \, \text{m}^3/\text{year}$$

   Continuation of similar calculations yields the population of the fourth year as $11{,}924 \times (1 + 0.04) = 12{,}401$ and water demand is,

   $$12{,}401 \times 120 \times 10^{-3} \times 365$$
   $$= 0.543 \times 10^6 \, \text{m}^3/\text{year}$$

Since this amount is bigger than the available groundwater volume $(0.543 \times 10^6 > 0.530 \times 10^6)$, after the fourth year, due to population increase, water stress and shortage will start.

3. Especially with the start of the fifth year, the local administrators should look for additional water supply alternatives for better management strategy. However, they should take precautions and make preparations under the light of the previous calculations, because the population growth indicates that within the fifth year water shortage will start. So, they can make their plans a few years earlier and start the most convenient additional water supply alternative(s) as mentioned in Example 6.2.

4. During a drought year with the same original population and 70 l/capita/day water consumption, the number of people, n, that can be supplied without water shortage is,

$$n = \frac{0.530 \times 10^6}{70 \times 10^{-3} \times 365} = 20{,}743 \, \text{capita}$$

6.8.2 Aquifer Features

The following aquifer features such as geometric, geologic, hydrogeological, and social properties are among the main factors for groundwater availability and management calculations in an area.

1. The geometry of aquifer boundaries, saturation and unsaturated thicknesses, and widths and areal extensions,
2. The hydrogeological properties of the aquifer material as for the storage coefficient, hydraulic conductivity, permeability, and transmissivity,
3. Aquifer material composition with reference to its homogeneity, isotropy, etc.
4. Piezometric (water table) level map and the direction of groundwater flow (hydraulic gradient),

5. Aquifer material type as porous, fractured, or karstic medium,
6. Groundwater flow regime as laminar or turbulent,
7. Water quality,
8. Recharge and/or spring locations,
9. Initial and boundary conditions,
10. Aquifer safe yield levels,
11. Data availability and reliability,
12. Individual or conjunctive management models,
13. Local legal rules and regulations,
14. Objective function and restrictions,
15. Political and transboundary conditions, if any.

6.9 HYDROGEOLOGICAL MANAGEMENT

In an effective groundwater management, it is necessary to link flow and storage managements in proportion for sustainable and harmonious exploitation alternatives. Groundwater exploitation and management processes go through a nonequilibrium phase, the length of which depends on the following points:

1. The rate of groundwater abstraction, and hence, the aquifer's hydrogeological features and parameters,
2. Boundary conditions and the rapidity of compensatory reactions, for example, reduction in the natural outflow and/or increase in the induced inflow.

According to exploitation strategies, the following management practices of the aquifer reserve are therefore conceivable and practical, but under certain restrictive constraints (Margat and Saad, 1983):

1. A maximum and lasting exploitation strategy of the renewable resources, in a regime of dynamic equilibrium with average abstraction, which is approximately equal to average recharge (possibly enhanced by boundary effects) without taking into account the season or even possible annual variations (increased abstraction in periods of drought). Thus, after a decrease in the initial phase of nonequilibrium, the stabilized reserve is used, usually, as a regulatory factor such as annual or multiannual, its natural regulatory function being amplified under the constraints of preserving a minimum flow rate at the outflow boundaries of the aquifer (springs, low water level in draining streams, etc.) or preserving the fresh-water/sea-water equilibrium in a coastal aquifer.

2. In a strategy of repeated exploitation of the storage and then through flow only, in a prolonged unbalanced regime, that may be "guided" or unintentional in the initial phase, and in which abstraction (increasing or stabilized) is greater than recharge (even when enhanced by boundary effects), a second phase involves reducing abstraction to restore the equilibrium. In this case, the depletion of the reserve contributes largely, sometimes predominantly, to the production of water, during an initial phase of limited medium- to long-term duration, which is limited either by external constraints (see strategy a below) or by a reduction in the productivity of wells (drawdown by deteriorating aquifer conditions and limited by the base of the aquifer). In the latter phase of possible reequilibration, the reserve may be either:
   a. stabilized, on average, provided that it has reached an equilibrium bringing abstraction close to recharge, or
   b. restored in part, by reducing the abstraction below average (repayment of the loan) and sometimes by artificially increasing recharge and then at a new equilibrium.

3. A strategy of mining or exhaustive exploitation with abstraction increases or otherwise, in operation greater than the

average recharge. In this case, depletion of the reserve provides most of the water produced, and the exploitation is in the long term more or less limited when drawdown becomes excessive without later returning to a regime of reequilibrium. The recovery of the reserve may be too slow and sometimes hindered by irreversible degradation of the original capacity of the reservoir due to subsidence.

6.9.1 Strategic Groundwater Storage Planning

In general, there are two strategic planning procedures in a region depending on a number of groundwater sources (aquifers). Herein, the meaning of a region is the area with at least two aquifers, so that they can compete with each other for joint strategic exploitation. Hence, in a regional planning, two complementary studies are:

1. Within aquifer (microplanning),
2. Interaquifer (macroplanning).

6.9.1.1 Within-Aquifer Management

All aquifers do not lend themselves indifferently to the methods of management as explained along the following strategies.

1. Strategy 1: This is appropriate for unconfined aquifers of small and medium capacity and of limited thickness, with a high rate of recharge (reserves of the order of one to 10 times the average annual volume of discharge), and includes the case where a confined aquifer is exploited near the point as it becomes unconfined. This strategy is used particularly in cases where constraints limit the possible drawdown (to preserve open water bodies or watercourses, to conserve hydraulic links with surface water supply sources to prevent the displacement of the fresh water/sea water interface, etc.), whatever the rate of recharge of the aquifer is.

2. Strategy 2: This is more convenient for unconfined and semiconfined aquifers with high capacity and a small-to-medium recharge rate (reserves of the order of 10 to a 100 times greater than the average volume of annual recharge), without appreciable constraints for the conservation of water levels. For example, this strategy is appropriate in an aquifer with outlets that are independent of water courses (areas of evaporation in arid regions).

3. Strategy 3: This is possible for aquifers of high capacity with little recharge (reserves in the order of 100 to 1000 times greater than the average annual volume of recharge). Development can take place in the unconfined and confined zones where transmissivity and storativity are high. This is the case for most of the deep aquifers in sedimentary basins, the most transmissive layers of which drain the less-permeable but higher-storage layers (aquitards).

The aquifer is considered as a whole for the rainfall occurrences and subsequent time-lagged infiltration (Chapter 1), which fills the voids in the unsaturated zones for strategic groundwater recharges and withdrawals (Chapters 3 and 4). Different parts of the aquifer must be identified with hydrogeological features that are meaningful in strategic planning and management. For instance, different parts of the main aquifer may provide competing alternatives. In a within-aquifer strategic planning study, the following points are significant for consideration:

1. An aquifer is considered as composed of different strategic parameter variations. Generally, any aquifer is considered as composed of three fuzzy parts ("low," "medium," and "high" potentialities), which are rather arbitrarily divided according to expert views (Chapters 3 and 4). On the other

hand, each part can be taken as a potential strategic planning unit.

2. An aquifer may also be considered as composed of different hydro-geo-stratigraphic units as aquifers and nonaquifers. Such a subdivision is useful from groundwater resources point of view, but it needs detailed information about the subsurface geological and hydrogeological information.

3. Division of any aquifer can be based on the groundwater quality zonation and such a division favors strategic groundwater resources planning for water quality satisfaction (Chapter 5).

4. Aquifers can be divided according to social, industrial, military, and agricultural uses. This is tantamount to saying that subdivision achievement can be obtained on the basis of priorities.

5. It is also possible to consider subdivision of an aquifer by recharge (infiltration and its augmentation) possibilities, because the recharge areas have the major significance for groundwater storage increments.

6. Strategic significance of an aquifer can be based also on the human activities, water shortages for certain periods, floods, and natural and social impacts. For instance, if an aquifer is far away from human activities, whatever the groundwater resources availability, it may not be included in strategic planning purposes in the short-term projections.

The rational and/or methodological comparisons of subdivisions will lead to the best management strategy. Prior to a regional study, strategic subaquifers must be depicted with all available data, field reconnaissance, knowledge, and information.

6.9.1.2 Interaquifer Management

This is a large-scale (macroplanning) strategic groundwater resources planning, including two or more aquifers with their competitive properties. In a large-scale strategic planning, the following points must be taken into consideration:

1. It is necessary that there is an actual or expected future possible cause for groundwater overexploitation. For instance, expansion of a city might lead to a steady strategic planning of surface water, and especially, groundwater resources in arid and semiarid regions. Population growth, land use, scarcity in the rainfall occurrences, and water quality deterioration due to mismanagement and uncontrolled pollution activities are among strategic groundwater planning and management ingredients.

2. It is preferable to interchange and to exploit groundwater resources jointly in nearby rather than far-away aquifers, which serve at times of emergency. If the regional hydrologic and subsurface groundwater balances require, it is possible to transport groundwater between adjacent aquifers.

3. The competitive divisions on a small scale (within aquifer) are considered first to compete among each other and additionally other small-scale features of different aquifers in the same region must be allowed to compete with the previous considerations. It may be feasible to transport and consume groundwater resource among nearby aquifers.

4. It is possible that each division from different aquifers may have different strategic planning facilities, and in such a situation, the priorities must be given according to local administration's or the central government's views.

Interaquifer planning should be effective after the completion of micro-planning strategies in each aquifer.

6.10 BASIC MANAGEMENT MODELS FOR AQUIFERS

In order to cope with future rational groundwater distribution and demand, there is an urgent need for efficient planning and management schemes taking into consideration the sustainability principles.

AM includes several concepts as supply, demand, wet and dry periods, economy, water quality, social affairs, etc. Merrit (2004) suggested that demand can have at least four quite distinctive implications such as the use of water, the consumption of water, the need for water, or the economic demand for water. He defined each one of these terms carefully in the context of regional hydrosocial balance. Water resource management is a form of "demand management," but such a concept has not been taken into consideration as one of the major active factors. Rather, the management strategies have dealt with increasing the water supply by new sources.

In potential aquifers, there is a real problem in groundwater resources exploitation including development and AM, which are almost nonexistent. Groundwater is not extracted by considering optimum benefit for the present and future generations, and especially, for possible critical dry periods. There is a more general point that AM is now widely seen principally as a form of "demand management."

Simple and effective AM rules and regulations must be put forward prior to the preparation of dynamic and adaptive management software. AM must be programmed in such a way that abstraction should be adjusted in accordance with the groundwater resources supply expectations within stipulated time periods in a sustainable and perennial supply manner. The renewable (recharge) character of potential aquifers must be considered with suitable aquifer pumping rates (discharges), well-field locations, and especially, priority of subunits within a single aquifer system as well as in the integrated and joint aquifer systems.

Balkhair (2002) has suggested the selection of five potential Wadis from arid regions of western Arabian Peninsula for the most optimum groundwater reservoir management through a multicriteria decision-making. He used 12 hydrologic and hydrogeological variables on a lump basis for each Quaternary deposit aquifer. Although his approach can be used as a preliminary step for solution, it includes many restrictive assumptions and subjective indexing of some variables.

Herein, generally, the basic decision variables for AM study are shown in Tables 6.1 and 6.2 for aquifers A and B, respectively. Although there are numerous management variables that may contribute to overall multivariable decision criteria for integrated AM system, herein, the most effective 18 numerical and 2 linguistic variables are considered. These tables include all types of numerical and verbal information needed for successful development and application of an effective groundwater management program. Each item is based on detailed field survey, office work, and the application of relevant techniques explained in the previous chapters.

Table 6.1 has four columns, whereas Table 6.2 has five columns each for different subaquifer within major aquifers A and B, respectively. Accordingly, in Table 6.1 there are $20 \times 4 = 80$ pieces of information and in Table 6.2, $20 \times 5 = 100$ inputs are available.

6.10.1 Logical and Rational Management Rules

Haphazard and heuristic techniques and readily available software are not employed, because they are with hidden and unclear logical and rational formulations. Prior to an AM system, it is necessary to consider the conceptual, rational, and logical basic model features first

TABLE 6.1   Strategic Groundwater Planning Management Variables in Aquifer A

| Aquifer Features | Aquifer A | | | |
| | Subaquifers | | | |
	A1	A2	A3	A4
Annual rainfall (mm/year)	189	192	173	182
Alluvium surface area ($\times 10^3$ m^2)	6612.1	5816.9	15494.1	1652
Direct rain water volume ($\times 10^6$ m^3/year)	1.25	1.12	2.68	0.3
Possible direct annual recharge ($\times 10^6$ m^3/year)	0.25	0.17	0.27	0.075
Water table area ($\times 10^3$ m^2)	5687.6	4449.1	12354.7	1236
Average saturated thickness (m)	31.6	57.2	36.6	18.7
Average unsaturated thickness (m)	10.3	19.6	27.3	14.8
Average transmissivity (m^2/day)	659.7	1510	282	557
Average hydraulic conductivity (m/day)	20.9	26.4	7.7	29.78
Hydraulic gradient	0.0126	0.008	0.005	0.00614
Subsurface flow rate (m^3/m/day)	8.31	12.08	1.41	3.42
Average storativity	0.09	0.128	0.045	0.068
Average effective porosity	0.25	0.25	0.2	0.3
Abstractable groundwater volume ($\times 10^6$ m^3)	16.17	32.57	20.35	1.57
Rechargeable direct groundwater volume ($\times 10^6$ m^3)	17.03	28.5	84.6	7.33
Abstractable fullness ratio	0.48	0.53	0.18	0.17
Present groundwater consumption ($\times 10^6$ m^3/year)	3.11	6.17	16.50	5.05
Average water quality (EC, micro-mhos)	1150	2245	2293	1020
Recharge possibility	High	Poor	Medium	Medium
Geological environment	Alluvium diorite granodiorite	Alluvium metabasalt volcano-clastic	Alluvium metabasalt volcano-clastic	Alluvium hornblende quartz-feldspar, marble

individually and then in an integrated manner. Such models provide information about the following points:

1. Definition of the phenomenon in terms of features recognizable by observations, experiences, and simple analyses or validated simulations. A management program should be based on various linguistic and simple conceptual principles and alternatives.

2. Description of the AM practices in terms of various alternatives, size, intensity, and accompanying groundwater conditions.

TABLE 6.2   Strategic Groundwater Planning Management Variables in Aquifer B

	Aquifer B				
	Subaquifers				
Aquifer Features	B1	B2	B3	B4	B5
Annual rainfall (mm/year)	357	357	166	166	166
Alluvium surface area ($\times 10^3 \, m^2$)	31895	8989	1514	3,027	35789
Direct rain water volume ($\times 10^6 \, m^3$/year)	11.39	3.21	0.25	0.50	5.94
Possible direct annual recharge ($\times 10^6 \, m^3$/year)	1.71	0.48	0.05	0.1	1.19
Water table area ($\times 10^3 \, m^2$)	29907	7678	9367	1873	22997
Average saturated thickness (m)	7.27	8.06	4.36	5.00	14.59
Average unsaturated thickness (m)	31.73	18.94	37.52	41.69	43.63
Average transmissivity (m^2/day)	187	14	63.8	659.7	431.8
Average hydraulic conductivity (m/day)	25.7	1.7	14.6	139.1	29.6
Hydraulic gradient	0.026	0.017	0.005	0.0086	0.011
Subsurface flow rate (m^3/m/day)	9.88	0.58	0.16	7.18	4.90
Average storativity	0.2476	0.1099	0.0694	0.1836	0.1391
Average effective porosity	0.25	0.25	0.20	0.30	0.25
Abstractable groundwater volume ($\times 10^6 \, m^3$)	53.83	6.80	2.83	1.72	56.83
Rechargeable direct groundwater volume ($\times 10^6 \, m^3$)	253.0	42.50	11.36	37.86	390.40
Abstractable fullness ratio	0.17	0.14	0.2	0.04	0.12
Present groundwater consumption ($\times 10^6 \, m^3$/year)	1.2	1.4	0.14	2.6	10.6
Average water quality (EC, micro-mhos)	977	838	1346	1063	1393
Recharge possibility	High	Moderate	Poor	Moderate	High
Geological environment	Alluvium biotic monzogranite	Alluvium diorite monzogranite	Alluvium diorite monzogranite	Alluvium diorite monzogranite	Alluvium diorite monzogranite

3. Logical statements about the physical processes, which enable understanding of the factors that dominate the mode and rate of groundwater recharges and consumptions. This corresponds to the derivation of basic and simple logical rules and regulations about the overall system performance.
4. Specification of the key hydrogeological fields demonstrating the main processes such as the recharge potential, saturated and unsaturated zone potentials for abstraction, and groundwater storage augmentation possibilities.
5. Guidance for hydrogeological conditions using diagnostic and prognostic stages that best discriminate between development and nondevelopment guidance for the AM.

The rational and logical principles provide decision makers with the following knowledge, which can be employed in any effective AM study:

1. Diagnosis: This is the preliminary step that helps in understanding the internal and external activities within the whole system under consideration.
2. Synthesis: Available verbal and numerical data and information must be synthesized for arriving at preliminary conclusions that help guide toward the optimum pattern of AM.
3. Logic: This provides a "mental picture" of the AM aspects within- and intraaquifer situations.
4. Isolation: The basic ingredients such as subaquifers and AM variables are isolated from each other so as to assess the individual effects.
5. Main pattern: This includes the identification of main and distinctive patterns within the whole complex system.
6. Numerical evaluation: Mental and logical aspects are executed numerically by considering fundamental formulations with individual and joint AM operations.

7. Numerical projections: Based on the previous steps, numerical predictions are produced as future projections under different scenarios.
8. Numerical products: Different tools are used for modification of numerical products.
9. Simple calculations: Rather than involved mathematical procedures, it is preferred to construct preliminary steps for a simple AM.
10. Data gaps: Suggested scenarios provide the possibility of filling in gaps in the data.

6.10.2 Basic Data

The joint and integrated groundwater management of aquifers in A and B provide an opportunity to enhance the joint water supply (Tables 6.1 and 6.2). The general configuration of joint management program for A−B system can be considered initially as in Figure 6.2.

The following points are important in the conceptualization of this system and they must be imbedded into the final AM program:

1. A and B do not have any direct interchange but their management will depend on what goes inside A−B block (joint management) diagram.

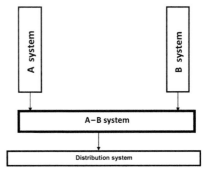

FIGURE 6.2 A−B overall groundwater management systems.

2. A–B block receives groundwater in different ratios from each aquifer but the total water amount remains equal to what monthly water demand is.

3. A–B block may be thought of as similar to surface water reservoirs (dams) where joint water withdrawals contribute in different proportions to overall demand.

4. In deciding on the water withdrawal ratios from each aquifer to A–B system one of the following three alternatives can be considered:

    a. Quantity-based ratio,
    b. Quality-based ratio,
    c. Quantity-quality-based ratios.

    In particular, for the initial development of the AM program, herein, the quantity-based ratio determination is used.

5. The groundwater supply from the joint aquifer (A–B) management goes directly into the water distribution system.

6.10.3 Conceptual Model of Aquifer (A–B) Systems

The conceptual model of subaquifers in each aquifer is presented in Figure 6.3 for the joint A–B management program. This is a block diagram that shows the natural groundwater flow from different subaquifers of each aquifer; A and B.

It is worth to notice that each subaquifer is connected either in parallel or in series. Parallel connection implies that the groundwater withdrawal from one of the subaquifers does not affect the other, and hence, their contribution to the A–B system is independent from each other. On the other hand, the serial connection implies

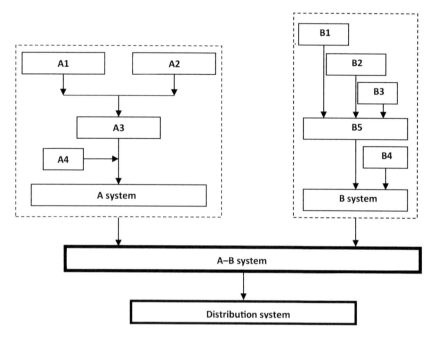

FIGURE 6.3 **A–B detailed groundwater management system.**

that the groundwater withdrawal is interconnected between the components. Depending on the aquifer transmissivity value, the greater the transmissivity the smaller is the required time of groundwater transportation. In the case of serial connection, the aquifer plays a similar role as that of a series of pipe flow. There are nine components (subaquifers) that may contribute to the overall A—B system in the AM stages.

6.10.4 Logical Management Principles

The joint operation between the subaquifers can be achieved by two logical connections: AND or OR. If the subaquifers are in serial connection then they must function for the overall performance of the system, and therefore, the subaquifers are connected by AND, whereas parallel units contribute to the overall performance of the system by OR connective. The statements including ANDs and ORs are referred to as the rule base. Consideration of these points provides possibility to extract the relevant rules for each conceptual model under consideration. Each rule indicates independent contribution to the overall demand. Let us consider the configuration A only in Figure 6.3. This leads to a set of rules as in the following frame, which gives the opportunity to assess the whole A basin subaquifer units in an exhaustive management program.

```
Rule 1. (A4)
              OR
Rule 2. (A1)AND(A3)
              OR
Rule 3. (A2)AND(A3)
              OR
Rule 4. (A1)AND(A2)AND(A3)
              OR
Rule 5. (A1)AND(A3)AND(A4)
              OR
Rule 6. (A2)AND(A3)AND(A4)
              OR
Rule 7. (A1)AND(A2)AND(A3)AND(A4)
```

Herein, there are seven independent logical rules (alternatives) for groundwater management in system A. The first rule is unique and very independent from other subaquifers. Other rules indicate joint operation of different subaquifer combinations which can be identified from the subaquifer connections as in Figure 6.3. The first two rules are for two of the upper subaquifers' joint operation for groundwater transfer to system A; the next three rules are for three subaquifers' joint operation, whereas the last rule is for the whole subaquifers' joint operation for water supply to system A. Similarly, consideration of subaquifer configuration for B yields the following logical rule chain:

```
Rule 1. (B4)
              OR
Rule 2.  (B1)AND(B5)
              OR
Rule 3.  (B2)AND(B5)
              OR
Rule 4.  (B3)AND(B5)
              OR
Rule 5.  (B1)AND(B2)AND(B5)
              OR
Rule 6.  (B1)AND(B3)AND(B5)
              OR
Rule 7.  (B2)OR(B3)AND(B5)
              OR
Rule 8.  (B1)AND(B2)AND(B3)AND(B5)
              OR
Rule 9.  (B1)AND(B2)AND(B3)AND(B4)AND(B5)
```

Hence, for B within-AM, there are nine independent logical management rules.

In order to supply groundwater from A—B system, there are many independent rules, and in total there are $7 \times 9 = 63$ alternatives. It is not necessary to write down all these rules, but the most logical ones are listed in the following frame:

1. (A1)AND(A2)AND(A3)AND(A4)AND(B1)AND(B2)AND(B3)AND(B5)AND(B4)
OR
2. (A1)AND(A2)AND(A3)AND(B1)AND(B2)AND(B3)AND(B5)
OR
3. (A1)AND(A2)AND(A3)AND (B1)AND(B2)AND(B5)
OR
4. (A1)AND(A3)AND(B4)AND(B5)
5. (A1)AND(A2)AND(A3)AND(A4)AND(B4)AND(B5)
OR
6. (A1)AND(A2)AND(A3)AND(B1)AND(B2)AND(B3)AND(B5)AND(B4)
OR
7. (A1)AND(A3)AND(B1)AND(B5)
OR
8. (A1)AND(A2)AND(A3)AND(B1)AND(B2)AND(B3)AND(B5)
OR
9. (A1)AND(A2)AND(A3)AND(A4)AND(B1)AND(B2)AND(B3)AND(B5)AND(B4)

It is a difficult task to identify the best and the most strategic joint AM rule among these logical rules. However, as the specific aspects are taken into consideration the picture becomes more obvious.

EXAMPLE 6.4 LOGICAL MANAGEMENT RULE

In an area (A) there are agricultural activities that need $15 \times 10^6 \, \text{m}^3$ groundwater per year. Adjacent groundwater resources with safe yields of $0.82 \times 10^6 \, \text{m}^3/\text{year}$, $11.2 \times 10^6 \, \text{m}^3/\text{year}$, and $0.56 \times 10^6 \, \text{m}^3/\text{year}$ are confined (C), unconfined aquifer (U), and leaky (L) aquifers, respectively.

1. Give a logical management plan for the agricultural activity possibilities with support of these aquifers by considering seasonality and generate few possible logical scenarios (Note: do not consider quantities).
2. Calculate whether these aquifers are necessary for sustainable yield satisfaction of the agricultural activities in this area.
3. Make suggestions, if needed.

Solution 6.4

It helps to consider a simple configuration of all the components in this problem, which is given in Figure 6.4.

Water supply to the agricultural area may be brought through a storage tank, which impounds different aquifer contributions according to monthly or seasonal activities.

1. In dry seasons, their joint use comes into practical application, and therefore, the logical management basis may have only one alternative as,

C(AND)U(AND)L.

This implies joint groundwater withdrawal from each aquifer. In wet seasons, depending on the circumstances, there may be various alternatives, two of which are given below and others may be generated by the reader.

If only one aquifer is enough for agricultural activity then the logical arrangement should be through ORing as,

C
(OR)
U
(OR)
L

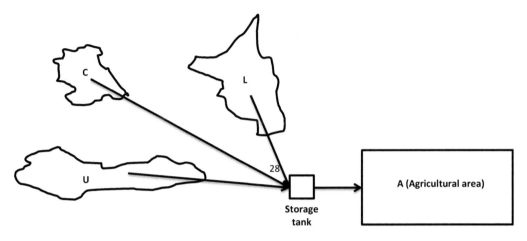

FIGURE 6.4 **Water resources and demand center configuration.**

If only two of the aquifers should cover the agricultural activities, then the following scenario is possible:

C(AND)U

OR

C(AND)L

OR

U(AND)L

2. Each aquifer has its safe yield, but sustainable yield for the agricultural activities is possible if the summation of these safe yields is more than the agricultural activity water demand. The total safe yield, S_T, of the system is,

$$S_T = 0.82 \times 10^6 + 11.2 \times 10^6 + 0.56 \times 10^6$$

$$= 12.58 \times 10^3 \, \text{m}^3/\text{year}$$

Since $12.58 \times 10^6 \, \text{m}^3/\text{year} < 15 \times 10^6 \, \text{m}^3/$ year, these groundwater resources cannot provide sustainable yield for the agricultural activity.

3. In order to maintain sustainable yield, it is necessary to import additional water from some other nearby aquifers or surface water impoundments. The amount of additional water is $15 \times 10^6 - 12.58 \times 10^6 = 2.42 \, \text{m}^3/\text{year}$.

EXAMPLE 6.5 LOGICAL GROUNDWATER MANAGEMENT

In an arid region, groundwater storage is available within and along the branches of a Quaternary basin as in Figure 6.5.

1. Write down all the possible logical alternatives for groundwater flow possibility to outlet, if only A, B, G, and H aquifers are functioning.
2. Write down all the possible logical alternatives for groundwater flow possibility to outlet, if only A, E, F, G, and H branch aquifers are active.
3. What are the logical possibilities if aquifers D, F, and A are depleted?

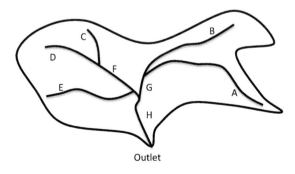

FIGURE 6.5 **Drainage basin with subbasins.**

Note: Assume that each alternative is sufficient for water demand at the outlet.

Solution 6.5

In the entire cases, one should keep in mind that the groundwater flow is from upstream toward the downstream and finally to the outlet.

1. The logical groundwater flow sequence toward outlet from branches A, B, G, and H can be written as follows,

   A(AND)G(AND)H
   (OR)
   B(AND)G(AND)H

EXAMPLE 6.6 RECHARGEABLE LOGICAL MANAGEMENT RULE

What are the alternatives of rechargeable water amount based on logical rules for the drainage basin A, as explained in Section 6.10.4. The necessary data are given in Table 6.1.

Solution 6.6

If the seven rules are valid as presented in Section 6.10.4 among the four subaquifers, then the quantitative rechargeable water volumes and the total of each rule are given in the following box:

Rule 1. $(7.33 \times 10^6 \text{ m}^3)$ – TOTAL : $7.33 \times 10^6 \text{ m}^3$
OR
Rule 2. $(17.03 \times 10^6 \text{ m}^3)AND(84.60 \times 10^6 \text{ m}^3)$ - TOTAL : $101.63 \times 10^6 \text{ m}^3$
OR
Rule 3. $(28.50 \times 10^6 \text{ m}^3)AND(84.60 \times 10^6 \text{ m}^3)$ - TOTAL : $113.10 \times 10^6 \text{ m}^3$
OR
Rule 4. $(17.03 \times 10^6 \text{ m}^3)AND(28.50 \times 10^6 \text{ m}^3)AND(84.60 \times 10^6 \text{ m}^3)$ - TOTAL : $130.13 \times 10^6 \text{ m}^3$
OR
Rule 5. $(17.03 \times 10^6 \text{ m}^3)AND(84.60 \times 10^6 \text{ m}^3)AND(7.33 \times 10^6 \text{ m}^3)$ - TOTAL : $108.96 \times 10^6 \text{ m}^3$
OR
Rule 6. $(28.50 \times 10^6 \text{ m}^3)AND(84.60 \times 10^6 \text{ m}^3)AND(7.33 \times 10^6 \text{ m}^3)$ - TOTAL : $120.43 \times 10^6 \text{ m}^3$
OR
Rule 7. $(17.03 \times 10^6 \text{ m}^3)AND(28.50 \times 10^6 \text{ m}^3)AND(84.60 \times 10^6 \text{ m}^3)AND(7.33 \times 10^6 \text{ m}^3)$ - TOTAL : $137.46 \times 10^6 \text{ m}^3$

---

2. A(AND)G(AND)H
   (OR)
   E(AND)H
   (OR)
   F(AND)H

3. Consideration of natural groundwater flow indicates that depletion in F implies depletion in C and D. Hence, the following logical management is possible

   B(AND)G(AND)H
   (OR)
   A(AND)G(AND)H
   E(AND)H

A first glance into this box yields quantitative management rule sequence according to ascending rechargeable groundwater volumes as 1, 2, 5, 3, 6, 4, and 7.

EXAMPLE 6.7 ABSTRACTABLE LOGICAL MANAGEMENT RULE

Table 6.2 includes five subbasins strategic groundwater resources management necessary data. Calculate the available abstractable groundwater volume and suggest the sequence of different scenarios.

Solution 6.7

The data values in Table 6.2 help convert the rule base for system B in Section 6.10.4 into the following quantitative abstractable water volume box:

used for evaluating the total uncertainty of lithological and aquifer parameter data assessment from water well records (Chapter 3). Randomness is used in the statistical sense in order to describe the groundwater phenomenon on a

---

Rule 1. $(1.72 \times 10^6 \text{ m}^3)$ – TOTAL : $1.72 \times 10^6 \text{ m}^3$

OR

Rule 2. $(53.83 \times 10^6 \text{ m}^3)AND(56.83 \times 10^6 \text{ m}^3)$ – TOTAL : $110.66 \times 10^6 \text{ m}^3$

OR

Rule 3. $(6.80 \times 10^6 \text{ m}^3)AND(56.83 \times 10^6 \text{ m}^3)$ – TOTAL : $63.63 \times 10^6 \text{ m}^3$

OR

Rule 4. $(2.83 \times 10^6 \text{ m}^3)AND(56.83 \times 10^6 \text{ m}^3)$ – TOTAL : $59.66 \times 10^6 \text{ m}^3$

OR

Rule 5. $(53.83 \times 10^6 \text{ m}^3)AND(6.80 \times 10^6 \text{ m}^3)AND(56.83 \times 10^6 \text{ m}^3)$ – TOTAL : $117.46 \times 10^6 \text{ m}^3$

OR

Rule 6. $(53.83 \times 10^6 \text{ m}^3)AND(2.83 \times 10^6 \text{ m}^3)AND56.83 \times 10^6 \text{ m}^3)$ – TOTAL : $113.49 \times 10^6 \text{ m}^3$

OR

Rule 7. $(6.80 \times 10^6 \text{ m}^3)OR(2.83 \times 10^6 \text{ m}^3)AND(56.83 \times 10^6 \text{ m}^3)$ – TOTAL : $66.46 \times 10^6 \text{ m}^3$

OR

Rule 8. $(53.83 \times 10^6 \text{ m}^3)AND(6.80 \times 10^6 \text{ m}^3)AND(2.83 \times 10^6 \text{ m}^3)AND(56.83 \times 10^6 \text{ m}^3)$ – TOTAL : $120.29 \times 10^6 \text{ m}^3$

OR

Rule 9. $(53.83 \times 10^6 \text{ m}^3)AND(6.80 \times 10^6 \text{ m}^3)AND(2.83 \times 10^6 \text{ m}^3)AND(1.72 \times 10^6 \text{ m}^3)AND(56.83 \times 10^6 \text{ m}^3)$ – TOTAL : $122.01 \times 10^6 \text{ m}^3$

The sequence of rules on the basis of ascending order for abstractable groundwater volume usages is as 1, 4, 3, 7, 2, 6, 8, and 9.

6.11 PROBABILISTIC RISK MANAGEMENT IN AN AQUIFER

Uncertainty within hydrogeological data and groundwater maps is rarely analyzed or discussed in practical applications. Rather, a set of simplifying assumptions are considered, and hence, the spatial variability in the aquifer parameters is ignored holistically. The increasing use of computers, for storing, retrieving, and serving large hydrogeological data sets made it easier to systematically analyze aquifer data. Error theory can be

regional basis (Seabear and Hollyday, 1966; Koch and Link, 1971; Davis, 1982).

Stochastic hydrogeology as suggested by Delhomme (1978) deals with uncertainty in groundwater resources evaluation and contaminant transport management problems. Although various geometrical (extent, areal coverage, thickness, slope, etc.), structural (fault, folds, fractures, fissures, joints), textural (porosity, specific surface, grain-size composition, sorting, etc.), hydraulic (hydraulic conductivity, storage coefficient, etc.), and geological properties are random in various degrees, hydrogeologists generally use deterministic approaches in problem solving. Deterministic approaches give on the average satisfactory single numerical results at the local scale, but on a regional scale (because of the heterogeneities, discontinuities, and anisotropies) the point values fail to provide reliable solutions especially in management studies. For instance, pumping tests provide representative and reliable aquifer parameters within the small depression cone influence areas, which is generally less than 300–400 m in confined aquifers and approximately 100 m in unconfined aquifers (Chapter 3). Deterministic results cannot represent large portions of the aquifers. Hence, the simplest way of regionalizing a variable or a parameter is possible through probabilistic and statistical techniques, which can then be incorporated into groundwater management programs.

Any measured data either on the field or in the laboratory represent quantities that are numerically different from each other. Randomness does not mean that the hydrogeological parameter values are spatially independent, but they are not predictable point wise even though their values might be available at a set of irregular sites. In order to obtain a representative and reliable estimate for the whole aquifer, the hydrogeologists often use a single deterministic value such as the arithmetic, geometric, or harmonic mean or the median or an expert value (Chapter 3). However, such a simplification brings the following disadvantages:

1. It is not possible to make risk assessments in groundwater problems.
2. The results cannot be related to actual surface or subsurface geology.
3. Theoretical probability distribution functions (pdfs) cannot be established.
4. Quantitative correlation analysis cannot be performed between lithological and hydrogeological parameters.

The probabilistic approach is but one technique used by geostatisticians to characterize spatial variability and to express a simple criterion for goodness of estimation (Journel, 1983). Such an approach needs at most a graphical representation of data without the need for computers, and hence, proves to be more practical.

In the classical groundwater storage volume calculations, simply the volume of the saturation zone is multiplied by the storage coefficient of the aquifer so as to find the amount of extractable groundwater volume similar to what has been explained in Chapter 4, Section 4.12.7. Groundwater storage volume, G_S, can be calculated by means of the following simple expression,

$$G_S = AhS \qquad (6.2)$$

where A is the water table area; h is the average aquifer thickness; and S is the average storage coefficient. For refining the estimation procedure, it is possible to define the specific groundwater storage capacity, g_S, as the volume of groundwater per unit water table area as

$$g_S = hS \qquad (6.3)$$

On the other hand, the specific subsurface flow rate, q, is equal to the volume of groundwater that passes through the whole section of the aquifer per unit width

$$q = Ti \qquad (6.4)$$

where T is the aquifer average transmissivity and i is the hydraulic gradient. Since these parameters (g_S and q) assume different values from one site to another, they are random. Their estimations can be achieved by probabilistic or statistical methods under a certain risk level. Hydrogeologists should be ready to accept a certain band of error ($\pm5\%$ or $\pm10\%$) in estimations.

6.11.1 Probabilistic Approach

Each one of the hydrogeological parameters varies spatially according to a certain pdf. For instance, from several thousand well logs in different aquifers, the specific yield is found to abide with log-normal pdf (Seabear and Hollyday,

FIGURE 6.6 Storage coefficient data.

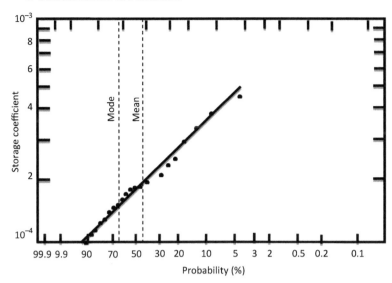

1966). The slope method in Chapter 3, Section 3.7, provides ability to identify the random characters in the hydrologic parameters even from a single aquifer test. A plot of storage coefficients obtained by the slope-matching procedure from Schultz (1973) data is shown on the logarithmic pdf paper in Figure 6.6.

In this plot, the probability, P_m, to each storage coefficient is attached by the classical empirical formula as

$$P_m = \frac{m}{n+1} \qquad (6.5)$$

where n is the number of data and m is the rank of the data value considered in the form of ascending sequence. Similarly, Davis (1969) concluded that for unconsolidated water-bearing formations, the hydraulic conductivity has a log-normal pdf. In addition, during a pumping test, heterogeneities in the transmissivity values have been quantified by the slope-matching method. The plotting of these transmissivity values on a logarithmic pdf paper gives a straight line indicating that the transmissivity values are also log normally distributed (see Figure 6.7). Krumbein and

Graybill (1965) and Way (1968) observed that the thickness of sedimentary beds is log normally distributed. Similar observations are confirmed by Gheorghe (1978).

As a result of all the aforementioned discussions, it is reasonable to assume that all the basic hydrogeological variables are generally distributed according to the logarithmic pdf, detailed information about which is presented by Aitcheson and Brown (1957).

Fortunately, the composite variables in Eqs (6.3) and (6.4) are expressed as simple multiplications of the relevant basic parameters. The multiplication property for the log-normally distributed variables makes the composite variable also to have a similar pdf. Hence, taking the logarithms of both sides in Eq. (6.3) leads to

$$\text{Lg}(g_S) = \text{Ln}(h) + \text{Ln}(S) \qquad (6.6)$$

The logarithmic transformation of logarithmically distributed random variables renders these variables into normally (Gaussian) distributed random variables. Hence, the arithmetic mean and standard deviation of the logs of the composite variable are sufficient for a complete

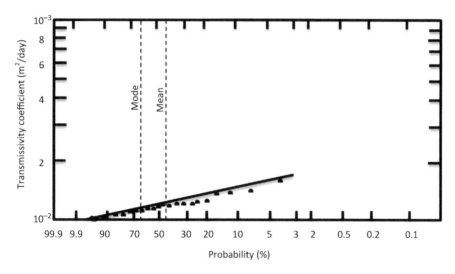

FIGURE 6.7 **Transmissivity (m^2/min) data.**

definition of the Gaussian pdf. For this purpose, the arithmetic mean, μ_{Lng_S}, and the standard deviation, σ_{Lng_S}, of the composite variable becomes after some algebraic manipulations as,

$$\mu_{Lng_S} = \mu_{Lnh} + \mu_{LnS} \qquad (6.7)$$

and

$$\sigma^2_{Lng_S} = \sigma^2_{Lnh} + \sigma^2_{LnS} + 2\overline{(Lnh)(LnS)} \qquad (6.8)$$

respectively. In these expressions, μ_{Lnh} and μ_{LnS} are the arithmetic averages and σ^2_{Lnh} and σ^2_{LnS} are the variances. Finally, $\overline{(Lnh)(LnS)}$ indicates the average of the cross-multiplication of Lnh and LnS values, which is known in the geostatistics literature as the cross-variance (covariance), that is representative of the cross-correlation between h and S in this case (Şen, 2009).

6.11.2 Statistical Management Model

Apart from the probabilistic approach, statistical methodologies consider errors involved in each aquifer parameter and their mutual interactions in forming the composite variable. The simplest treatment is possible by the perturbation theory where each basic variable is composed of random fluctuations around their respective arithmetic averages. For instance, a random variable, say x, can be written in two parts, namely, its average value $\bar{x}$ and fluctuations e_x corresponding to errors as $x = \bar{x} + e_x$. In such a definition, the error term e_x has zero arithmetic average with standard deviation equivalent to the standard deviation of the original random variable, x.

Similarly, according to the perturbation theory, each basic hydrogeological variable such as thickness, h, storativity, S, transmissivity, T, and hydraulic gradient, i, can be written as $h = \bar{h} + e_h$, $S = \bar{S} + e_S$, $T = \bar{T} + e_T$, and $i = \bar{i} + e_i$, respectively. Substitution of these expressions into Eqs (6.2) and (6.3) leads to the following averages of the composite variables,

$$\overline{g_S} = \overline{hS} + \overline{e_h e_S} \qquad (6.9)$$

and,

$$\overline{q} = \overline{Ti} + \overline{e_T e_i} \qquad (6.10)$$

where $\overline{e_h e_S}$ and $\overline{e_T e_i}$ show the cross-covariance between the two basic hydrogeological parameters considered and they may assume positive or negative values corresponding to over- and underestimation, respectively.

It is obvious from these two expressions that errors in terms of deviations from the arithmetic mean play a significant role when they are dependent on each other. These expressions imply that over- or underestimations appear in the specific groundwater storage capacity and the subsurface flow rates due to the $\overline{e_h e_S}$ and $\overline{e_T e_i}$ terms, respectively. However, their contribution might be equal to zero under the following circumstances:

1. If the storativity is independent of aquifer thickness, although this is valid in a single well, but when multitude of wells are considered in a region then the storage coefficient may be dependent upon the thickness, which may be different in different well locations. Hence, in an area of rough bedrock topography the thicknesses may vary significantly in the region. In an ideal (theoretical) case, when the thickness of the aquifer is assumed as constant over the entire aquifer area then $\overline{e_h e_S} = 0$.
2. If the aquifer reservoir material composition is homogeneous and isotropic, it implies that $\overline{e_S} = 0$. Contrary to the previous case, the aquifer storage coefficient is constant due to the material composition, but the aquifer thickness may vary due to the piezometric level and subsurface bedrock topographies. Areal constancy of the storage coefficient causes $\overline{e_h e_S} = 0$.
3. If both the aquifer saturation thickness and the storage coefficient values are areally constant, then again $\overline{e_h e_S} = 0$.

Unfortunately, none of the aforementioned three cases exist in actual field studies. Similar arguments may be stated for $\overline{e_T e_i}$ being equal to zero. Since neither the aquifer material is homogeneous and isotropic nor the saturation thickness is uniform, there will always be additional error terms in the form of cross-covariance as in Eqs (6.9) and (6.10).

Fitting a straight line on probabilistic papers as in the previous subsection is equivalent to obtaining the population of the concerned variable. In order to obtain the corresponding pdf, the necessary calculations are given step by step in Chapter 4, Section 4.14.2.

The plots of the resulting pdfs are shown in Figures 6.8 and 6.9 for the specific storage and subsurface flow rate, respectively. Deterministically, the total volume of available groundwater within the saturated zone of the Quaternary deposits is given as $3.1 \times 10^6 \, \text{m}^3$ provided that the average thickness and the storage coefficient are used within a $3.4 \, \text{km}^2$ area. This volume corresponds to $0.91 \, \text{m}$ of specific groundwater storage. However, the most frequently occurring specific groundwater storage and the total

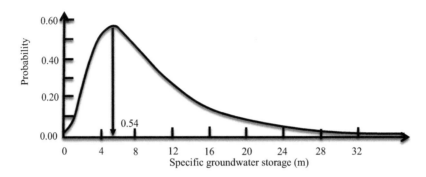

FIGURE 6.8　**Specific groundwater storage pdf.**

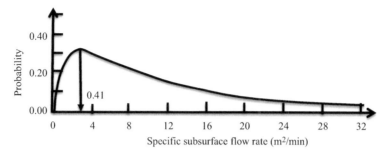

FIGURE 6.9 **Specific subsurface flow rate pdf.**

groundwater storage volume are 0.54 m (see Figure 6.8) and 1.8×10^6 m^3, respectively. Hence, the relative error is almost 73% between the deterministic and statistical approaches. This is an extremely large overestimation which may give rise to a multitude of problems such as water shortages for agricultural activities or urban water supply, overpumping and mining of the aquifer, and economic losses due to extra wells, pipes, pumps, etc.

In Figure 6.9, the probability is on the vertical and the specific subsurface flow (according to Eq. (6.4)) is on the horizontal axis. For instance, the risk associated with the classical estimation of the groundwater storage volume is 0.22. However, the most frequently occurring estimate has a risk value of 0.41. As a result of using the latter estimate, the risk decreases as the cost of reliability increases.

In general, the increase in the groundwater storage volume gives rise to an increase in the risk component. Practical questions arise as "how to estimate the groundwater resources volume in a given area?" and "what risk level should be accepted in such a determination?" Once the estimation of the available groundwater storage volume is determined, all of the subsequent activities of exploitation, distribution, management, well number and location, development, and budget to be invested for such activities are dependent primarily on this estimate. In the absence of any further information, such as economic and political conditions, the best estimate

corresponds to the maximum frequency as explained above (Figures 6.8 and 6.9).

If the risk level is predetermined, then the cumulative pdf such as in Figure 6.10 provides an estimate of the groundwater storage volume. For instance, for a given risk values of 0.36 (the reliability is 0.64) the corresponding specific groundwater storage value is 8 m.

There is always some risk associated with the estimate and for zero risk the groundwater storage volume is also equal to zero. As for the specific subsurface flow, the most frequently occurring rate under unit hydraulic gradient is 44 m^2/min with a risk level equal to 0.70 (see Figure 6.11) and 0.85 m^2/min from the classical approach, which has the risk value as 0.22. Hence, the relative error is 91%, which is a large difference, and it affects dangerously the water balance equation for the study area, because it implies more output than likely.

Risk levels are comparatively small compared to the case of specific groundwater storage. The aforementioned methodology has produced the following important implications and advantages over commonly used classical deterministic techniques:

1. It is a quantitative way of assessing hydrogeological risk for groundwater resources evaluation.
2. It is a flexible procedure that yields various alternative estimates of composite decision variables.

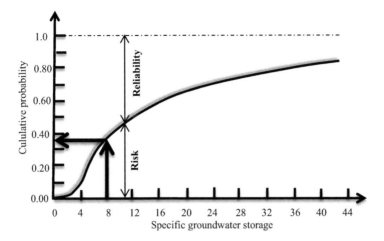

FIGURE 6.10   **Cumulative pdf of the specific groundwater storage.**

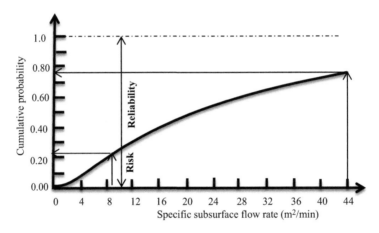

FIGURE 6.11   **Cumulative pdf of the specific subsurface flow rate.**

3. It is a probabilistic approach for determining the best estimate based on most frequently occurrence (model) value.
4. Finding the pdf of the composite variables provides meaningful information with regard to the population behavior of these variables, hence reducing more data necessity for meaningful results.
5. It is a way of incorporating the risk concept in groundwater development and management studies.

6.12 GROUNDWATER LEVEL RISE MODELING IN CITIES

Due to mismanaged human activities, leakages, and infiltrations from different sources (water supply pipes, water supply storage tanks, and some local sewerages), the natural groundwater balance is overturned into an unbalanced situation where the input to water table is comparatively more than the outgoing groundwater flow toward the depression areas, rivers, or seas. Depending on the season of the year, the groundwater table not only fluctuates but there is also a steady increase in the form of steady groundwater rise. For instance, the studies carried out in Jeddah City, Kingdom of Saudi Arabia, indicated that on the average there was about 0.10 cm groundwater table rise in each year (Al-Sefry and Şen, 2006). Among the main reasons are leakages from the city water supply pipes, insufficient tanker water haulages and their use for irrigation purposes, sewage water leakages, leakages from water storage tanks, garden watering, rainfall, and natural groundwater flow.

The changes in recharge caused by urbanization influence the groundwater levels and flow regime in underlying aquifers. As suggested by Van Stempvoort and Simpson (1995), this can take considerable time because the response constants of aquifers are normally the largest of all components of the urban hydrological cycle. Hydrogeologists speak of urbanization impact on groundwater in terms of large-scale temporal changes, while the geotechnologists tend to think in terms of the impact on urban structures in a site-specific steady state sense (Foster et al., 1994). In urban areas, subsurface infiltration rates increase and the effects on groundwater table level are dependent on the following points:

1. The single-most significant aquifer parameter for groundwater regime and flow is hydraulic conductivity that changes laterally and vertically. Hence, the lateral and vertical hydraulic conductivity changes must be determined as much as possible (Chapter 3). Otherwise, theoretically isotropic and homogeneous aquifer assumptions cannot lead to reliable management studies.
2. The aquifers are heavily exploited in urban areas for water supply. The groundwater levels decline or rise if malfunctioning water distribution and wastewater collector systems exist.
3. Impermeabilization of Earth's surface in urban areas by asphalt roads, roofs, airports, etc. reduces the natural groundwater recharge from direct rainfall or runoff.

Groundwater regime and levels beneath urban areas vary with urbanization development and such a variation may cause different problems including stability, integrity, and operation of subsurface installations and engineering structures. The following are the side effects of groundwater level rise:

1. Dampness, seepage, and flooding take place in low lands, depressions, and basements of buildings.
2. Groundwater rise may show damages on engineering structures such as foundations, bridge piers, etc.
3. In groundwater rise cases, effective stress and shear strength reduction may lead to soil-bearing-capacity reductions and effective soil stress decrease causes instability of the slope that may cause instability in structures (roads, buildings, bridges, etc.).
4. Groundwater rise coupled with chemical composition may cause corrosion in foundations. This happens if the groundwater has high SO_4^{2-}, Cl^-, and Mg^{2+} concentrations.
5. Groundwater level rise may lead to uplift of structures such as water storage tanks, cesspools, septic tanks, building basements, tunnels, and drainage channels.
6. When wastewater collectors are beneath the groundwater table, then major seepage may occur into these collectors and this extra water volume raises pumping costs.
7. High groundwater levels may lead to leakage in water supply pipe lines, and hence, the water supply network distributes polluted supply to house taps.

6.12.1 Groundwater Rise Calculations

The simple risk, R, can be defined as the probability of a variable, V, to be greater than the critical level, C_L, (nationally or internationally allowable levels) at least once over the stipulated time period, T_r, (return period). If the sequence of future likely occurrence of V is $V_1, V_2,..., V_n$ then the joint probability of n event nonoccurrence, $P(n)$, is defined as, (Şen, 1999a)

$$P(n) = P(V \leq C_L) = P(V_1 \leq C_L, V_2$$
$$\leq C_L, ..., V_n \leq C_L) \qquad (6.11)$$

Hence, the simple risk, R, the probability of occurrence as a complementary event is defined as,

$$R = 1 - P(n) = 1 - P(V_1 \leq C_L, V_2$$
$$\leq C_L, ..., V_n \leq C_L) \qquad (6.12)$$

The calculation of the multivariate probability term on the right-hand side is dependent on the data variability and theoretically it can be calculated by multiple integration of the multivariate pdf through tetrachoric series expansion (Saldarriaga and Yevjevich, 1970). However, in the case of a simple dependence structure such as the first-order Markov process, the right-hand side factorizes into various terms, which are explained by Şen (1976, 2009).

In the risk assessment of any variable, it is necessary to decide first on the frequency of critical-level exceedance, which is related to T_r after which it is then possible to determine the magnitude of the design variable based on the most suitable pdf. The return period is defined as the average length of time over which V will exceed only once. The random variable, T_r, specifies the time between any two successive exceedances. Its distribution in the case of independent discrete observations is given by Feller (1967) as,

$$P(T_r > i) = q^{j-1} \qquad (6.13)$$

or

$$P(T_r = j) = pq^{j-1} \qquad (6.14)$$

where j is discrete duration of nonexceedence, and p and q are the risk and safety probabilities, respectively. The return period can be obtained as,

$$E(T_r) = T_r = \sum_{j=1}^{\infty} jP(T_r = j)$$
$$= p\sum_{j=1}^{\infty} jq^{j-1} = \frac{1}{p} \qquad (6.15)$$

where

$$p = P(V > C_L) \qquad (6.16)$$

which is the probability of exceedance. If the theoretical pdf of the variable concerned is denoted by $f(V)$ then Eq. (6.16) can be rewritten as,

$$p = \int_{C_L}^{\infty} f(V)dV \qquad (6.17)$$

In order to apply Eq. (6.17), it is necessary to specify the theoretical pdf of the variable concerned from the available data. For the case study, 118 groundwater-sampling points are scattered over the whole of Jeddah, Kingdom of Saudi Arabia, and the water samples from these locations are analyzed hydrochemically in the laboratories for major ionic concentrations. Herein, only two quality variables, namely, chloride (Cl^-) and sulfate (SO_4^{2-}) are considered with actual groundwater level measurement (G_L). The theoretical logarithmic-normal, Cauchy and Weibull pdfs are found suitable for Cl^-, SO_4^{2-}, and G_L, respectively. The theoretical pdfs are obtained from the experimental histograms as shown in Figures 6.12−6.14.

In risk calculation, the basic function is the cumulative pdf for each variable, which shows the probability of exceedance.

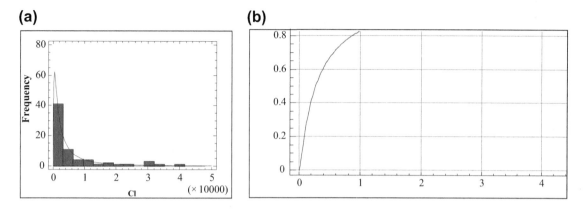

FIGURE 6.12   (a) Histogram (b) log-normal cumulative pdf for chloride.

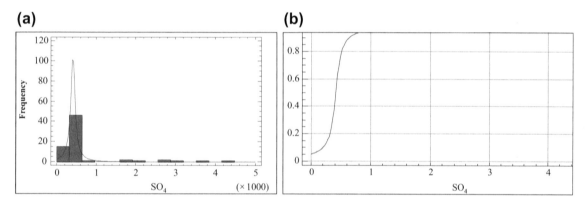

FIGURE 6.13   (a) Histogram (b) Cauchy cumulative pdf for sulfate.

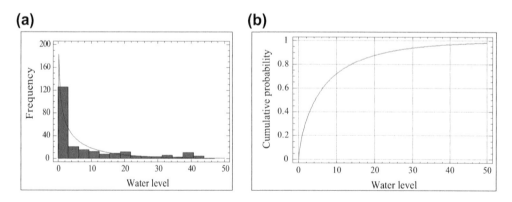

FIGURE 6.14   (a) Histogram (b) Weibull cumulative pdf for groundwater level.

6.13 AQUIFER UNCERTAINTIES AND STRATEGIC PLANNING

Aquifer test assumptions collectively are considered as valid conceptually and deterministically for models without random components for the aquifer system, and hence, they do not allow risk assessments (Chapters 3 and 4). The following major concepts are significant in any groundwater assessment, planning and management:

1. For water availability, aquifer thickness of saturation zone at a set of cross-sections must be explored by geophysical methods for their cross-section dimension determinations. In calculations, rather than depending on average depths, the possible uncertainties in the saturation thickness should be taken into consideration by considering a convenient pdf of depths in the study area. Another source of uncertainty for the availability calculation is the aquifer parameters that vary spatially. Herein, groundwater availability will be calculated based on risk levels of both the saturation depth and aquifer parameter as already explained in Chapter 4.

2. The average thickness of the unsaturated zone should be taken into account in calculating the possible strategic additional groundwater augmentation in the future. In such a calculation, there is little uncertainty in the unsaturated thickness, after all it is measured directly on the field prior to aquifer test as static water level, but the uncertainty in the calculation of possible future additional water storage is attached with the occurrence time, the amount of storm, rainfall, and the antecedent soil conditions. These uncertainties are calculated from the past rainfall records, and accordingly, risk levels are adopted (Chapters 1 and 4).

3. Potentiality of any aquifer is related to its response to yielding water supply through wells, in other words, the easiness of

groundwater withdrawal. This is dependent on aquifer transmissibility values, hence the uncertainties in transmissibility are accounted by simple probabilistic calculation, and finally, the abstractable groundwater volumes are provided in a graphical form for different risk levels and time durations (Chapter 4, Section 4.14.4).

6.13.1 Risk Calculations for Various Hydrogeological Parameters

Storativity (S), transmissivity (T), saturation thickness (S_t), unsaturated thickness, (U_t), and the porosity (p) variables are basic parameters in any strategic groundwater resources evaluation for aquifer potentiality (Chapter 4). For risk calculations, it is assumed that their regional variations are represented by logarithmic-normal pdf and the necessary steps are given already in Chapter 4, Section 4.14.2.

For the strategic groundwater resources calculations, it is useful to prepare a table (Table 6.3), which shows the variation of each aquifer parameters with a given set of risk levels (5%, 10%, 25%, 50%, 75%, 95%, and 99%).

Fitting a straight line on the semilogarithmic paper is equivalent to obtaining the regional variable, and hence, the probability statements can

TABLE 6.3   Hydrogeological Parameter Risk Levels

Risk (%)	S	T (m²/day)	p	D_S (m)	D_U (m)
5	0.8	2800	0.65	28	48
10	0.62	1950	0.58	24	42
25	0.28	620	0.43	18	30
50	0.075	310	0.32	13	18
75	0.019	58	0.23	10	9
95	0.0071	16	0.17	7	6
99	0.0088	10	0.11	4	5

Note: S = Storativity; T = Transmissivity (m²/day); p = Porosity; D_S = Saturation thickness (m); D_U = Unsaturated thickness (m).

be made including risk amounts corresponding to any value. For instance, the probability of the storage coefficient being equal to or greater than, say, 5.8×10^{-2} is approximately as 0.89 (Figure 4.38). Furthermore, for any valid transmissivity and saturation thickness corresponding risks can be read from Figures 4.39 to 4.40, respectively. The amount of groundwater storage underneath unit square meter of area can be calculated as (Al-Sefry and Şen, 2006),

$$V_u = \overline{D_S S} + \rho_{D_S S} \sigma_{D_S} \sigma_S \qquad (6.18)$$

where $\overline{D_S S}$ is the average of storativity and saturation thickness multiplication, $\rho_{D_S S}$ represents the correlation coefficient between storativity and saturation thickness, σ_{D_S} and σ_S are standard deviations of saturation thickness and storativity, respectively.

EXAMPLE 6.8 GROUNDWATER PROBABILISTIC STORAGE CALCULATION

Table 4.12 presents the random behavior of some hydrogeological variables in an area. Calculate the groundwater storage volume per square meter by considering random variability of each parameter.

Solution 6.8

The groundwater storage volume per square meter has already been given by Eq. (6.18). The relevant parameters from this table are, $\overline{D_S} = 8.75$ m, $\overline{S} = 1.556 \times 10^{-1}$, $\rho_{D_S S} = -0.203$, $\sigma_{D_S} = 8.12$ m, and $\sigma_S = 0.0914$ m. Hence, the substitutions of these values into the relevant equation yields,

$$V_U = 8.75 \times 1.556 \times 10^{-1} + (-0.203) \times 8.12$$
$$\times 0.0914 = 1.21 \text{ m}$$

In order to calculate the whole available groundwater in the aquifer, this value must be multiplied with the area of the water table.

6.14 OPTIMUM YIELD AND MANAGEMENT IN A WELL FIELD

Optimum yield is the quantity of water that can be pumped out from a well with no damage either to the aquifer or to the well itself. In management studies, pumping discharges must be kept below the optimum yield. In order to avoid extra costs and unnecessary drawdowns, the interference between the wells must be avoided through an effective management program. In many areas, the wells are already drilled, and hence, the distances between them are known. These distances can be considered as constraints in the evaluation of optimum discharge such that there is no interference among adjacent wells. Optimum discharge is neither unique nor constant, but its value at any time depends on the spacing between wells and the aquifer material properties in the vicinity of each well. In any groundwater management, the question is what amount of optimum discharges should be withdrawn from each well. The answer must be such that:

1. There is no damage to the well-aquifer interactions in individual well locations.
2. No interference should occur among the multiple well locations.

These two conditions should be satisfied simultaneously during a simple management program. It is apparent from the radial flow concept that as the groundwater approaches a well its velocity increases. Exceedance of velocity over a critical limit gives rise to the transportation of fine particles from the aquifer into the gravel pack. Subsequently, these fine particles cause plugs in the filters, which might lead to extra head losses, and consequently, drawdown increments. Approach velocity attains its maximum value at the well circumference and filters where the transport of fine materials is most likely to occur. The groundwater velocity at any point within the aquifer is equivalent to the specific discharge, q, (Darcy velocity) which has been given in Chapter 2 by Eq. (2.24) and also in this chapter by Eq. (6.4). This velocity should not be confused with the real flow velocity for which the flow area is the intergranular pore space. The real velocity, v_r, is estimated after dividing the specific discharge by the porosity, p, as,

$$v_r = \frac{Q}{A_n} = \frac{q}{p} \qquad (6.19)$$

in which A_n is the circumferential pore area of the well interface within the whole aquifer. Physically, the approach velocity is dependent on the grain-size distribution of the aquifer material and on the hydraulic conductivity more than any other aquifer parameter. It is not possible to establish such dependence through analytical methods. By observing a number of operating wells, Sichard (1927) proposed an empirical relationship between the hydraulic conductivity of aquifer, K, and approach velocity, v_{aq}, which is supported by factor as (Huismann, 1972),

$$v_{aq} = \frac{\sqrt{K}}{60} \qquad (6.20)$$

This relationship provides a valid design criterion where conductivity is determined preferably by in situ pumping tests or from the specific discharge measurements within each well, if possible. The second alternative is employed in calculating the transmissivity values, which are then converted into hydraulic conductivities by dividing these values by the aquifer thickness at each well location. It is possible to rewrite the optimum discharge, Q_{oi}, for the i-th well by considering the approach velocity in Eq. (6.20) (Holting, 1980) as,

$$Q_{oi} = 2\pi r_{wi} m_i \frac{\sqrt{K_i}}{60} \quad (i = 1, 2, 3, ..., n) \quad (6.21)$$

where r_{wi} and K_i are the well radius and hydraulic conductivity, respectively, m_i is the saturation thickness, and n is the number of wells.

The second stage in a simple management program is the condition that there will not appear any interference between the wells. Two or more wells pumping from the same aquifer will interfere with each other if the distance between them is more than the summation of their radii of influence values. Several interfering pumping wells impose addition of the drawdowns at any point within the aquifer. For confined aquifer in the steady state flow case, the relevant discharge expression can be obtained from Eq. (3.17) in Chapter 3 for i-th well by considering the radius of influence, R_i, and the well drawdown, s_{wi} as,

$$Q_{oi} = \frac{2\pi T_i s_{wi}}{\log\left(\frac{R_i}{r_{wi}}\right)} \quad (i = 1, 2, 3, ..., n) \quad (6.22)$$

in which r_{wi} is the well radius and T_i is the transmissivity at the same well location. In practical studies, one can adopt the radius of influence from an empirical formulation suggested by Holting (1980) as,

$$R_i = 3000 s_{wi} \sqrt{K_i} \qquad (6.23)$$

The substitution of this expression into Eq. (6.22) leads to,

$$Q_{oi} = \frac{2\pi T_i s_{wi}}{\log\left(\frac{3000 s_{wi} \sqrt{K_i}}{r_{wi}}\right)} \quad (i = 1, 2, 3, ..., n)$$

$$(6.24)$$

In order to satisfy simultaneously the aforementioned two stages in the same management program, Eqs (6.21) and (6.24) are set equal to each other and after some algebra one can obtain the optimum drawdown expression as,

$$s_{wi} = \frac{r_w \frac{\sqrt{K_i}}{60} \log\left(\frac{3000 s_{wi} \sqrt{K_i}}{r_w}\right)}{K_i} \quad (i = 1, 2, 3, ..., n)$$

$$(6.25)$$

In this expression, all the quantities are known except the drawdown s_{wi} within the i-th well. For each well site, the optimum drawdown can be calculated from this expression by trial-and-error approximation. These solutions do not take into consideration the distances between the existing wells. They are converted into the sequence or radius of influences, R_{wi}, through Eq. (6.23). In any existing well field, the distances among the adjacent wells can be found directly from the well location map. Let D_{ij} indicate the distance between two adjacent wells at sites i and j. If there are n existing wells, there will be $n(n-1)/2$ different distances. For a reliable management program, the summation of the two adjacent well radius of influence should be smaller than the actual distance between them, D_{ij},

$$D_{ij} < R_i + R_j \quad (i = 1, 2, 3, ..., n) \qquad (6.26)$$

This is the distance constraint in the management program. Equation (6.26) should be checked for each pair of adjacent wells. If the constraint is not satisfied, then the radius of influence in one or two of the wells will be reduced conveniently until the acceptable inequality is obtained. Reduction in the radius of influence implies reduction in the pump discharges. According to new values of the radius of influences, it is possible to calculate the corresponding drawdown in concerned wells from Eq. (6.23) as,

$$s_{wi} = \frac{R_i}{3000\sqrt{K_i}} \quad (i = 1, 2, 3, ..., n) \qquad (6.27)$$

With adjusted values at hand, the optimum discharge for each well can be recalculated from Eq. (6.24). Finally, corresponding groundwater velocities, v_{oi}, can be obtained from Eq. (6.26) at the well surface as,

$$v_{oi} = \frac{Q_{oi}}{2\pi m_i r_{wi}} \quad (i = 1, 2, 3, ..., n) \qquad (6.28)$$

The following points help achieve optimum groundwater management program and they are important for inclusion in any management program in the future:

1. The regional optimum water levels can be estimated by the simple management rules and formulations presented above.
2. Interference among adjacent wells must be avoided during any groundwater management program.

EXAMPLE 6.9 OPTIMUM YIELD MANAGEMENT

Figure 6.15 indicates well field configuration and Table 6.4 includes necessary information about the wells and aquifer parameter values. Find the optimum well yields such that there is no inference between the wells.

Solution 6.9

The first step is to decide about the radius of influence for each well. This can be decided as in Figure 6.16 by considering the distances between neighboring wells such that there will not be interference between their depression cones.

This is the most ideal design, which gives the radius of influence from considerations of W1−W2 and W3−W4 pairs. Hence, from the distances between W1 and W2, the radius of influence for each well is $3.5/2 = 1.75$ km and for

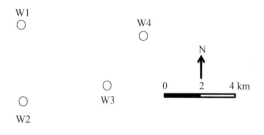

FIGURE 6.15 **Well-field configuration.**

TABLE 6.4 Well-Field Configuration and Hydrogeological Quantities

Well number		W1	W2	W3	W4
Well radius, r_i (m)		1.20	0.95	1.10	1.23
Hydraulic conductivity, K_i (m/s)		3.5×10^{-4}	5×10^{-4}	1.3×10^{-4}	3.5×10^{-4}
Saturation thickness, m (m)		56.8	87.5	14.9	43.6
Distances between wells, D_{ij} (km)	W1	0	2.1	3.5	3.1
	W2		0	2.6	3.8
	W3			0	1.9

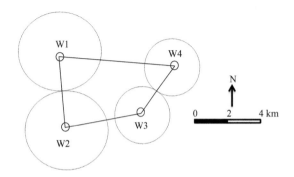

FIGURE 6.16 **Radius of influence design.**

TABLE 6.5 Calculation Results

Well number	W1	W2	W3	W4
Well radius, r_i (m)	1.20	0.95	1.10	1.23
Drawdown (m) from Eq. (6.26)	0.62	0.90	0.23	0.62
Optimum discharge (m³/s) from Eq. (6.23)	0.0396	0.0845	0.0038	0.0305
Optimum discharge (l/s)	39.62	84.47	3.81	30.51

1. Let us first determine the optimum drawdown values from Eq. (6.27) for each well. The substitution of the radius of influence and the hydraulic conductivity values into this expression yields the third line in Table 6.5.

2. The inference among the adjacent wells is avoided according to the aforementioned calculations and accordingly the optimum discharge values can be obtained from Eq. (6.24) by substitution of the relevant values and the results are given in the fourth and fifth rows in m³/s and l/s, respectively.

6.14.1 Optimum Aquifer Yield

The safe yield is already defined in Section 6.6.1. Any withdrawal in excess of safe yield is an overdraft. The term overdraft may include not only depletion of the groundwater reserves, but also the intrusion of undesirable water quality (Domenico, 1972; Kazmann, 1972).

For the explanation of optimum aquifer yield, Al-Kharj area close to Al-Riyadh, capital of Saudi Arabia, is chosen for explanation (Figure 6.17).

This area is an important center of growth and agricultural activity. The available water comes from four productive multiple aquifers, namely,

wells W3 and W4 it is $1.9/2 = 0.95$ km. These are the maximum allowable radii of influences. Hence, there are plenty of alternatives provided that the radius of influences remains less than these values.

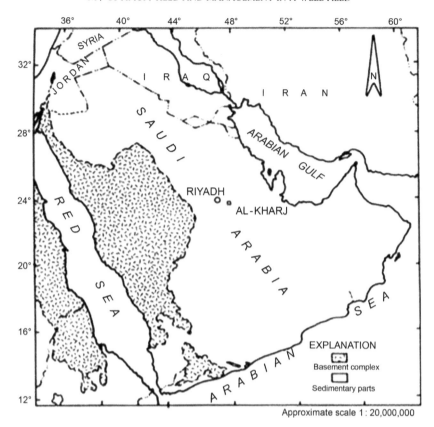

FIGURE 6.17 **Location map.**

Wasia, Biyadh, Sulaiy, and Arab sedimentary formations, which should be exploited with an efficient management plan. The Wasia and Biyadh formations are clastic productive aquifers, whereas the Sulaiy and Arab formations include nonclastic aquifers. The Sulaiy formation has undergone considerable underground solutions to form a karstic topography over some parts of Al-Kharj area. The Arab formation is characterized by joints, fractures, and solution openings up to 1 m in diameter (Bazuhair, 1989).

Due to difficulty in finding satisfactory observation wells, only 10 wells are tested during field work. The pumping test data are analyzed using different analytical methods as Theis (1935), Cooper and Jacob (1946), and Boulton (1963). The specific capacity method of Walton (1970) is applied to estimate the transmissivity of the productive aquifers (Chapter 3). The sieve analysis of the clastic samples are plotted to provide cumulative frequency curves (Chapter 4). From these curves, different parameters necessary for the detection of the permeability are obtained and substituted in the equations provided by Hazen (1893) and Masch and Denny (1966).

The permeability values for these productive aquifers are determined according to different methods presented in Table 6.6 where the big range of Biyadh and Wasia aquifers is due to the differences in grain size, lateral variation, cementation, and/or secondary permeability.

TABLE 6.6　Summary of Results

Aquifer Name	Cooper and Jacob (Chapter 3)	Recovery Straight Line (Chapter 3)	Theis Type Curve (Chapter 3)	Boulton Type Curve (Chapter 4)	Specific Capacity (Chapter 2)	Grain-Size Method (Chapter 2)	
						Hazen Method K (m/day)	Mean Diameter (mm)
	T (m²/day)						
Wasia	2756.2	–	–	1500–2300	1500–2200	13.5–42.3	14
Biyadh	108.9–604.8	34.6–388.8	96.9–817.7	–	31–1950	2.88–25.92	7
Sulaiy	–	34.6	–	–	28–54	–	–
Arab	2877.12	–	–	666.65	2000–2950	–	–

TABLE 6.7　Parameters Used for Calculation of Storativity

Aquifer Name	T (m²/day)	(day)	S_{max} (m)	Q Practical Discharge (m³/day)	r_w (m)	S (−)
Wasia	2758.10	0.20000	1.17	1507.16	0.16	2.30×10^{-7}
Biyadh	664.00	0.404000	9.19	5443.2	0.16	3.22×10^{-2}
Sulaiy	34.56	0.29166	7.00	276.5	0.16	2.65×10^{-2}
Arab	2873.90	0.186100	3.05	5024.0	0.16	1.5×10^{-5}

Şen (1983) derived a solution to determine the storativity in the absence of observation well as already explained in Chapter 3 (Eq. (3.89)). His equation is applicable to confined, and after drawdown correction (Eq. (4.9)) to unconfined aquifers. The parameters required for the application of his formula as well as the calculated S values are shown in Table 6.7.

In areas, like Al-Kharj, where the groundwater is the only source for supply, the increase in demand will mean abstraction of more groundwater with increase in the population (Example 6.2). In order to preserve the aquifer safe yield, it is very convenient to have proper plan, design, management, and operation rules. The optimum aquifer yield determination is one of the significant factors in managing the available groundwater in each aquifer of Al-Kharj area. For this purpose, Eq. (6.21) can be used for optimum well discharge calculations. On the other hand, Eqs (6.23) and (6.24) are ready for the empirical calculations of the maximum radius of influence, R_{max}, and the optimum well discharge, Q_{opt}. The calculations from Eq. (6.25) are given in Table 6.8 for optimum aquifer yield.

In the study area, the maximum radius of influence is useful for calculating the allowable maximum number of wells for production without interference between the adjacent wells. Table 6.9 gives all the necessary numerical data for this purpose.

The number of wells is equal to the division of total aquifer area, by the influenced area of the

TABLE 6.8   Parameters Used to Determine R_{max} and the Optimum Aquifer Yield

Aquifer Name	Average Thickness, h (m)	K (m/day)	$\sqrt{K}$ ($\sqrt{(m/s)}$)	r_w (m)	s_{max} (Practical) (m)	R_{max} (m)	πR^2 (m²)	Q_{opt} (m³/hour)
Wasia	235	40.0	0.0220	0.16	1.17	77.2	18713.9	1296.0
Biyadh	39	17.0	0.0160	0.16	9.19	441.1	610947.3	151.2
Sulaiy	30	1.8	0.0046	0.16	7.00	96.6	29301.1	33.5
Arab	164	18	0.0144	0.16	3.05	131.8	54545.7	575.0

TABLE 6.9   Parameter Values

Parameters	Quantity
Total aquifer area, A_a	25,000,000 m²
Irrigated area, A_r	23,000,000 m³
Biyadh average saturation thickness (m)	65 m
Transmissivity (T)	664 m²/day (well no.4)
Hydraulic conductivity (K)	10.22 m/day = 0.000118 m/s
Well radius (r_w)	0.16 m
Maximum drawdown (s_{max})	9.19 m
Radius of influence (R_{max})	303.27 m
Optimal discharge (Q_{opt})	2900 m³/day
Area of influence (AI)	288,794 m³
Existing daily discharge	198,995 m³
Required daily discharge	250,720 m³
Existing wells	55

well. Hence, the number of wells is $A_a/\pi R_{max}^2 = 79$ wells. Accordingly, at least one standby well must be ready for use in case of any well malfunction, and therefore, the total well number should be taken as 80.

There is a logical relationship between the aquifer water abstraction discharge, number of wells, and time as in Figure 6.18 for Wasia aquifer.

Such charts are achieved by assuming various discharge values quantity charges from which the yearly consumption can be calculated for a given number of wells. The number of years for depletion is calculated by dividing the total amount of water by the yearly consumption. However, the daily discharges are considered as summer and winter for 10−24 hours, respectively. The average saturation thickness helps calculate the total volume of the aquifer by considering the aquifer material as homogeneous and isotropic. The total volume of water can be calculated from Eq. (6.2).

EXAMPLE 6.10 WELL NUMBER, DEPLETION-YEAR DURATION, AND DISCHARGE CALCULATIONS

In Al-Kharj area, optimum groundwater yield planning is thought such that the aquifer will be mined.

1. If optimum well discharge alternative planning is considered by means of 20 wells with different set of continuous discharges as 5 m³/hour or 10 m³/hour, then provide a few alternatives for aquifer safe yield depletion during 25 years.

2. In this area, the annual water demand is 0.5×10^6 m³, for which one of the alternatives in the previous step is the most suitable.

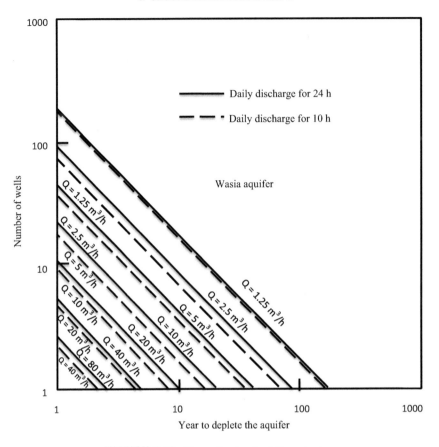

FIGURE 6.18 **Groundwater depletion charts.**

Explain your interpretation of the final decision.

Solution 6.10

Optimum groundwater resources planning require optimum discharge variation with a number of wells and years as in Figure 6.18.

1. There are various alternatives for the solution because each well from 20 can be arranged such that either 5 m^3/hour or 10 m^3/hour can be selected. According to the circumstances, one can decide which wells should pump one of these alternatives. The reader can appreciate that there are numerous alternatives. However, herein a few of them will be explained so that the reader can make practice accordingly with other alternatives.

Since it is mentioned that there is continuous discharge, it means that the pumps will work 24 hours and the stipulated time horizon for aquifer is 25 years. The discharge-year-well number relationship is already provided for the study area (Al-Kharj) in Figure 6.18. Since the safe aquifer capacity is expected to deplete in a 25-year period, the vertical line in Figure 6.19 indicates the relationship between the discharges and well numbers.

It is obvious that 20 wells can deplete the aquifer safe capacity with discharge much less than 1.25 m^3/hour during 25 years. The possible

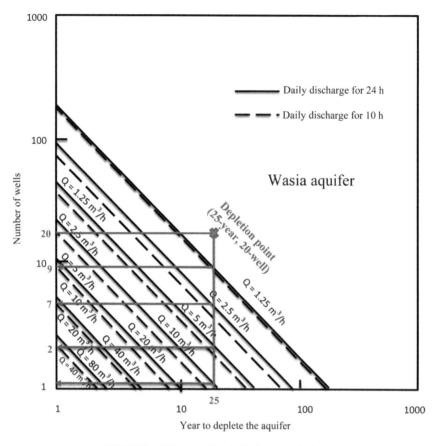

FIGURE 6.19 **Aquifer depletion calculations.**

discharges for 24 hour operation for 25-year duration are nine wells with 1.25 m³/hour, seven wells with 2.5 m³/hour discharge, two wells with 5 m³/hour, and one well with 10 m³/hour. Under these circumstances, the combination of 20 wells can be achieved through the following different alternatives:

a. Two 9-well groups each with 1.25 m³/hour discharge and two wells with 5 m³/hour. Hence, the total discharge becomes
$2 \times 9 \times 1.25 + 2 \times 5 = 32.500$ m³/hour.

b. Two 9-well groups each with 1.25 m³/hour discharge and two 1-well groups with 10 m³/hour. The total discharge adds up to $2 \times 9 \times 1.25 + 2 \times 1 \times 10 = 42.500$ m³/hour.

c. One 9-well group with 1.25 m³/hour, one 7-well group each with 2.5 m³/hour, and two 2-well groups with 5 m³/hour. This makes totally $9 \times 1.25 + 7 \times 2.5 + 2 \times 5 = 48.750$ m³/hour.

d. Two 7-well groups each with 2.5 m³/hour discharge and three 2-well groups with 5 m³/hour. The total discharge is $2 \times 7 \times 2.5 + 3 \times 2 \times 5 = 65.000$ m³/hour.

e. There are many other combinations that are left for the reader to make similar deductions and calculations.

2. In order to make a decision one should know what is the total annual water amount from each alternative.

    a. The total annual yield from the first alternative is

        $32.500 \times 24 \times 365 = 0.285 \times 10^6 \, \text{m}^3$,

    b. The same calculation for this step is

        $42.500 \times 24 \times 365 = 0.373 \times 10^6 \, \text{m}^3$,

    c. For the this alternative the total annual water volume is

        $48.750 \times 24 \times 365 = 0.427 \times 10^6 \, \text{m}^3$, and

    d. Finally, $65.000 \times 24 \times 365 = 0.569 \times 10^6 \, \text{m}^3$.

Among these alternatives the last one is sufficient for the necessary water demand, because $0.569 \times 10^6 > 0.500 \times 10^6$. This alternative implies from the previous step that two 7-well groups (14 wells) and three 2-well groups (6 wells) are necessary. So, it is up to the planner to decide which 14 wells and 6 wells should be distributed among 20 wells.

6.15 WATER STORAGE VOLUME RISK CALCULATION

In any strategic groundwater reservoir planning, the main concern is not with the individual haphazard changes of hydrogeological parameters only, but also their effects on possible storage volume calculations. It has already been mentioned and the procedure is presented fully in Section 6.10 that the single-most significant composite variable is the specific groundwater storage capacity, g_s, which is defined as the multiplication of the storage coefficient, S, by the saturation thickness (Eq. (6.3)). According to Figure 4.37 in Chapter 4, abstractable groundwater volume is calculated as $71 \times 10^6 \, \text{m}^3$. By considering 50 and 70 l/person/day as mentioned in Section 6.4, various scenarios can be thought as shown in Table 6.10.

It is possible to draw population–duration curves by considering 50, 70, 100, and 150 l/day/person consumption rates as shown in Figure 6.20.

Practical questions are how to estimate and manage the groundwater resources volume in a given area and what risk level should be accepted in such a management. Once the estimation of the available groundwater storage volume is determined, all of the subsequent main activities of exploitation, distribution, management, well number and location, development and economy for such activities are dependent primarily on these estimates. It is possible to extend such management studies in the following directions:

1. For regional modeling and management purposes, the pdfs of different hydrogeological variables can be incorporated within the framework of geostatistical prediction (Şen, 2009).
2. For economic purposes, various risk levels can be combined with their consequences in terms of, say, dollars and an economically optimum prediction can then be obtained.
3. The pdfs and associated risks can be used in conjunction with opinions of experts so as to unite the theory, field observations, and professional experience in arriving at reasonable estimates.

EXAMPLE 6.11 PROBABILISTIC RISK MANAGEMENT

The application of the probabilistic risk and statistical formulations developed in previous sections is demonstrated for the Wadi Qudaid unconfined aquifer near the city of Jeddah, Kingdom of Saudi Arabia (Şen, 1999a). The Wadi alluvium is composed of coarse to medium Quaternary deposits and surficial sand dunes. The bedrock topography and the alluvial thickness are very irregular in addition to the heterogeneity of the aquifer material. These features are very obvious from Table 6.11 where the basic hydrogeological parameters, as found in each of the seven wells in the area, exhibit regional variations.

TABLE 6.10 Different Water Supply Scenarios

Consumption 50 l/capita/day			Consumption 70 l/capita/day		
Population (×10⁶)	Time		Population (×10⁶)	Time	
	Day	Month		Day	Month
0.5	7584	252	0.5	5417	180
1.0	3792	126	1.0	2708	90
1.5	2528	84	1.5	1805	60
2.0	1896	63	2.0	1354	45
2.5	1516	50	2.5	1083	36
3.0	1264	42	3.0	902	30
3.5	1083	36	3.5	773	25
4.0	948	31	4.0	677	22
4.5	842	28	4.5	601	20
5.0	758	25	5.0	541	18
5.5	689	22	5.5	492	16
6.0	632	21	6.0	451	15
8.0	474	15	8.0	338	11
10.0	379	12	10.0	257	8

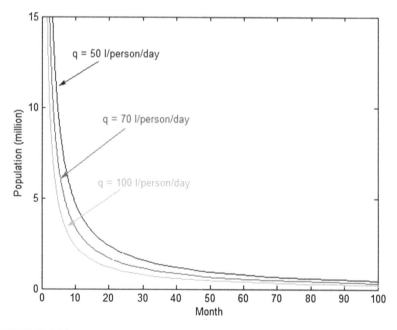

FIGURE 6.20 Per capita per day water demand-population-duration relationships.

TABLE 6.11   Pumping-Test Results

Well No.	Storativity S (−)	Error (%)	Transmissivity (m²/min) T	Error (%)	Saturation thickness m (m)	Error (%)
1	3.3×10^{-2}	-3.6×10^{-2}	1.8×10^{-1}	-7.1×10^{-1}	13.40	0.34
2	3.5×10^{-2}	-3.4×10^{-2}	1.2×10^{-1}	-7.7×10^{-1}	12.10	−0.96
3	2.0×10^{-2}	-4.9×10^{-2}	4.1×10^{-1}	-4.8×10^{-1}	10.50	−2.56
4	8.5×10^{-2}	1.6×10^{-2}	2.3×10^{-1}	-6.6×10^{-1}	12.20	−0.86
5	1.9×10^{-1}	1.2×10^{-1}	4.2×10^{-0}	3.3×10^{0}	10.30	−2.76
6	5.2×10^{-2}	-1.7×10^{-2}	1.2×10^{-1}	-7.7×10^{-1}	14.95	1.89
7	7.0×10^{-2}	1.0×10^{-2}	9.6×10^{-1}	-7.0×10^{-2}	18.00	4.94
Averages	6.9×10^{-2}	≈ 0.0	8.9×10^{-1}	≈ 0.0	13.06	≈ 0.0
Variances	13.2×10^{-2}	7.4×10^{-2}	10.98×10^{-1}	16.9×10^{-1}	35.19	6.6
Cross-variances	$\overline{e_S e_T} = 7.45$		$\overline{e_S e_h} = 1.66 \times 10^{-2}$		$\overline{e_T e_h} = -1.23$	

TABLE 6.12   Risk Levels of Various Hydrogeological Variables

Risk (%)	Saturated Layer Thickness (m)	Unsaturated Layer Thickness (m)	Porosity (%)	Storativity (−)	Transmissivity (m²/day)	Specific Groundwater Storage (m)	Specific Recharge Storage (m)
0.95	28.12	49.87	68.05	0.6602	210	18.56	33.93
0.90	23.80	38.59	57.78	0.2201	199	5.24	22.30
0.75	18.00	26.00	43.97	0.1852	184	3.33	11.43
0.50	13.20	16.78	32.46	0.0765	167	1.01	5.45
0.25	9.67	10.82	23.96	0.0316	153	0.31	2.59
0.10	7.32	7.29	18.23	0.0143	140	0.10	1.33
0.05	6.19	5.76	15.48	0.0089	134	0.06	0.89

Solution 6.11

The parameters are obtained from a large-diameter well aquifer test according to Papadopulos and Cooper (1967) type curves. Comparison of each parameter value at individual well with arithmetic averages indicates rather random behavior. The error terms are also presented in Table 6.12. For the probabilistic risk calculations, first storativity, transmissivity, and thickness values are plotted separately on logarithmic pdf papers with the most suitable straight lines. These plots are presented in Figures 6.21−6.23 where the scatter of points appears along straight lines which confirm the assumption that the parameters abide by a logarithmic-normal pdf.

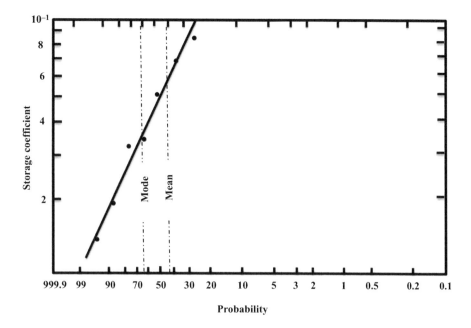

FIGURE 6.21   Storage coefficient data from Wadi Qudaid.

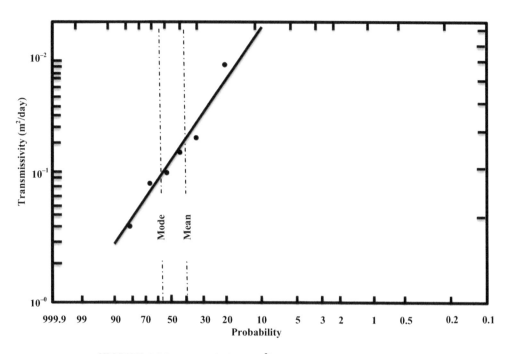

FIGURE 6.22   Transmissivity (m²/min) data from Wadi Qudaid.

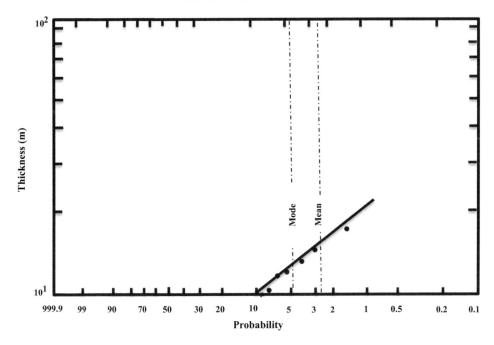

FIGURE 6.23   **Alluvium thicknesses in Wadi Qudaid.**

6.16 AQUIFER UNCERTAINTIES AND MANAGEMENT

Availability and potentiality of groundwater resources are two significant concepts for strategic planning. Geometric dimensions with hydrogeological parameter estimations help assess these two concepts numerically.

Large error in the deterministic method affects the water balance equation. The methodology described in this section has the following important implications and advantages over commonly used classical deterministic techniques as already explained in Section 6.11.2.

In the previous sections, arithmetic average formulations are used because there is an implied assumption that all the deviations have Gaussian (normal) pdf. However, the pdf of many hydrogeological variables accord with logarithmic-normal pdf (Freeze and Cherry, 1979). After the decision on the logarithmic-

normal pdf as the best representative model, it is possible to calculate numerical risks attached with the parameter values. Preliminary steps are presented by Şen (1999b) for risk assessment in groundwater calculations. In order to make probabilistic risk assessments of various possible scenarios for groundwater strategic availability and recharge possibilities, the following hydrogeological variables should be attached with risk levels.

1. The storativity (S) risk model: It is natural that the storativity value changes from well location to other within the geological formations in general, and within the Wadi aquifer in particular. Ten aquifer tests are performed on the field and additionally seven tests are also given by Şen (1999b), hence totally 17 well-test results exist. Their logarithmic-normal pdf model is shown in Figure 6.24.

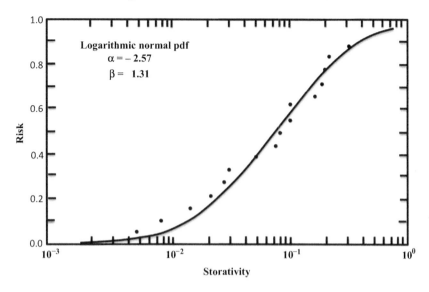

FIGURE 6.24 Storativity-risk models.

2. Transmissivity(T)-risk model: The same number of data is available for this parameter also from two sources. It is well established in the literature that hydraulic conductivity or transmissivity has logarithmic normally distributed values within geological formations. This statement validates the use of logarithmic-normal pdf for transmissivity risk assessments. The transmissivity-risk model is presented in Figure 6.25.

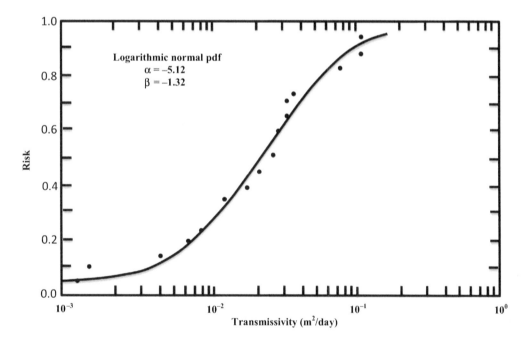

FIGURE 6.25 Transmissivity-risk models.

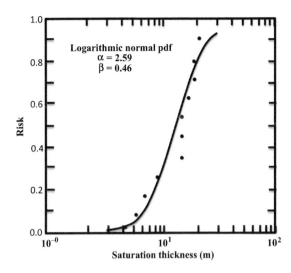

FIGURE 6.26   **Saturation-thickness-risk model.**

3. Saturation-thickness(D_S)-risk model: Within the aquifer area, the saturation thickness changes according to the alluvium bed topography. This is the reason why different saturation thicknesses are valid at each well-test location. In order to generalize the saturation thickness within the Wadi, again logarithmic-normal pdf model is used, and consequent saturation thickness-risk relationship is obtained as shown in Figure 6.26, where only 10 well depths are available. It is clear that there are sharp risk variations within small differences in the saturation thickness values.

4. Unsaturated thickness(D_U)-risk model: Almost all the groundwater reservoirs in the study area appear as unconfined aquifers, and therefore, the groundwater table is expected, in general, to follow the surface topographic features. However, due to heavy pumping in

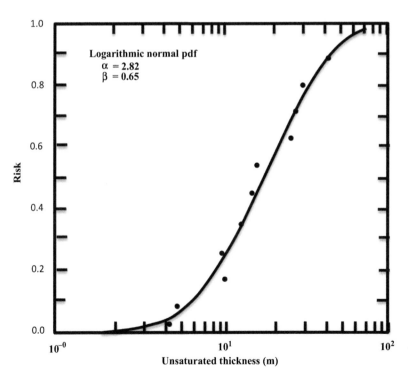

FIGURE 6.27   **Unsaturated-thickness-risk models.**

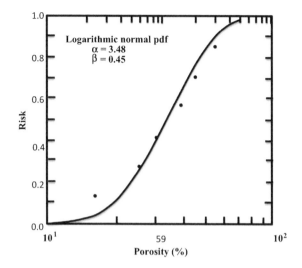

FIGURE 6.28 Porosity-risk numbers.

the area, this is not the case, and consequently, there are big variations in the unsaturated thickness. The risk model is again provided through the suitable logarithmic-normal pdf as in Figure 6.27.

Comparison of the last two figures indicates actually that the unsaturated thickness has a wider domain of variation than saturation thickness.

5. Porosity(p)-risk model: In order to be able to calculate the possible amount of recharge and additional groundwater rise over the present groundwater table, it is necessary that the porosity distribution or its model should also be taken into consideration in the calculations. For this purpose, the possible porosity values obtained in the alluvium aquifers are considered with its logarithmic-normal model as shown in Figure 6.28. It is obvious that the matching of measured points to this model shows a good agreement.

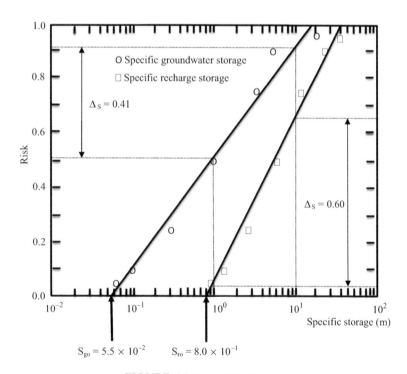

FIGURE 6.29 Specific storages.

Table 6.12 gives risk levels at a set of levels for hydrogeological parameters. The last two columns include the risky specific groundwater and the specific recharge storage values. The specific recharge value is the amount of water that the aquifer can take from unit surface area, and hence, its calculation is a result of porosity multiplication by unsaturated layer thickness. The plots of risk, R, versus specific groundwater and recharge storages yield straight lines on a semilogarithmic paper as in Figure 6.29.

The specific groundwater, G_S, and recharge, R_S, storage equations can be found from the slopes and intercepts on the horizontal axis as,

$$G_S = 0.055 + 0.41 \log (R) \qquad (6.29)$$

and,

$$R_S = 0.8 + 0.60 \log (R) \qquad (6.30)$$

For an expert who is ready to work under 10% risk, the specific groundwater storage appears as 0.36 m per square meter. For the sake of argument, if an inexperienced hydrogeologist is ready to accept 50% risk then his/her estimate will be 0.44. However, as his/her experience increases the risk level will become smaller, and accordingly, s/he will arrive at more conservative estimates.

References

Aitcheson, J., Brown, J.A.C., 1957. The Lognormal Distribution with Special Reference to its Uses in Economics. Cambridge University Press, 176 pp.

Al-Sefry, S.A., Şen, Z., 2006. Groundwater rise problem and risk evaluation in major cities of arid lands-Jedddah case in Kingdom of Saudi Arabia. Water Resour. Manage. 20, 91–108.

Aubrecht, C., Levy, M., De Sherbinin, A., Yetman, G., Jaite, M., Steinnocher, K., Metzler, S., 2010. Refinement of regionally modeled coastal zone population data enabling more accurate vulnerability and exposure assessments. In: Proceedings of International

Disaster and Risk Conference IDRC 2010. Davos, Switzerland.

Balkhair, K.S., 2002. Outranking strategic groundwater basins in Western Saudi Arabia using multi-criterion decision making techniques. In: Proceedings of the International Conference on Water Resources Management in Arid Regions, March 23–27, Kuwait.

Bazuhair, S.A., 1989. Optimum aquifer yield of four aquifers in Al-Kharj area, Saudi Arabia. Jour. King Abdulaziz Univ. 2, 37–39.

Boulton, N.S., 1963. Analysis of data from non-equilibrium pumping test allowing for delayed yield from storage. Proc. Inst. Civil Engineers 26, 469–482.

Cooper, H.H., Jacob, C.E., 1946. A generalized graphical method for evaluating formation constants and summarizing well field history. Trans. Am. Geophys. Union 27, 526–534.

Davis, S.N., 1969. Porosity and permeability of natural materials. In: DeWeist, R.J.H. (Ed.), Flow through Porous Media. Academic Press, New York, pp. 54–89.

Davis, G., 1982. Prospect risk analysis applied to groundwater reservoir evaluation. Groundwater 20, 657–662.

Delhomme, J.P., 1978. Kriging in the hydro-sciences. Adv. Water Resour. 1, 251–266.

Domenico, P.A., 1972. Concepts and Models in Groundwater Hydrology. McGraw-Hill, New York, 405 p.

Dottridge, J., Jaber, N.A., 1999. Groundwater resources and quality in Northeastern Jordan: safe yield and sustainability. J. Appl. Geog. 19, 313–323.

Feller, W., 1967. An Introduction to Probability Theory and Its Application, vol. 1. John Wiley and Sons, Inc., New York, 509 pp.

Foster, S.S.D., Morris, B.L., Lawrance, A.R., 1994. Effects of Urbanization on Groundwater Recharge. Groundwater Problems in Urban Areas. Institution of Civil Engineers. Thomas Telford, London, 453.

Freeze, R.A., Cherry, J.A., 1979. Groundwater. Prentice-Hall, p. 604.

Gheorghe, A., 1978. Processing and Synthesis of Hydrogeological Data. Abacus Press, Kent, 390 pp.

Hall, O., Duit, A., Caballero, L., 2008. World poverty, environmental vulnerability and population at risk for natural hazards. J. Maps, 151–160.

Hazen, A., 1893. Some Physical Properties of Sands and Gravels, Mass. State Board of Health, 24th Annual Report.

Heath, R.C., Spruill, R.K., 2003. Cretaceous aquifers in North Carolina: analysis of safe yield based on historical data. Hydrogeol. J. 11, 249–258.

Holting, B., 1980. Hydrogeologie. Einfuhrung in die Allgemeine und Angewandte Hydrogeologie.

Huismann, L., 1972. Groundwater Recovery. Winchester Press, New York.

Journel, A.G., 1983. Non-parametric estimation of spatial distributions. Math. Geol. 15, 445—468.

Kazmann, R.G., 1972. Modern Hydrology, second ed. Harper and Row, New York. 365 p.

Koch, S., Link, R.F., 1971. Statistical Analysis of Geological Data. Dover, New York, p. 375.

Krumbein, W.C., Graybill, F.A., 1965. An Introduction to Statistical Models in Geology. McGraw-Hill Book Co., New York, 475 pp.

Margat, J., Saad, K.F., 1983. Concepts for the utilization of non-renewable groundwater resources in regional development. Nat. Resour. Forum. United Nations. New York 7 (4).

Masch, F.D., Denny, K.J., 1966. Grain size distribution and its effect on the permeability of unconsolidated sands. Water Resour. Res. 2, 665—677.

Meinzer, O.E., 1934. History and development of ground-water hydrology. Wash. Acad. Sci. J. 24 (1), 6.

Merrit, M., 2004. Estimating Hydraulic Properties of the Floridian Aquifer System by Analysis of Earth-Tide, Ocean-Tide, and Barometric Effects, Collier and Hendry Counties, Florida. U.S. Geological Survey, Water-Resources Investigations Report, Florida, Tallahassee, Florida, 64 pp.

Nian-Feng, L., Jie, T., Feng-Xiang, H., 2001. Eco-environmental problems and effective utilization of water resources in the Kashi plain, Western Terim basin, China. Hydrogeol. J. 9, 202—207.

Papadopulos, I.S., Cooper, H.H., 1967. Drawdown in a well of large diameter. Water Resour. Res. 3, 241—244.

Saldarriaga, J., Yevjevich, V., 1970. Application of Run-lengths to Hydrologic Series. Hydrology Paper 40. Colorado State University, Fort Collins, Colorado.

Schneiderbauer, S., 2007. Risk and Vulnerability to Natural Disasters — From Broad View to Focused Perspective: Theoretical Background and Applied Methods for the Identification of the Most Endangered Populations in Two Case Studies at Different Scales (Ph.D. dissertation). Free University of Berlin, Germany.

Schultz, E.F., 1973. Problems in Applied Hydrology. Water Resources Publications. Colorado State University, 510 pp.

Seabear, P.R., Hollyday, E.F., 1966. Statistical analysis of regional aquifers. Amer. Geophys. Union Meeting, San Francisco.

Şen, Z., 1976. Wet and dry periods of annual flow series. J. Hydraul. Div. ASCE 102 (HY10), 1503—1514. Proc. Paper 12457.

Şen, Z., 1983. Large Diameter Well Evaluations and Applications for Arabian Shield. In: Symposium on Water Resources in the Kingdom of Saudi Arabia Management, Treatment and Utilization, vol. 1. A-257—A-270.

Şen, Z., 1999a. Simple risk calculations in dependent hydrological series. Hydrol. Sci. Jour. 44 (6), 871—878.

Şen, Z., 1999b. Simple probabilistic and statistical risk calculations in an aquifer. Groundwater 37 (5), 748—754.

Şen, Z., 2008. Wadi Hydrology. Taylor and Francis Group, CRC Press, Boca Raton, 347 pp.

Şen, Z., 2009. Spatial Modeling Principles in Earth Sciences. Springer, New York, 351 pp.

Şen, Z., 2010. Fuzzy Logic and Hydrological Modeling. Taylor and Francis Group, CRC Press, Boca Raton, 340 pp.

Şen, Z., 2013. Philosophical, Logical and Scientific Perspectives in Engineering. Springer, 247 pp.

Şen, Z., Al-Sefry, S.A., Al-Ghamdi, S.A., Ashi, W.A., Bardi, W.A., 2013. Strategic groundwater resources planning in arid regions. Arab. J. Geosci. 8 (6), 4363—4375.

Sichard, W., 1927. Das Fassungsvermogen von Bohrbrunnen und seine Bedeutung fur die grundwassesabsenkung inbesondere fur grossere Absenktiefen (dissertation). Technische Hochshule, Berlin.

Van Stempvoort, D.R., Simpson, M., 1995. Hydrogeology of the Southeast Aquifer Management Plan Area. Saskatchewan Research Council, Publication R-1220-9-E-94, Saskatoon.

Sweitzer, J., Langaas, S., 1995. Modelling population density in the Baltic sea states using the digital chart of the world and other small scale datasets. In: Gudelis, V., Povilanskas, R., Roepstorff, A. (Eds.), Coastal Conservation and Management in the Baltic Region. Proceedings of the EUCC-WWF Conference, 2—8 May 1994, Riga-Klaipeda-Kaliningrad, pp. 257—267.

Tatem, A.J., Linard, C., 2011. Population mapping of poor countries. Nature 474, 36.

Theis, C.V., 1935. The relation between lowering of the piezometric surface and the rate and duration of discharge of a well using ground water storage. Trans. Am. Geophys. Union, 519—524, 16th Annual Meeting, Part.2.

Todd, D.K., 1980. Groundwater Hydrology. John Wiley and Sons, New York, 535 pp.

Walton, W.C., 1970. Groundwater Resource Evaluation. McGraw-Hill Book Co., New York.

Way, J.H., 1968. Bed thickness analysis of some carboniferous fluvial sedimentary rocks near Joggin, Nova Scotia. J. Sediment. Petrol. 83, 424—435.

Index

Note: Page numbers followed by f indicate figures; t, tables; b, boxes.

Printed and bound by CPI Group (UK) Ltd, Croydon, CR0 4YY

08/05/2025

01864906-0001